The Science and Culture of Surfing

David M. Kennedy
Editor

The Science and Culture of Surfing

Editor
David M. Kennedy
School of Geography, Earth and Atmospheric Sciences
The University of Melbourne
Parkville, VIC, Australia

ISBN 978-3-031-80978-1   ISBN 978-3-031-80979-8   (eBook)
https://doi.org/10.1007/978-3-031-80979-8

© The Editor(s) (if applicable) and The Author(s), under exclusive license to Springer Nature Switzerland AG 2025

This work is subject to copyright. All rights are solely and exclusively licensed by the Publisher, whether the whole or part of the material is concerned, specifically the rights of translation, reprinting, reuse of illustrations, recitation, broadcasting, reproduction on microfilms or in any other physical way, and transmission or information storage and retrieval, electronic adaptation, computer software, or by similar or dissimilar methodology now known or hereafter developed.
The use of general descriptive names, registered names, trademarks, service marks, etc. in this publication does not imply, even in the absence of a specific statement, that such names are exempt from the relevant protective laws and regulations and therefore free for general use.
The publisher, the authors and the editors are safe to assume that the advice and information in this book are believed to be true and accurate at the date of publication. Neither the publisher nor the authors or the editors give a warranty, expressed or implied, with respect to the material contained herein or for any errors or omissions that may have been made. The publisher remains neutral with regard to jurisdictional claims in published maps and institutional affiliations.

This Springer imprint is published by the registered company Springer Nature Switzerland AG
The registered company address is: Gewerbestrasse 11, 6330 Cham, Switzerland

If disposing of this product, please recycle the paper.

Contents

1   **Introduction to the Science and Culture of Surfing** 1
David M. Kennedy

2   **Surfing and Indigeneity: Heʻe Nalu, Discourse, and Empire** 9
Hunter H. Fine

3   **Why Are There Waves?** 27
Alessandro Toffoli

4   **What Forms My Break?** 51
Javier X. Leon and Tom D. Shand

5   **Beach Safety and Surf Hazards** 75
David M. Kennedy

6   **Lifeguards in the Lineup: Surfers as Rescuers** 93
Robert W. Brander, William A. Koon, and Amy E. Peden

7   **Surf Medicine and Health** 117
James Furness

8   **SheShaka: Surfing Is a Feminist Practice That Promotes Spatial Justice** 157
Gemma Tarpey-Brown, Fran Edmonds, Natalie Galea, Georgina Sutherland, Cathy Vaughan, and Karen Block

9   **Surfing Economics: Understanding, Managing and Protecting the Value of Surfing Ecosystems** 179
Ana Manero

10   **Surf Tourism** 203
Danny O'Brien

11   **The Surf Industry** 223
Craig Sims and Danny O'Brien

12 **The Surf Media** 241
   Craig Sims

13 **Sonic Waves and Acid Screens: Surf Culture
   and the Long 1970s** 259
   Sean Lowry, Danny Butt, and Jason Beech

14 **Conclusions and the Future of Surfing** 275
   David M. Kennedy

Index 281

Chapter 1
Introduction to the Science and Culture of Surfing

David M. Kennedy

1.1 Introduction

Surfing is an instant experience: to slide down the face of a breaking wave and ride it toward the beach provides an irreplaceable moment of pleasure and action (Fig. 1.1). No two rides are the same; each trip to the beach brings a new experience and a different perspective on the power of nature in the endless battle between sea and land. To surf is to experience nature in its purest and rawest form.

That instant moment of wave breaking—the focus of the surf experience—is the culmination of energy dynamics spanning the globe. It drives erosion of the coast and over long time periods, millennial in scale, shapes the shoreline that we enjoy today. The power of waves is related to the instantaneous circulation of the atmosphere across the planet. During storms, winds create waves, and these spread out through the ocean basins. Your local wave almost certainly derives part of its story from a storm thousands of kilometers away.

While riding the wave is the pinnacle of the surfing experience, it only accounts for 5% of the time in the water (Fig. 1.2). This means in a typical 1–2-hour session, only 180–360 seconds is spent surfing a wave (Meir et al., 2015; O'Neill et al., 2021) (Chap. 7). When one considers the effort taken in travelling to a break, whether it be your local or a holiday destination, the investment in equipment and health, and the industry required to produce the products required for surfing, it is clear that surfing is worth the effort.

The impact of surfing is felt way beyond the waves themselves. An identifiable culture surrounds and emanates from surfing. From the classic "surfer dude," captured in the media and movies of the mid-twentieth century to the ecofeminist radical surfers of today, surfing has been a shaper of, and responding to, cultural shifts

---

D. M. Kennedy (✉)
The University of Melbourne, Parkville, VIC, Australia
e-mail: davidmk@unimelb.edu.au

© The Author(s), under exclusive license to Springer Nature Switzerland AG 2025
D. M. Kennedy (ed.), *The Science and Culture of Surfing*,
https://doi.org/10.1007/978-3-031-80979-8_1

Fig. 1.1 The joy of riding a wave, always different, always exhilarating. (Photo: Peter Jovic)

Fig. 1.2 Sitting, waiting for a good wave, is just one of a multitude of experiences that comprises a surf session. Surfing also requires an industrial economy to provide the equipment needed to surf from wetsuits to boards. (Photo: Peter Jovic)

1 Introduction to the Science and Culture of Surfing

of the time. Through fashion, music, and personal character, a surfer can often be identified across cultural boundaries. Such is the appeal of the lifestyle that even people who are far from the water seek a surfer persona through sporting brands and outlets. The power of surfing is undeniable.

1.2 Surfing Science and Culture

In this book, we explore the multifaceted nature of surfing and the impact it has on society from culture to economics. The book starts, as everything does, with history. To fully understand where we are and where we are going, we have to know where we have been. Surfing started with the seafaring Austronesian people as they ventured into the then unpopulated Pacific Islands thousands of years ago. Riding waves has therefore been present throughout Polynesia, and in Hawai'i, this activity was called *he'e nalu,* meaning the art of surfing or literally translated as "wave sliding" (Chap. 2). Surfing was first presented to "Western" eyes by John Webber in 1778 in his painting depicting Cook's arrival in Kealakekua Bay, Owyhee (Hawai'i). Earlier to this, surfing (of tidal bores) was also known in dynastic China (Zanella, 2019). While represented, especially in the surfing movies and magazines of the mid-twentieth century, as the pursuit of youth in California and Australia, its origins are deep in the culture of the Pacific. Surfing is a gift from these peoples to modern society to which we are eternally grateful.

The breaks that we ride are the end of a global-scale energy conveyor belt that starts with waves generated in high latitude storm belts and tropical storms and ends with them reaching the coast. The physics of energy transfer between the air (through wind) and into the ocean is highly complex, as is the change in the form of this energy as it travels through the ocean (Chap. 3). Once a wave is formed, it loses little energy before it hits the shore; hence, large waves occur in areas which are calm—the perfect conditions for surfing. While the physics of fluid dynamics is incredibly complex—and not even fully understood by scientists—the basic relationship between the strength of the wind, how long it blows, and how much water it blows across is straightforward enough to produce highly accurate models of wave height (Chap. 3). The revolution in computational understanding and environmental monitoring now means the surfer can simply use an app-based forecast to know when and where to go.

For a wave to be surfable, it must break in the right form. Here the geomorphology, the shape of the landscape both above and below the sea, is critical (Chap. 4). The shape of the seafloor determines where and when the wave will break and also the form that it will take. A reef break is a different ride to that on a beach bar. The canyons of Nazaré, Spain, give a very different surfing experience to the lefts and rights in the sandy bay of Zarautz, Spain. It is the landscape that determines this diversity in experience (Chap. 4). As the seabed affects the break, the waves also change the shape of the geomorphology. This can happen at a range of scales from seconds to minutes when moving loose sand, to thousands of years when eroding hard rock. The key here is the feedback between energy input from the waves and the ability of the geomorphology to resist this attack (Chap. 5).

The impact of waves on the people who ride them is significant both in a mental and physical sense. Surfing is fun—that is why people do it! In fact, it is estimated that the global mental health benefit of surfing is worth US$0.38–1.30 trillion per year (Buckley & Cooper, 2023). It is also a recognized therapy under Australia's National Disability Insurance Scheme (NDIS, 2022). Surfing is also dangerous as the rolling breakers, and the currents they create can cause physical harm and even death (Chap. 5). On average, recreational surfers will experience one to two injuries per 1000 hours of activity (Chap. 7). As breaking waves are highly unstable environments, excellent balance, posture, and body position is critical for performing maneuverers. In fact, people who have surfed for a long time have far better postural control and balance than otherwise fit people who only cycle, walk, or swim (Chap. 7).

As surfers are not the only people who enjoy the beach, they also act as life savers (Chap. 6). Surfers carry a flotation device and have innate knowledge of surf conditions at their location perfectly placing themselves as informal lifeguards. A recent survey indicated that 80% of surfers would readily help someone in distress in the water (Attard et al., 2015). Apart from saving lives and the positive impact this creates socially, the economic result is also huge. A study in New Zealand estimated that if just 1% of surfers conducted one rescue that saved a life per year, it would prevent an economic cost of NZD $6.4 billion (Mead et al., 2024).

It is estimated that there are up to 35 million surfers across the globe (Lawes et al., 2023), around equal to the UN estimates of the total population of Malaysia or Uzbekistan. A US$4 billion surf industry (Statista, 2022) revolves around these surfers whose travels generate between US$32 billion and US$64 billion of direct expenditure (Chaps. 9 and 10). Surfing is certainly big business, reaching the sporting realms of Olympic level in 2020. This size creates challenges to the images of surfing, in how does the authenticity of a surfing subculture get diminished when it enters the mainstream (Chap. 11). This is especially the case when companies to maximize profits seek to engage an even wider customer base that just its participants. This is succinctly articulated by Jarratt (2010) (p. 10) who observed that "Surfing's biggest brands (were able to) cross the billion-dollar threshold by thinking big and staying cool … and that's a hell of a balancing act" (Chap. 11).

For a sport that is rooted in youthful rebellion, the modern surfer most often is more likely middle aged and probably less rebellious than they would consider themselves. In its early days, the average age of a surfer was around 18 years; today, it is around 30, with peak participation being in the 45–54 age category for men and in the 35–44 age category for women. Surfing can no longer be legitimately positioned as a youth sport tied to youth culture (Sims & Scott, 2022) (Chap. 11).

The "coolness" of surfing can be tracked through time in it its representation in film, television, and print, through to today's social media (Chaps. 12 and 13) as well as art (Fig. 1.3). "Surfing is at once a physical activity without inherent

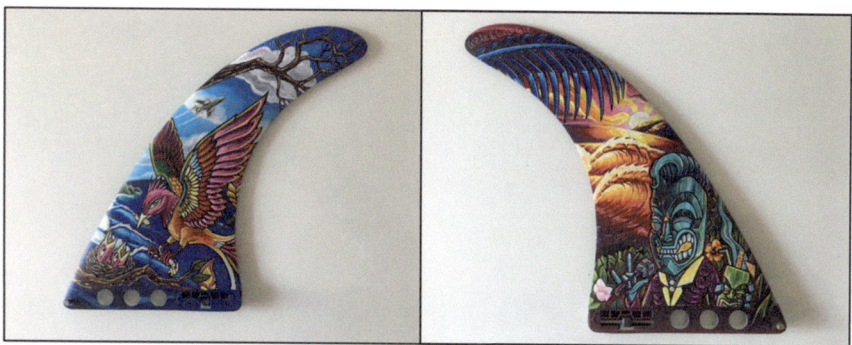

Fig. 1.3 Fin art showing different aspects of surfing culture. (From the collection of Craig Sims)

meaning and a performative act richly infused with social and cultural significance…. It has long functioned as a meeting point and catalyst for social development" (Chap. 13). The classic surfing films of the 1960s and 1970s drove the representation of the sport to a global audience while providing, sometimes subtle other times overt, social critiques of society (Chap. 13). The framing of this medium has rapidly changed as the transferal of knowledge systems has evolved into today's mass media forms (Chap. 12). Originally surfing only occurred through a personal, often familial knowledge exchange between elders to their direct or extended whānau (Māori for wider family). In traditional Hawai'i, *he'e nalu* was the realm of all where prowess in the surf was celebrated as social status (Chap. 2). This personal culture of surfing was carried as the sport spread across the world, though the original indigeneity was lost for a while. As more sought surfing, knowledge exchange moved into the realms of media both for skill development and eventually simple sport based entertainment. This landscape is now much more complex and diverse than in the mid-twentieth century as surfing's individualistic radical roots intersect with commercial economic interests of the modern era (Chap. 12).

The power of surfing is now evolving onto the political stage. It was declared the official sport of the State of California. The US State Department even uses surfing as a form of soft diplomacy through their "Sports Envoy Program" which funds high profile surfers from Hawaii to tour communities in Papua New Guinea, run surf clinics, and host discussions about gender-based violence and equitable access to education and health (Chap. 10).

The simple act of catching a wave, the hedonistic lifestyle that is created around this pursuit, is powerful, global reaching, and culturally rich. Surfing has and will continue to make the world a better place. Its intimate link to nature underpins its importance in recognizing the power of the Earth and protecting the natural world on which we all depend (Fig. 1.4).

Fig. 1.4 "The act of surfing on a wave is, therefore, an ephemeral event; an activity bound by space and time that leaves no trace of its existence (Ford & Brown, 2005)" (Chap. 11). (Photo: Peter Jovic)

References

Attard, A., Brander, R. W., & Shaw, W. S. (2015). Rescues conducted by surfers on Australian beaches. *Accident Analysis & Prevention, 82*, 70–78.

Buckley, R. C., & Cooper, M.-A. (2023). Mental health contribution to economic value of surfing ecosystem services. *npj Ocean Sustainability, 2*(1). https://doi.org/10.1038/s44183-023-00027-5

Ford, N. J., & Brown, D. (2005). *Surfing and social theory: Experience, embodiment and narrative of the dream glide*. Routledge.

Jarratt, P. (2010). *Salts and suits*. Hardie Grant Publishing.

Lawes, J. C., Koon, W., Berg, I., van de Schoot, D., & Peden, A. E. (2023). The epidemiology, risk factors and impact of exposure on unintentional surfer and bodyboarder deaths. *PLoS One, 18*(5), e0285928. https://doi.org/10.1371/journal.pone.0285928

Mead, J., Le Dé, L., & Moylan, M. (2024). The unexplored role of surfers in drowning prevention: Aotearoa, New Zealand as a case study. *Environmental Hazards, 23*(2), 150–166.

Meir, R., Duncan, B., Crowley-McHattan, Z., Gorrie, C., & Sheppard, J. (2015). Water, water, everywhere, nor any drop to drink: Fluid loss in australian recreational surfers. *Journal of Australian Strength & Conditioning, 23*(6).

NDIS. (2022). *National Disability Insurance Scheme: What do we mean by social and recreation support?* Australian Government. Retrieved 28/8/24 from https://ourguidelines.ndis.gov.au/supports-you-can-access-menu/social-and-community-participation/social-and-recreation-support/what-do-we-mean-social-and-recreation-support

O'Neill, B., Leon, E., Furness, J., Schram, B. E. N., & Kemp-Smith, K. (2021). The effects of a 2-hour surfing session on the hydration status of male recreational surfers. *International Journal of Exercise Science, 14*(6), 1388–1399. https://search.ebscohost.com/login.aspx?direct=true&db=s3h&AN=173328987&site=ehost-live&scope=site

Sims, C., & Scott, O. (2022). *The nature of Gen-Z's Influence on the future of printed Surf Magazines*. Bond University.

Statista. (2022). *Global surfing industry market size 2022–2027*. Statista Research Department. Retrieved 28/8/24 from https://www.statista.com/statistics/1327319/surfing-market-size-worldwide/

Zanella, N. (2019). *Children of the tide: An exploration of surfing in Dynastic China*. Cordee Ltd.

Chapter 2
Surfing and Indigeneity: He'e Nalu, Discourse, and Empire

Hunter H. Fine

2.1 Introduction

Developed in Oceania as the result of the migration of people across the open ocean of the Pacific, the practice of riding waves was present throughout Polynesia. In Hawai'i, this activity was called *he'e nalu*, meaning the "art of surfing" or literally translated as "wave sliding." Surfing in the present day builds on these origins. In popular discourse on the subject, surfing is largely described as an endeavor that is revived, progressed, and mastered by Western practitioners. Subsequently, the role of Indigenous wave riders, notably *Kānaka Maoli* or native Hawaiians, is often vastly understated (Walker, 2011; Ingersoll, 2016). As the activity was popularized throughout the Western world, particularly in the USA, this ignoring of cultural knowledge initially served colonial ends, and even today this legacy still persists. This chapter attends to some of this obfuscated territory and functions to recuperate ways of knowing that have been intentionally destroyed or casually omitted. Any representation of surfing references the histories of Indigenous wave riding, *Kānaka Maoli he'e nalu*, and Austronesian seafaring.

In Matavai Bay, Tahiti 1777, William Anderson, a surgeon on Captain James Cook's *Resolution*, described a local practitioner performing the basics of paddling a board, waiting and watching for a swell, and finally riding a wave along and toward the shore. Over a 100 years before this description, in the 1600s along the west coast of Africa in what is now Ghana, practitioners rode waves in "prone," kneeling, sitting, and standing positions in "one-person canoes" (Dawson, 2017). These Indigenous Africans developed independent wave riding practices before colonial contact creating and riding both canoes and boards in surf zones.

H. H. Fine (✉)
University of Guam, Mangilao, Guam, USA
e-mail: fineh@triton.uog.edu

© The Author(s), under exclusive license to Springer Nature Switzerland AG 2025
D. M. Kennedy (ed.), *The Science and Culture of Surfing*,
https://doi.org/10.1007/978-3-031-80979-8_2

Relatedly, the *caballito*, a reed canoe-like vessel, was used before Western contact by Indigenous Peruvians to slide down the face of a breaking wave when returning to shore (Warshaw, 2010). It therefore is easily stated according to scholars such as Matt Warshaw (2010), Isaiah Heikunihi Walker 2011, 2017), Scott Laderman (2014), Kevin Dawson (2017), Colleen McGloin (2017), Dexter Zavalza Hough-Snee and Alexander Sotelo Eastman (2017), and others that contemporary surfing is grounded in Indigenous knowledge that predates colonization.

Surfing is evidenced to have been practiced in a general manner in locations such as Peru, Senegal, and the Ivory Coast of Africa; however, it became particularly evident in Oceania. In Hawai'i notably, riding waves was known specifically as *he'e* or "to slide" and *nalu* or "wave" (Pukui & Elbert, 1986). Wave sliding was more than a physical practice—it functioned as a significant custom within the complex governing system of *Kapu*, which was a set of rules that administered daily life as a complete societal regulatory system. *Kapu*, or similar, was present throughout many islands in the Pacific (Warshaw 23). Secondary definitions of wave riding provide a glimpse into its conceptual nature as *he'e* was also defined as "to put to flight" and *nalu* "to ponder, meditate, reflect, mull over" and "speculate." These words are illustrative of an Indigenous knowledge set and are indicative of a much larger body of knowledge. To slide along the crest of a wave was to put one's body and mind into flight bringing about a host of cerebral, environmental, and physical negotiations in the process (Chap. 7).

John R.K. Clark (2011), working from Hawaiian sources, produced a Hawaiian and English language reference work containing a glossary of over 1000 Hawaiian terms relating to the practice of *he'e nalu*. This surviving knowledge represents a fraction of the meaning and importance placed on the practice as over a 1000 years ago *Kānaka* rode waves on *papa* or boards in meaningful performances of social identity and individual expression (Finney and Houston 106, 24; Warshaw 22). Children rode a small *pupo* or body board in the shore break or *po'ina nalu*, while *ali'i* or royalty rode the large buoyant *olo* board on a *ha'i, ha'i maika'i* or perfectly breaking wave. Such sought-after surf zones, materials, and boards were often protected sites under *Kapu* as there is ample evidence of wave riding as a physical, conceptual, and political act.

In each instance of Indigenous wave riding, particularly in Oceania, the connections between the practice of riding canoes and boards on waves are evident. These moments indicate a shared culture that consisted of a series of unique practices including material harvest, tool fabrication, carving, and vessel navigation initially honed by Indigenous Austronesian cultures throughout Oceania. What we refer to as surfing, which is not used as a term until the twentieth century, is an encounter with the practice of *he'e nalu* and the larger culture of Austronesian seafaring. Surf scholarship and the ongoing critique of colonial legacies have created a valuable conversation concerning the social, cultural, and political roles of *he'e nalu* and contemporary surfing; however, seldom is the connection made between surfing and seafaring in their Indigenous forms.

When Indigenous spaces are colonized, locations such as the surf zone, the wave riding practices themselves, and the ways in which we discuss and represent them become sites of tension. In turn, the marking of these connections to Indigenous cultural roots and structures is political as they also comment on current forms of structural domination. McGloin and Walker have noted that surfing has functioned a form "of resistance to the enduringness of colonialism" (McGloin 2017). Today surfing is a borderland, a physically and symbolically contentious site. Walker (2011) defines the *ka poʻina nalu* as a space, where social differences converge and "unique social and cultural identities are formed." He claims that as a result, "state-sanctioned authority is often absent from" these zones. To surf is to reference both Indigenous agency and a colonial history as themes of cultural seizure are enmeshed in surfing narratives and practices (Chap. 13).

Walker notes that the *ka poʻina nalu* or surf zone is a site of resistance due to the sustained presence of Hawaiian surfers. Whether it is the fluidity of the zone itself or the radical nature of the movement employed in its negotiation, the surf zone creates a liminal space. He adds that due to the preservation of these spaces through surfing, "colonial powers were less able to conquer" (Walker, 2017). Dawson (2017) adds that a continual wave riding presence and culture "challenged these incursions as twentieth-century coastscapes were embattled political spaces" which remained "Indigenous culturalscapes" (144). Surf zones all around the globe remain a physical and intellectual location where different narratives of the world compete.

Many Indigenous scholars such as Elizabeth Lynn-Cook (2012), Epeli Hauʻofa (2008), Marissa Muñoz (2019), and Vincent Diaz (2016), among others, have discussed the ways that Indigenous paradigms are revealed when paying attention to geographical and socio-spatial contexts. Many others have noted the social productions of space in general, such as Henri Lefebvre (2007) and Marc Augé (1995); the interplay between physical places and more ideological spaces such as Yi-Fu Tuan (2007) and Tim Cresswell (2004); the ways individuals and groups embody, reify, and resist spatial and social productions through everyday practices such as Michel de Certeau (1984) and Pierre Bourdieu (1990); and the intersections that are created between geographical borders and boundaries of identities such as Gloria Anzaldua's (1985) work on borderlands.

Today the *ka poʻina nalu* continues to function as such a space, which is transferred to the identities of surfers (Fine 2018; Walker, 2011; Modesti, 2008). From these symbolic and physical intersections, a production occurs, which Sonja Modesti (2008) refers to as a "material consequentiality" or the ways in which discourse can produce material affects. The surf zone, the narratives consumed, and the riders themselves become indicative of the spatial tensions they embody. Such performances mix with popular narratives, and in turn, surf scholarship has observed that surfing has performed many different, often contradictory, roles throughout history. The practice, its spaces, and its discussion remain important points of tension that continue to unfold.

McGloin notes that riding waves is a form of learning, a pedagogy involving indigeneity. In this regard, she notes that when we recognize Indigenous aspects of surfing, this "recenters indigenous knowledges, epistemes, histories, and spatiotemporalities" (197). The act of riding waves is entangled with the symbolic modes it has performed throughout history as surfing has also been used to colonize Indigenous land and sea spaces according to scholars such as Laderman and Dawson (Dawson 144). Laderman (2014) argues in *Empires in Waves: A Political History of Surfing* that the Americanization of riding waves fueled imperial expansion. Similarly, Dawson (2017) states that as surfing was Westernized, it was concurrently used to further encroach on Indigenous spaces previously unavailable to European and American settlers.

Karin Amimoto Ingersoll in *Waves of Knowing: A Seascape Epistemology* (2016) states that on the Hawaiian Islands, surf tourism and the surf industry have continued to confine and limit the movements of Indigenous Hawaiians as outsiders crowded, "surf breaks, channels, and beaches, leaving no 'space' for autonomous Kanaka movement." Krista Comer (2010), in "Surfeminism, Critical Regionalism, and Public Scholarship," notes that the gendering of surfing reflects global dimensions that can further sexist constructions while acting as a cover for, "Western imperial land grabs in surfing's new emergent markets." Today surfing is caught between numerous borderlands. In *Surfer Girls in the New World Order* (2010), Comer adds to her analysis by stating that "critical femininities" can also be developed alongside "critical sensibilities," arising from an awareness of the ways in which surf zones function. She emphasizes the connections between surfing sites and identity formations surrounding gender, race, and nationality.

In contemporary contexts, the *ka poʻina nalu* continues to function as a potential space of recuperation. Comer mentions how the contemporary surf zone can be harnessed to breach patriarchal structures. Relatedly, on the Hawaiian Islands *Kānaka* enjoyed *heʻe nalu* in mixed male and female spaces. This contrasted with other social areas such as dining that were often divided based on sex. Meanwhile both male and female practitioners had access to the same surfing zones outside of royalty. It was a space of temporary equality where all had access to forms of cultural capital. Princes and princesses alike were often highly celebrated for their surfing ability, adding to their societal status through displays of surfing prowess. These examples illustrate the divergent and similar qualities of Indigenous wave riding and contemporary surfing as sites of transgression and cultural capital.

Any wave capable of being ridden is both a contemporary resource that remains accessible according to the conditions of land and a historical symbol of Indigenous knowledge. Everyday acts such as riding waves continue to play large roles in defining the identity of a community and nation. In this regard, the reordering of *heʻe nalu* along Eurocentric lines transgresses the various contemporary frameworks in which it is often placed, such as exercise, escape, and sport. In this chapter, I examine the discursive intersections between Indigenous wave riding and its Western manifestation, surfing. I focus on the symbolic transformations involved as *heʻe nalu* is intentionally transformed into surfing. To examine its symbolic representation is to examine its material performance. I argue that these historical tensions continue in contemporary discourse and manifest themselves in material forms having much to do with larger colonial and postcolonial structures. I first discuss

Indigenous surfing as an extension of Austronesian seafaring culture before illustrating how Western surfing was developed from Indigenous Hawaiian *heʻe nalu*. Finally, I examine points of intersection in early discourse on wave riding between nation, identity, and culture.

2.2 Austronesian Seafaring and Surfing

Austronesian refers to the Indigenous inhabitants of Taiwan, maritime Southeast Asia, Micronesia, coastal New Guinea, Melanesia, and Polynesia, as well as Madagascar who are linked through a common language and ancestry. Stemming from their seafaring cultural awareness, knowledge, and practices, *heʻe nalu* developed on the precolonial Hawaiian shores. The acts of riding waves and carving boards are developed through the practices required to develop a sustained network of open sea navigation and canoe building (Finney & Houston, 1966, p.22). From the moment of European interaction by Ferdinand Magellan with CHamoru and Filipinos and Cook in Hawaiʻi with *Kānaka* and Indigenous Tahitians, the practices of canoe building, seafaring, and wave sliding were symbolized according to distinct narratives of authors of the times that reflect their various political contexts. What is seldom discussed is the place riding waves occupies in relation to the practices of seafaring culture unique to Oceania.

Riding waves utilized a hand carved board that created a unique relationship with the intensity of the ocean, which was first established through traditions of Austronesian seafaring. Indigenous seafaring and surfing share similar epistemological underpinnings, which are part of a much larger historical and widespread network of Indigenous knowledge stretching back over 4000 years from the Philippine and Indonesian archipelagos to Madagascar in the east, Hawaiʻi and *Rapa Nui* to the north and west, and *Aotearoa* (New Zealand) and Australia to the south. Each site within the vast region covering much of the globe has a shared and unique seafaring tradition recognized most visibly by the outrigger canoe. On the Hawaiian Islands and elsewhere in Polynesia, this tradition is known to also produce boards and wave riding.

The first Western illustration of a surfboard in 1778 by John Webber on Cook's arrival in Kealakekua Bay, Owyhee, or Hawaiʻi depicts a local paddling out to sea among numerous traditional outrigger canoes (Finney & Houston, 1966). The HMS *Resolution* and the HMS *Discovery* sit in the background of the image as many *Kānaka* seafaring practitioners make their way to them from all directions (Fig. 2.1). This person, their posture, board, and accompaniment of outrigger canoes, alludes to the shared qualities of these crafts in the water. The vessels depicted looked and moved through the water similarly. They were undoubtedly carved according to equivalent traditions and out of the same materials. Once a central hull was carved, planks or boards were carved to build up the sides of larger canoes. These boards were rudimentarily shaped much like boards for riding waves. Austronesian seafaring is part of a system of physical practices and concepts, a series of interconnected

Fig. 2.1 Painting by John Webber in 1778 on Cook's arrival in Kealakekua Bay, Owyhee, or Hawai'i showing Indigenous canoes and board paddlers in the surf

methods, terms, and ideas whether they manifest themselves in the Marianas in Micronesia, Hawai'i and Tahiti in Polynesia, or the Philippines in Southeast Asia.

Such practices were based on local identification and harvesting of specific trees and plants, carving canoes, and tool construction, which effectively lead to the same identifying, harvesting, and carving practices that produced the *olo* and the *alaia* boards of *he'e nalu* (Clark, 2011) (Fig. 2.2). This connection remains an undeniable link between surfing and traditional seafaring as Indigenous networks connects not only the islands and sea spaces of Oceania but also to contemporary practices such as surfing. Riding waves and its physical and conceptual precursor, Austronesian seafaring practices are part of a shared body of Indigenous knowledge.

The ancient Austronesian art of carving, with adze, blade, and fire, is a learned trade that produces these material objects (D'arcy, 2006). All of which are made possible by a community familiar with a network of shared practices passed down through a formal oral tradition of teaching. Along with the construction, riding waves on boards developed as an endeavor complete within an incalculable number of social meanings and hierarchies. Most of these mirrors those related to seafaring culture. The significance placed on *he'e nalu* on the Hawaiian Islands is akin to the importance of seafaring within community structures throughout Oceania. Finally, the practices of seafaring and surfing exhibit a similar approach to the negotiation of dynamic seascapes, a complete familiarity with the ocean, and an equally complete subjectivity that arises from the social, physical, and intellectual practice. Seafaring and surfing share not only a particular relationship with the sea but more specifically a way of knowing based on calculated responses to constantly changing environments.

Austronesian seafaring culture constitutes a network of learning and information. It is comprised by an almost infinite series of interconnected practices, terms, and ideas that result in the creation of large material objects such as canoes to the

Fig. 2.2 One of the earliest known photographs of a surfer, taken on Waikīkī Beach, Hawai'i, in the late 1800s (most likely 1890). The surfer is carrying an *alaia* style board while gazing out at the waves. Diamond Head is in the background. (Source: ANMM Collection, https://collections.sea.museum/en/collections)

grand acts of travel such as open sea voyages. This is most visible in the voyaging achievements of the Polynesian Voyaging Society and the *Hōkūle'a*, a double hulled craft fabricated according to traditional Polynesian designs. In 1976, Grandmaster navigator Pius Mau Piailug navigated over 3000 miles of open ocean from Hawai'i to Tahiti by traditional navigational means taught and practiced in Micronesia. During the journey he utilized and taught one of its central tenets, the Carolinian star compass, which works in unison with other environmental navigation techniques. He shared the knowledge of his ancestors, not only with his sons grandmaster navigator Sesario Sewralur and master navigator Antonio Piailug and others on his home island of Satawal but also on the *Hōkūle'a*. In the geopolitical contexts, this monumental journey became a deliberate act of social advocacy.

Since this maiden voyage, numerous seafaring traditions throughout Oceania have been revived by local practitioners. The various forms of traditional seafaring practices throughout Oceania and the world exemplified by the Canoe Federation CNMI, Polynesian Voyaging Society, Tahitian Voyaging Society, and the Micronesian Voyaging Society, among many others, represent a practice that promotes Indigenous optics. Meanwhile Indigenous wave riding, particularly Hawaiian surfing and surfers, are often recognized as a sovereign cultural and national identity often unobtainable in other arenas (Walker 79). Surfing is an aspect of seafaring, a larger, more dispersed, and arguably recognized form of Indigenous social advocacy, yet Indigenous surfing particularly in Hawai'i as many scholars have noted has functioned as salient site of resistance.

Acknowledging that surfing is part of this larger cultural constellation reminds us that surfing as well is composed by a larger network of meaning. Within Western discourse, the acts of surfing and seafaring are separated as are the various acts and roles of surfing within surfing culture; however, each practice and element such as carving boards and riding waves or navigating social and physical spaces is closely connected. Hawaiian newspapers from the nineteenth century indicate these links that form this shared cultural structure. The extensive undertaking of moving through oceans on fabricated crafts can only occur after a complex series of practices that involves the creation of canoes for paddling and sailing and boards for surfing from trees. This larger structure constitutes a set of knowledge that provides the foundation for riding waves as well as an indication of their ordering within a system:

> He [Kamehameha] chose kahunas who were makers of double canoes (wa'a kaulua), war canoes (wa'a peleleu), single canoe (wa'a kaukahi), sailing canoes (wa'a kialoa)— either one masted canoes (kiakahi) or two-masted (kialua); and kahunas who were makers of holua sleds and [alaia] surfboards (papa he'e nalu). (Clark, 2011).

Kamehameha, royalty, chose specialists to manufacture various canoes, boards, and sleds to be used to traverse the sea. The excerpt above indicates a shared maker of numerous crafts that all are constructed, designed, and used in similar ways. As both canoe and board building and use were discouraged or even banned, during colonial times, the connections between the art and science of surfing and its wider cultural important were also severed. As the endeavors reemerge in differing forms, they became aligned with mainstream depictions as individual sports of daring, leisure, and privilege, rather than everyday pursuits of life-sustaining spatial engagement, stemming from a personal and often intimate fabrication.

Ingersoll (2016) states that like surfing, "voyaging becomes an ideology," which constitutes a way of knowing. Both endeavors engage in unique situational contexts as frameworks for action, navigating according to a variety of signs and directing a surfboard through water to catch and ride swells (Lewis, 1972). Within the cultural constellation of traditional Polynesian seafaring, the Hawaiian outrigger canoe, the *wa'a*, for example, has a shallow hull designed to navigate potentially shallow waters, which often surround islands, while the more specific design, the *wa'a pā.kā.kā nalu* is described as a canoe specifically intended for surfing waves (Clark, 2011). Surfing is an aspect of seafaring, and both vessels function today as forms of Indigenous agency. Indigenous surfing is an extension of Austronesian and Oceanic traditions of seafaring and canoe building, entailing a constellation of meaningful practices. The ways in which these tensions play out has much to say about our cultural, social, and political identities.

When the *Hōkūle'a* made its second voyage, most surfers with connections to Hawai'i instantly saw the connections between seafaring cultural heritage in

Oceania and surfing. Many were eager to support and participate given the obvious shared cultural and operational territory. The navigation of the second voyage did not include Micronesian navigator Mau Piailug and centered around a Hawaiian centric revival. Surfers were some of the most focused and eager to contribute possessing a rich history of involvement in battling intense aquatic spaces. Surfing for decades had also already been embattled as forms of social resistance. Like the shortboard revolution in surfing that appears after Indigenous forms, there is a reversal of Indigenous knowledge in settler-colonial contexts. Surfing awareness and the ability to charge into what are, for many, intimidating aquatic environments made epistemological sense when riding waves intersected with the precursory cultural knowledge of traditional seafaring. Eddie Aikau came into the scenario as a legendary surfer, water person, and lifeguard moving Kānaka agency on a global scale much as Kahanamoku had earlier. The second voyage of the *Hōkūleʻa* capsized in the waters between Oʻahu Duke and Molokaʻi in 30 knot trade winds and 6–10 foot swells. The crew was stranded on a capsized vessel with no support boat, 5 hours after they departed Ala Wai Harbor. The next day, Aikau lost his life attempting to save the lives of others as he had so many times before. He ventured out on a board and was tragically never seen again. The rest of the *Hōkūleʻa* crew was eventually rescued and renewed attention to safety and the larger cultural underpinnings insured that the *Hōkūleʻa* would go on to unparalleled success, partially in memory of Aikau. The connections remain indelible as boards and canoes were carved together in the same spaces using identical materials, tools, and techniques. They eventually were paddled out alongside each other and came to constitute important acts of cultural meaning and resistance in contemporary contexts.

2.3 *Heʻe Nalu* to Surfing

There exists, centering around its development on the Hawaiian Islands, a sustained interpretation of riding waves in the English language writings of Hiram Bingham, Henry T. Cheever, William Ellis, James Cook, Mark Twain, and Jack London. Other Western writers have added to the narratives concerning surfing including Isabella Bird, James Michener, Thomas Wolfe, Timothy Leary, and Kem Nunn (Comer, 2010, 9). Along with these popular English language texts, there exists a large body of information pertaining to *heʻe nalu* contained in Hawaiian language newspapers written before 1900. Much of the information drawn from these newspapers beginning in 1834 are available due to the work of Hawaiian scholars such as John Papa ʻIʻI, Samuel Kamakau, Zephrin Keauokalani, David Malo, Mary Kawena Pukui, Samuel H. Elbert, and others (Clark 2–3). Through the examination of the discourse concerning *heʻe nalu* and surfing, we can better understand what surfing is and how it functions.

On the Hawaiian Islands prior to colonial contact, the practice of riding waves for most began with the shaping of a *papa* or board, typically the *alaia* (Clark,

2011). A Hawaiian language newspaper in 1865 describes the widespread practice of riding waves.

> Surfing is a very popular sport in Hawai'i from the chiefs to the commoners. This is how you do it. The board is created ahead of time out of koa, kukui, 'ohe, wiliwili, or other woods that are good for making boards. (Clark, 2011, 406).

The description of riding waves begins with the harvesting of material before producing an object that is now a commodity of an industrial market. The same processes and tools used for carving a canoe are now discussed solely for the purposes of carving boards, which serve similar aquatic functions. The directions start with the searching for "woods that are good for making boards or plant identification, harvesting, and fabrication intended for specific use (Clark, 2011). To this day, boards are referred to as being shaped rather than made and often are products of handmade labor. Within surfing discourse, the questions of where the designs or ideas for surf boards initially came from or how the stances and gestures of paddling and standing developed are seldom asked. The answers to these questions are contained in the lived memories of Indigenous practitioners, descriptions such as these of *Kānaka he'e nalu*, as well as in the objects and practices themselves.

The design of the *alaia*, for example, is specific while generally adaptable to the rider (Fig. 2.2). It is the outcome of an evolution of riding the *waha* or steep section near the inside of a curling wave (Clark, 2011). This is still the most sought-after posture and position in surfing, crouching low as if riding an *alaia* as one is required to do to maintain control on a thin finless tomb-shaped board with long straight edges, tucking one's torso under the lip of a hollow breaking wave. There are many such waves on the Hawaiian Islands but as surfing is developed from these Indigenous wave riding practices, the more gradually breaking waves and use of the large royal *olo* become popular. The thin narrow *alaia* functioned as the board of the people commonly used to ride waves in Hawai'i and is typically constructed by the rider themselves rather than an expert carver as was often the case with royalty. The harvesting of material was a sacred process protected heavily by rules of the *Kapu* system, and to cut down a tree useful for a community must be a social affair. Thus, the use of larger trees was often off limits to common practitioners, and those that rode waves were left to work with smaller portions.

The modern short board, which dominates contemporary surfing, represents an innovation, a revival, and an obfuscation of a previous cultural gesture that predates the longboard, typically placed before the shortboard in Western surfing narratives. The longboard was an adaptation of the *olo* and was initially in Western surfing discourse merely referred to as a surfboard before the supposed short board revolution recaptured the ethos of the *alaia*. Once again, Western culture presents existing Indigenous territory as a discovery. Dawson (2017) and Laderman (2014) agree that "post-annexation (1898) Hawaiians planted 'the roots of global surf culture' before considering how surfboards became imperial implements" (Dawson, 2017). The *alaia* and the contemporary shortboard as well as the gestures required for their navigation are symbolic reminders of this history.

The act of riding particular boards at specific sites, by certain practitioners, indicates status and identity achieved through meaningful movement much as seafaring did within wider community structures. As *he'e nalu* morphed into surfing, so did its enjoyment and participation. It changed from an activity of the "many" to one that is enjoyed primarily by the unique or special. Waves can be considered part of land, breaking due to the interaction of water with the seabed (Chaps. 3 and 4). They therefore can also be considered part of the social constructions of both spaces. Surf zones continue to function as sites of dialog due to the Indigenous practices and colonial pasts involved in the practices being performed. None perhaps are more significant than those along the North Shore of Oahu, notably Pipeline. The names of surf zones along this famous 7-mile shoreline from Haleiwa in the south to Sunset in the north indicate a symbolic regrouping, Gas Chambers, Off the Wall, Log Cabins, Leftovers, and Velzyland, named after Southern California surfer Dale Velzy by John Severson after visiting for a film production (Chap. 13). As they become renamed, U.S. surfers are credited repeatedly as the first to ride the waves at these long-revered sites.

Ben Finney and James Houston compile a list of over 100 identified surf zones throughout the Hawaiian Islands (pp. 28–32). Clark (2011) notes that *Kapuni* on Waikīkī Beach is the first to turn into Canoes—subsequently all the surf zones were given English names by the early 1900s. Just as European explorers claimed to discover unknown lands, provide religious salvation, and introduce scientific knowledge, U.S. surfers claimed to recover the abandoned art of riding waves. Brown narrated in *The Endless Summer* (1966) that in the 1950s, intrepid surfing explorers such as John Kelly, George Downing, Gregg Noll, Pat Curren, Peter Cole, and Fred Van Dyke were the first to ride Waimea Bay, perpetuating similar narratives of Western discovery (Chap. 13). Ingersoll affirms, based on Hawaiian periodicals and oral histories or *mo'olelo*, that Indigenous Hawaiians were the first who "rode and named the waves at Waimea Bay and elsewhere" (Ingersoll, 2016). Along with the sites of the surf zones, the knowledge associated with the practices of riding waves becomes adapted according to new meanings. Clark (2011), working from Hawaiian sources previously mentioned, produced a Hawaiian and English language reference work containing a glossary of over 1000 Hawaiian terms related to the practice of *he'e nalu*. This surviving knowledge represents a fraction of the meaning and importance placed on the Indigenous practice.

2.4 Surfing and Settler Contexts

Surfing establishes a space where competing conceptions of culture, community, and identity play out. Walker notes that riding waves creates contested social spaces where cultural, identity, and social production can be challenged and affirmed. The practices of riding the waves continue to function symbolically according to the geopolitical structures of the land settlements nearby. For example, surfing is marked as a distinctly Californian endeavor consisting of a series of innovations including the wetsuit and leash. Assembly Bill 1782, which establishes surfing as the official sport of the State of California, declared surfing "an iconic California sport" and listed many aspects of surfing culture and technology that were pioneered in California (Moser 2022). California state identity therefore is in part funneled through an Indigenous activity of riding waves developed in the central and western Pacific Ocean. The symbolization noted here is drawn from a constellation of attitudes and values that shape the material reality of the *ka poʻina nalu*.

By 1895, Nathaniel Emerson lamented the decline of this vibrant cultural past time, writing that he "cannot but mourn its decline," as "today it is hard to find a surfboard outside of our museums and private collections" (Warshaw, 2010, 34). The practices of riding waves in Indigenous forms are celebrated while being simultaneously removed through discourse. Museum displays, imagery, and narrative replace materiality. The theme of absence within the historiography of riding waves is not neutral but indicative of the ways in which colonial and imperial contexts control. Indigenous perspectives expressed through grand and everyday practices reflect an untold loss, and *heʻe nalu* is but one example; however, the practice continued, like many others including weaving, dance, language, and seafaring.

Most agree, such as Richard Kenvin (2014), Laderman (2014), Walker (2011, 2017), Ingersoll (2016), and others, that surfing was introduced to North America when visiting sovereign princes of Hawaiʻi, Jonah Kuhio Kalanianaʻole, David Laʻamea Kahalepouli Kawananakoa, and Edward Abnel Keliʻiahonui, rode the waves at the mouth of the San Lorenzo River in Santa Cruz County, California, around 1885 (Fig. 2.3). As sovereign *aliʻi* visitors, they had a royal *olo* board shaped from a local timber mill for the undertaking (Kenvin, 2014). Surfing in California began with sovereign Hawaiian royalty riding traditionally noble boards; this moment is a cultural instance of importance and an interaction between nations.

The board and practice of riding functioned as an exchange of global politics. The use of boards in waves became a vehicle of diplomatic exchange. A plaque remains in commemoration at Steamer Lane, a popular surf spot now at the heart of Northern California surfing culture. A similar yet contradicting plaque in Redondo Beach, California, indicates that George Freeth was the "first surfer in the USA" in 1907. Freeth was employed to bring surfing from Hawaiʻi to California specifically for the purposes of promoting the domestic resources of water and waves. The practice of surfing functions primarily as a royal and sport endeavor, leaving out much of its larger societal functions and epistemological associations. The aspects of plant harvesting, board carving, tool construction, and the close fabrication of a board for one's own use were also severed from the larger cultural network of meaning in these narratives.

Today both Huntington Beach and Santa Cruz compete for the moniker of Surf City USA as surfing remains a form of economic and cultural capital that plays a role in the creation of a national identity. National recognition and the political and cultural status of Hawaiʻi regarding surfing on the global sporting stage are a contested area. Surfing is firmly situated as a sport pursued by individuals seeking excellence or proficiency, yet this takes away from the communal aspects of

Fig. 2.3 Jonah Kuhio Kalaniana'ole, David La'amea Kahalepouli Kawananakoa, and Edward Abnel Keli'iahonui the sovereign princes of Hawai'i pictured in 1886 in uniform while attending military school in San Mateo County, California. (Source: Hawaii State Archives Call Number: PP-97-2-011)

seafaring and surfing; however, in and out of national and sports frameworks, surfing functions as a site of cultural capital and social resistance. Hawaiian surfer Carissa Moore has won the women's Surf World Championships and an Olympic gold medal, yet like other Hawaiian competitors, she surfs under the Hawaiian flag in the World Surf League but not the Olympics, as in the latter people can only compete under a national, UN-recognized country flag. This is also true in the International Surfing Association and the National Scholastic Surfing Association, which "define Hawai'i as an independent region" (Walker, 2017). These recognitions fall within the sport of king's framework and unfold as a series of tensions involving Indigenous agency and the legacies of colonialism.

As the accomplishments of Freeth and Duke Kahanamoku are framed according to national and sport optics, they help propel surfing, swimming, Hawai'i, and the USA onto the global stage. They both gave surfing demonstrations around the world, and Kahanamoku won Olympic gold medals for the USA. Walker remarks that Hawaiian representation as semiautonomous in international competitions, outside of the Olympics, enables "Hawaiian surfers to develop unique and empowering identities," which have functioned in the context "of longstanding cultural, political, and economic struggle evolving from a 150-year history of colonization" (2017). Along with Hawai'i, the World Surf League also recognizes Guam as an independent region which, according to the United Nations Special Committee on Decolonization, is 1 of 17 entities that remain under control by UN member states, including the USA, the largest territorial power (Statham, 2002). As the sports centric framework unfolds, so do notions of national identity and sovereignty, which furthers the sport of kings on the global stage.

The dialog between nations was performed in the water and continued through discourse as the act of riding waves was transformed through symbols to fit within the interests of nation-states and sports centric productions. The tensions between *he'e nalu* and its constructions as surfing are present in these early and more contemporary moments. Western chroniclers cataloged a certain way of looking at the practice that rewrites its corporeality, yet this rearrangement often entailed a symbolic razing. *He'e nalu* was in practice in 1866 when Mark Twain visited Hawai'i and observes practitioners riding waves on a *papa he'e nalu* or "board for wave-sliding" (Warshaw, 2010, 32). The practice is so unfamiliar to Twain that he remarked, "none but the natives will ever master the art" (Winchester, 2015). Three years later, in 1898, Hawai'i was forcefully annexed by the USA, and surfing reemerged as a series of gestures with new cultural and national connotations.

South Carolina transplant, Alexander Hume Ford, sought to establish a European and U.S. vision of surfing along with statehood. The two unfolded together with each serving the constructions of the other. Ford, like many others endowed with a similar colonial disposition, believed that foreigners could learn all the cultural secrets of the Hawaiian surfer. He formed organizations and publications that served the interests of settler-colonial communities. Surfing was placed on resort, airline, and tourist brochures serving the spread of a national identity from the mainland to the islands. Many years later, popular films such as *The North Shore* (1987) echoed this narrative as a surfer from landlocked Arizona masters the unruly surf zones of O'ahu's North Shore. Surfing subsequently spread around the world with Europeans and Americans as inheritors of an abandoned Hawaiian practice, who go on to become experts of the craft (Walker, 2011, 95; 32). As surfing moves from its cultural center in Hawai'i to California, it subsequently becomes a literally floating signifier of U.S. imperial culture.

London, on a trip through the Pacific Ocean, stopped in Hawai'i in 1908. After meeting Ford and Freeth, London experienced and cataloged his experience riding waves. His attitude toward the practice, due to the encouragement of Ford, was decidedly different from Twain's; as London declared in *The Cruise of the Snark* (2017), "The Snark shall not sail from Honolulu until I, too, wing my heels with swiftness of the sea," referring to the act of riding waves. The article in which London described his attempts to ride waves on boards was originally titled, "A Royal Sport: Riding the South Sea Surf." Here he noted the royal conception of the practice and perhaps paralleling the loss of sovereignty with the perceived loss of a prominent cultural practice (Winchester, 2015).

London and other Westerners, as Dawson (2017) states, "perceived surfing as people's attempt to conquer nature." London (2017) was indebted to Ford as the possessor of Indigenous wisdom, writing, "I am always humble when confronted by knowledge. Ford knew." Furthermore, his knowledge is gained without a teacher as London (2017) embellishes, "he had no one to teach him, and all that he had laboriously learned in several weeks he communicated to me in half an hour." London was actively searching for new experiences and cultural perspectives for readers to absorb. On his way to the Hawaiian Islands, he wrote a similar statement after learning bits of information about celestial navigation, an artform and science at the heart of Austronesian seafaring. These narratives seize not just land but also the social spaces that stemmed from deeply engrained aquatic relationships and knowledge.

Sam Low (2013) notes that Ford intended to turn Hawai'i into a "white man's state" creating a "beckoning paradise for the growing number of Pacific tourists," establishing the islands "as a crucial outpost of American global power." Ford formed the *Outrigger Canoe Club* and intended to develop "the great sport of surfing in Hawai'i." The name of this club reflects the English name change of the nearby surf zones such as canoes. This all-white club is commonly recognized as surfing's first organization, outside of the marginalized origins of *he'e nalu* (Warshaw, 2010). The group promoted not only surfing as they saw it, but Western segregation. As Franz Fanon (1963) concludes in *Wretched of the Earth*, no colonial system draws its justification from the fact that the territories it dominates are culturally nonexistent. Ingersoll (2016) states that the organization functioned as a means of exclusion, which provided a space where only some could enter and "engage in a paradoxical act of negation and appropriation" by participating in an Indigenous practice while excluding an Indigenous population. The dominant narrative, like many colonial tales of discovery and mastery, indicated that riding waves was all but lost, if it were not for the revival of it by Ford, London, and other Westerners. There is a dwindling of many local pursuits, yet if there is a revival of the practice after colonial and imperial encounters, it is already underway by Indigenous practitioners of *he'e nalu*.

In 1911, a group of Hawaiian surfers, notably Kahanamoku, formed the *Hui Nalu Club* or *Surfing Riding Club* as an organization of Indigenous agency. The Hawaiian language name of this organization refers to the early signifiers of the practice discussed in this text. This collective was a political response to what was becoming the racist infrastructure of surfing, which marks surfing as a form of social advocacy. It is a form of empowering optics that responds to the segregation and subordination suffered at the hands of what was becoming a society based on Hawaiian oppression. Today the physical spaces and bodies within mainstream contemporary surfing remain "phenotypically White" as Brenda Wheaton reminds us (Wheaton, 2017). Such homogeneity according to structural forces of power that involve subject positionality is created intentionally and is direct reflection of the histories discussed in this chapter.

James Baldwin notes that, "the great force of history comes from the fact that we carry it within us, are unconsciously controlled by it in many ways, and history is literally present in all that we do" (Giroux, 2021). Hui Nalu members find an identity that is linked to a precolonial past in a threatened cultural space. Later in 1976, the Hui O He'e Nalu or Club of Wave Sliders was formed as a revival of the original Hui Nalu Club in order to respond to the increasing numbers of U.S. and Australian surfers brought to the shores of Hawai'i by the growing global surf and sport industry. The *ka po'ina nalu* was situated as an endangered native Hawaiian space due to the incursion of the modern surf industry, which many positioned as a continuation of colonial seizure (Walker, 2011). The group continues to champion *Kānaka* agency by outlining the club's objectives to preserve Indigenous Hawaiian influence over the sacred waves of the North Shore.

2.5 Conclusion

He'e nalu developed from traditional seafaring and canoe building practices present throughout maritime Southeast Asia and Oceania, which resulted in the initial settling of Polynesia. The relationship with the ocean and the unique movement on boards in surf zones embodied by the aforementioned Hawaiian princes, Freeth, and Kahanamoku are indicative of an Indigenous epistemology associated with *he'e nalu* and traditional Austronesian seafaring cultures embodied by grandmaster navigator Mau Piailug and his sons. Recognizing the function of riding waves in historical contexts reveals some of a much larger body of Indigenous knowledge. It also marks a shared territory of obfuscation and functions as forms of social advocacy within contemporary contexts. From the Hawaiian Islands' precolonial contact, I turned to the symbolic transformations involved as *he'e nalu* was interpreted as surfing. I pay particular attention to the ways in which notions of political, cultural, and social identity are appropriated, produced, and maintained in this process.

Such tensions within the history of surfing discourse are felt and experienced in the contemporary practices of surfing around the globe as the *ka po'ina nalu* continues to function as a borderland. This space is then transferred to the identities of the practitioners themselves through material consequentiality. The practice of riding waves functions as a very visible and popular form of cultural representation and for its ability to mark loss, a form of advocacy. Ultimately, we find in surfing a dialectical tension that reflects forces of cultural repression and forms of Indigenous wisdom ingrained in the practice of riding waves on boards.

References

Anzaldúa, G. (1985). *Borderlands/La Frontera: The new mestiza* (4th ed.). Aunt Lute Books.
Augé, M. (1995). *Non-places: Introduction to an anthropology of supermodernity.* (John Howe, Trans.). Verso.

Bourdieu, P. (1990). *The logic of practice*. Polity Press.
Certeau, Michel de. The practice of everyday life. Trans. Steven Rendall. : U of California P, 1984.
Clark, J. R. K. (2011). *Hawaii surfing: Traditions from the past*. U of Hawai'i P.
Comer, K. (2010). *Surfer girls in the New World order*. Duke UP.
Cook-Lynn, E. (2012). *A separate country: Postcoloniality and American Indian nations*. Texas Tech UP.
Cresswell, T. (2004). *Place: A short introduction*. Wiley Blackwell.
Dawson, K. (2017). Surfing beyond racial and colonial imperatives in early modern Atlantic Africa and Oceania. In D. Z. Hough-Snee & A. S. Eastman (Eds.), *The critical surf studies reader* (pp. 135–154). Duke UP.
Diaz, Vincent M. Critical Indigenous Studies. "In the Wake of Matāʻpang's Canoe: The Cultural and Political Possibilities of Indigenous Discursive Flourish.": U of Arizona P, 2016. 119–138.
Fanon, F. (1963). *The wretched of the earth*. (Constance Farrington, Trans.). Grove Press.
Fine, H. H. (2018). *Surfing, street skateboarding, performance, and space: On board motility*. Lexington.
Finney, B. R., & Houston, J. D. (1966). *Surfing: The sport of Hawaiian kings*. Charles E. Tuttle.
Giroux, H. A. (2021). *Race, politics, and pandemic pedagogy: Education in a time of crisis*. Bloomsbury Academic.
Hauʻofa, Epeli. (2008). *We are the ocean: Selected works*. U of Hawaiʻi.
Hough-Snee Zavalza, Dexter and Alexander Sotelo Eastman (2017). "Introduction." The Critical Surf Studies Reader. Ed. Dexter. Durham: Duke UP.
Ingersoll, K. A. (2016). *Waves of knowing: A seascape epistemology*. Duke UP.
Kenvin, R. (2014). *Surf craft: Design and the culture of board riding*. MIT Press.
Laderman, S. (2014). *Empire in waves: A political history of surfing*. U of California P.
Lefebvre, H. (2007). *The production of space*. Blackwell.
Lewis, D. (1972). *We, the navigators: The ancient art of landfinding in the Pacific*. UP Hawaiʻi.
London, J. (2017). *The cruise of the Snark*. Seawolf Press.
Low, S. (2013). *Hawaiki rising: Hōkūleʻa, Nainoa Thompson, and the Hawaiian renaissance*. Island Heritage.
McGloin, C. (2017). Indigenous surfing: Pedagogy, pleasure, and decolonial practice. In D. Z. Hough-Snee & A. S. Eastman (Eds.), *The critical surf studies reader* (pp. 196–213). Duke UP.
Modesti, S. (2008). Home sweet home: Tattoo Parlors as postmodern spaces of agency. *Western Journal of Communication, 72*(3), 197–212.
Moser, P. (2022). *Surf and rescue: George Freet and the birth of California Beach culture*. U of Illinois P.
Muñoz, M. (2019). River as lifeblood, river as border: The irreconcilable discrepancies of colonial occupation from/with/on/of the Frontera. *PUB PRESS*, 62–81.
Statham, E., Jr., & Robert. (2002). *Colonial constitutionalism: The tyranny of United States' offshore territorial policy and relations*. Lexington.
Phelps, W., & (Dir.). (1987). *The north shore*. Universal Pictures.
Pukui, M. K., & Elbert, S. H. (1986). *Hawaiian dictionary: Hawaiian-English, English-Hawaiian*. U of Hawaiʻi P.
Tuan, Y.-F. (2007). *Space and place: The perspective of experience*. U of Minnesota P.
Walker, I. H. (2011). *Waves of resistance: Surfing and history in twentieth-century Hawaiʻi*. U of Hawaiʻi P.
Walker, I. H. (2017). Kai Ea: Rising waves of national and ethnic Hawaiian identities. In D. Z. Hough-Snee & A. S. Eastman (Eds.), *The critical surf studies reader* (pp. 62–83). Duke UP.
Warshaw, M. (2010). *The history of surfing*. Chronicle Books.
Wheaton, B. (2017). Space invaders in Surfing's white tribe: Exploring surfing, race, and identity. In *The critical surf studies reader* (pp. 177–195). Duke University Press.
Winchester, S. (2015). *Pacific: Silicon chips and surfboards, coral reefs, and atom bombs, brutal dictators and fading empires*. Harper.

Chapter 3
Why Are There Waves?

Alessandro Toffoli

3.1 Introduction

Waves are rhythmic oscillations, or disturbances, of the water surface that can be observed in any water body, from rivers, lakes, seas, to oceans. Ancient mariners and coastal dwellers often attributed them to the whims of gods or mystical forces, as vividly depicted in Homer's epic poem "The Odyssey." In reality, waves exist because an initial state (e.g., the still water surface) is perturbed by an external disturbance and restored by a compensating force (Toffoli & Bitner-Gregersen, 2017). For the waves that most surfers ride, wind is the primary perturbation, though for the more adventurous individuals, those waves formed by gravitation interactions of the Earth, Moon, and Sun (i.e., tides) can also break in the right circumstances and therefore be ridden. Normally, these forces are compensated by gravity. However, surface tension plays a more important role for very short waves (which we will learn to call capillary waves), while very long waves (tides) tend to be restored primarily by Coriolis force (which is related to the Earth's spin). The different origins and nature of surface oscillations affect the form of a wave in terms of its height and length, and this results in the large variety of waves found on the coast.

Waves are a key research topic in earth science and engineering. They are a considerable source of stress for any structures in the sea such as ships, offshore platforms, pipelines, moorings, renewable energy installations, and coastal defenses (e.g., Faltinsen, 1993; Reeve et al., 2018; Bitner-Gregersen and Toffoli, 2014). Furthermore, motion induced by waves below the ocean surface induces turbulence throughout the water column, contributing to expanding or contracting water layers at depth, thereby regulating biological processes in the upper ocean, exchanging

---

A. Toffoli (✉)
The University of Melbourne, Melbourne, VIC, Australia
e-mail: alessandro.toffoli@unimelb.edu.au

© The Author(s), under exclusive license to Springer Nature Switzerland AG 2025
D. M. Kennedy (ed.), *The Science and Culture of Surfing*,
https://doi.org/10.1007/978-3-031-80979-8_3

heat and gases at the air-sea interface, and, hence, controlling the global climate (Babanin, 2011; Csanady, 2001). In coastal regions, waves also regulate erosion, transport, and deposition of sediment (Reeve et al., 2018; Chaps. 4 and 5).

For surfers, waves are not just a natural phenomenon, but the essence of their sport. The Polynesians were among the first to harness this energy, developing the art of surfing over a thousand years ago (see, e.g., Finney et al., 2007; Chap. 2). Understanding why waves exist and how they form is crucial for anyone interested in surfing, as it not only enhances the experience but also underscores the profound connection between riding a wave and the environment.

This chapter provides a comprehensive overview of ocean waves, covering the basic science of their generation, propagation, and transformation as they approach the coastline and eventually become the breaking waves that surfers ride. These principles are important as they underlie, and are the engine of, all surf forecasting apps. Finally, this chapter explores the global wave patterns and how climate change will affect these into the future.

3.2 Basic Wave Properties

Water waves are fluctuations of the water surface. In their simplest form, waves are sinusoidal in shape (see Fig. 3.1), and this form is used to define their most basic properties.

3.2.1 The Basic Wave Form and Related Properties

At a specific point in space x and instant in time t, the position of the water surface is described by its elevation $\eta(x, t)$ relative to a reference level (the mean water level). A wave (or a wave cycle) is the oscillatory profile of this elevation between two successive upward crossings of the mean surface level (Fig. 3.1). Likewise, a

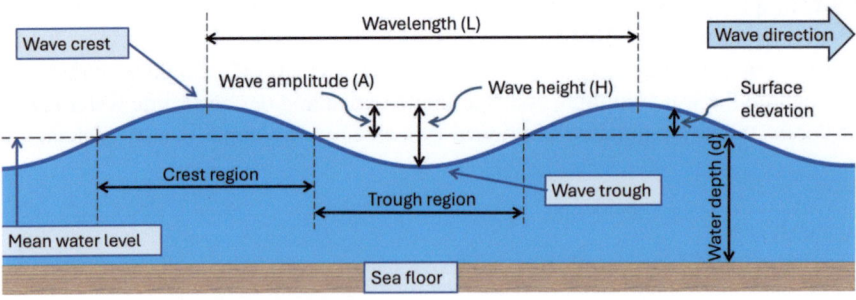

Fig. 3.1 The basic wave form showing each of the components used to describe its shape

3 Why Are There Waves?

wave can also be identified as the profile between two consecutive downward crossings of the mean surface level. The distance of the reference level to the sea floor is the water depth d. Essentially, think of a wave as an "S" shape, stretched and lying on its side!

The portion of the oscillation above the mean water level is defined as the crest region, and its highest elevation is known as the crest height. The portion below the mean water level is the trough region, and its lowest depression is the trough depth. The vertical distance between the lowest depression and the highest elevation of a wave is the wave height H (Fig. 3.1). Since a sinusoidal wave is vertically symmetric, crests and troughs are identical. In this framework, crest height and trough depth are also known as wave amplitude A, and the wave height can be described as $H = 2A$.

The distance over which the wave pattern repeats itself is the wavelength L. The time required for a wave to complete a full cycle is the wave period T. Typically, the wave period is expressed as the wave frequency, which is inversely proportional to the wave period.

The overall shape of the wave form can be formulated into the ratio of wave height to wavelength, known as the wave steepness ($\varepsilon = H/L$).

3.2.2 The Shape of Ocean Waves

When observing ocean waves from a beach, or floating on a board, one immediately realizes that waves do not look alike and their form is not exactly sinusoidal. An example of what might be observed is shown in Fig. 3.2. Realistic ocean waves are irregular, as if shorter waves ride on top of longer ones. While this description may seem like a visual impression, it does not differ from reality. Indeed, ocean waves are a superposition, or combination, of many basic sinusoidal waves with different heights (or amplitudes), wavelengths, and directions of propagation (Holthuijsen, 2010). However, while the shape of a sinusoidal wave passing through a specific point is constant over time, it varies in irregular sea states. Therefore, basic

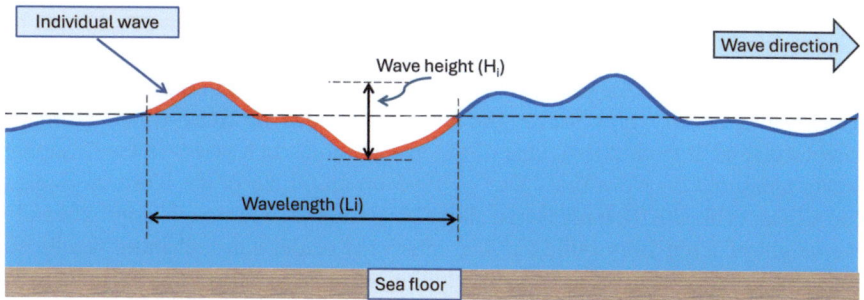

Fig. 3.2 The basic wave form which is most commonly not a perfect sinusoidal shape

properties such as height, period, and length that describe an irregular sea state have to be defined as average values over a determined period of time (e.g., 20 minutes). This is because when describing waves, we need to capture all those impacting the coast. A break therefore is often formed by the interaction of four or five different waves at the same time.

3.2.2.1 Wave Height

If one monitored waves passing through a single point over a fixed period of time, they would record a number of waves, each with its own specific height. To calculate the average (mean) height, we simply sum the total number of heights measured and divide by the number of waves observed. Mathematically, this can be expressed as follows:

$$\bar{H} = \frac{1}{N}\sum_{i=1}^{N} H_i \qquad (3.1)$$

where $\bar{H}$ is the average wave height, N is the number of waves measured, H is the wave height, and the subscript i indicates a generic (individual) wave.

To an observed or even an experienced mariner, waves always look bigger than the average. This is because we are naturally biased toward noticing large waves. Therefore, when describing waves, often the significant wave height (Hs) is used which is simply the average of the highest one-third of waves. The significant wave height is the basic parameter describing the characteristic wave height in the wave field and has become a standard product of the marine forecasts.

For certain applications, such as the design of coastal defenses (e.g., Reeve et al., 2018) or to be extra cautious when making decisions, the average of the highest one-tenth of waves is used to obtain a representative value of the largest waves. It should be noted that even this measure is still an average representation of large waves in a record and does not necessarily represent the absolute largest wave that can be encountered.

In general, the maximum wave height that can be encountered in a sea state is approximately twice that of the significant wave height ($H_{max} = 2Hs$) (Holthuijsen, 2010). For example, if one surfs waves with $Hs = 1$ m, the largest waves that can be encountered would be 2 m in height. As all these measurements deal with averages, it is certainly possible for some waves to be higher than the relationships described. These exceptionally high waves are known as rogue or freak waves (Bitnerr-Gregersen and Toffoli, 2014). One of the most famous such waves is the Draupner wave, measured on New Years Day in 1995 on an oil rig in the North Sea (e.g., McAllister et al., 2019). It measured about 26 m from crest to trough, approximately the height of a ten-story building. The interested reader can find more details on rogue waves in Dudley et al. (2019), Holthuijsen (2010), Massel (1996), Onorato et al. (2013), and Toffoli et al. (2024).

3.2.2.2 Wave Period and Wavelength

Similar to investigating height, wave period and wavelength are also expressed and analyzed as averages. Mathematically the mean of wave period is

$$\bar{T} = \frac{1}{N}\sum_{i=1}^{N} T_i \tag{3.2}$$

and, likewise, the mean wavelength is

$$\bar{L} = \frac{1}{N}\sum_{i=1}^{N} L_i \tag{3.3}$$

where T is the period, L the wavelength, and i the reading of a particular wave. Period and length, as with height, are also expressed as maximums and significant values when forecasting developing predictive models and of course for forecasting apps.

3.2.2.3 Wave Spectrum

In many areas of physics and engineering, it is common to represent the superposition of many different waves using a wave energy spectrum. This refers to how the total energy recorded at a break is comprised of many different wave types. In simple terms, it is like breaking down a piece of music into its individual notes and seeing how much energy each note contributes to the overall sound.

The superposition of waves on the water surface elevation mentioned in Sect. 3.2.2 at a specific point in space and over time can be expressed mathematically as follows:

$$\eta(x=0,t) = \sum_{i=1}^{N} A_i \sin(\omega_i\, t, \phi_i) \tag{3.4}$$

where $\omega_i = 2\pi/T_i$ is an expression of the wave frequency and ϕ_i is a generic phase associated to the generic frequency. The latter information indicates the position within the wave cycle at a specific point in space and time (i.e., it describes whether a wave is at its peak, trough, or any point in between) and can assume values between 0 and 2π. A schematic representing this superposition is shown in Fig. 3.3.

The energy associated with each individual sinusoidal wave is proportional to its amplitude. The energy spectrum (ω) represents the distribution of energy as a function of frequency. The typical spectrum for ocean waves displays a bell shape (see left panel in Fig. 3.3), indicating that the wave field is dominated by a few wave components that carry most of the energy (those at the spectral peak).

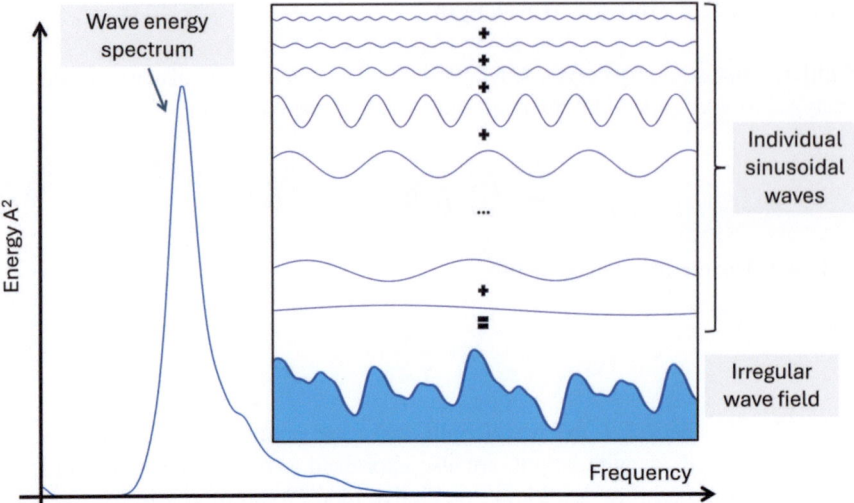

Fig. 3.3 Schematic of the superposition of various individual waves with amplitude A_i and frequency ω_i forming an irregular wave field (right) and concurrent wave energy spectrum (left)

Height can be estimated from the energy spectrum by calculating the area underneath the curve. Mathematically, the significant wave height can be expressed as four times the square root of the variance m_0 of the spectrum:

$$H_s = 4\sqrt{m_0} = 4\sum S(\omega)\Delta\omega \tag{3.5}$$

where $\Delta\omega$ is the spectral bandwidth of a single frequency, which is often constant. The significant wave height extracted from a wave spectrum according to Eq. 3.5 is approximately 5% higher than one obtained as the average of the highest one-third of the waves in a record.

The relationship above is the core component for marine forecasting, and, hence, it is of critical importance for surfers.

3.3 Types of Ocean Waves

Numerous types of waves exist on the ocean surface, from tides and tsunamis to wind-generated waves. If we were to view all wave types in the ocean, the energy spectra would be dominated by wind waves—the ones we surf (Fig. 3.4). Understanding these distinct types is essential for understanding oceanic dynamics. Here, a brief overview is presented. For more details, the interested reader is referred to Toffoli and Bitner-Gregersen (2017).

Tides are the longest waves on Earth, generated by the gravitational interactions among the Earth, Moon, and Sun. Their periods range from a few hours to over a

3 Why Are There Waves?

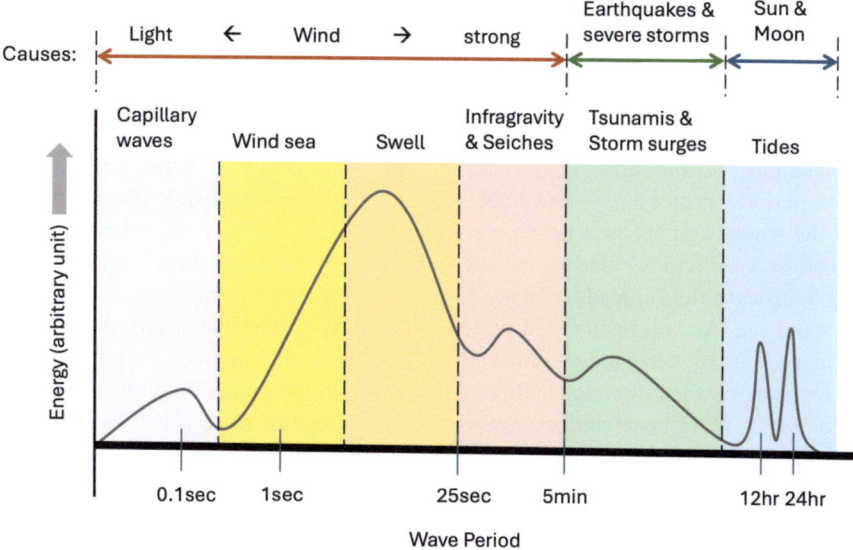

Fig. 3.4 Diagram illustrating types of surface waves, their causes, and the amount of energy they contain

day, and their wavelengths span from a few hundred to a few thousand kilometers. Due to their significant length, tides do not exhibit a typical oscillatory profile observable to coastal observers. Instead, they are recognized by the rise and fall of water levels. The peak of the tide is commonly referred to as high tide, while the lowest point is known as low tide.

Storm surges have slightly shorter wavelengths than tides and typically occur during severe storms characterized by low atmospheric pressure and strong winds. These conditions disturb the ocean surface over vast areas, spanning hundreds of kilometers and lasting for 1 or 2 days. Similar to tides, storm surges manifest as variations in water level. They can occur rapidly and pose a significant risk of severe coastal flooding, especially when they coincide with high tides.

Tsunamis are waves generated by any massive disturbance to the ocean—most commonly this is caused by underwater landslides, earthquakes, or volcanic eruptions. Despite being shorter than storm surges, these waves are still extremely long with a small amplitude and are barely noticeable in deep ocean waters. However, tsunamis can cause extensive damage upon reaching coastal areas, where their wave height rapidly increases (see Sect. 3.5).

Seiches are standing waves that resonate within a basin, such as harbors or bays, matching the natural frequency of the basin. They are typically triggered by strong winds that push water from one end of the basin to the other. When the wind subsides, the water rebounds to the opposite side of the enclosed area, inducing an oscillation that persists back and forth for hours or even days.

Infragravity waves are generated by groups of wind-generated waves, characterized by periods of a few minutes and very small amplitudes. Despite their low energy content, infragravity waves pose challenges in ports by causing excessive motion for moored ships. They are also very important for distributing wind-wave energy on beaches leading to the development of rips (Chap. 5).

The most visible form of ocean waves is the wind-generated wave. These waves have periods shorter than 30 seconds (or wavelengths shorter than 1500 m)—nearly all the waves you see at a beach are wind-generated waves. Being directly influenced by local winds, wind-generated waves are irregular and short crested, meaning their wave field spreads in many directions. These waves are commonly known as wind sea. As wind-generated waves move away from their origin, they are no longer forced by wind and organize into more regular and long-crested patterns that move along distinct directions. This transition marks the transformation of wind sea into swell. Wind-generated waves with periods shorter than 1/4 second (wavelengths shorter than about 10 cm) are affected more by surface tension than gravity and are called capillary waves.

3.4 Observation Techniques

Measurement of waves is essential. Even though models can provide some of the answers without data on what is actually happening at a particular place, it is impossible to know what is reality unless it is directly measured. There are many ways to measure waves. Two broad techniques are used, namely, in situ, where the wave is measured at its location, or remotely where a wave is measured from afar (such as from space).

3.4.1 In Situ Techniques

In situ instruments are deployed either at or below the sea surface. These instruments primarily measure time records of the vertical motion of the water surface at specific locations. That is, they don't directly measure the height but measure some other physical factor (such as water pressure or velocity) that is affected, or caused, by changes in wave height.

Wave buoys (Fig. 3.5a) are the most common in situ devices, designed to closely track water particle motion by floating on the surface. The basic method involves measuring vertical acceleration with an onboard accelerometer. These buoys typically feature radio and satellite communication to transmit data to onshore receiving stations. Techniques have also been developed to derive wave direction through measurements of the sea surface slope or the buoy's horizontal motion.

Wave poles are installed on offshore platforms and consist of suspending wires vertically from above the water surface to below it. They measure water surface

Fig. 3.5 (**a**) Wave-rider buoys (SOFAR Spotter types) being prepared for deployment as part of the Vic Waves monitoring network (vicwaves.com.au), (**b**) a pressure sensor (RBR Solo) deployed at 10 m depth off Australia's Victorian coast to investigate wave hazards. (Photos: David M. Kennedy)

movement along the wire, often by measuring wire length using electrical resistance or capacitance techniques. However, these methods do not provide directional wave information unless multiple poles are used in an array configuration, such as three poles forming a triangle to estimate surface slope (Young, 1994).

Other in situ techniques include instruments installed below the water surface, such as the inverted echo sounder, pressure transducers, and current meters (Holthuijsen, 2010). The inverted echo sounder measures water surface position with an upward-looking sonic beam from beneath the sea. Pressure transducers (Fig. 3.5b) detect wave-induced pressure fluctuations at depth to estimate wave characteristics. Current meters, also deployed below the surface, measure wave-induced orbital motion, from which wave parameters can be extrapolated.

3.4.2 Remote Sensing Techniques

Remote sensing techniques allow for the measurement of the ocean surface without direct contact with water, using instruments on towers, ships, airplanes, or satellites. These techniques involve measuring reflections from the sea surface of visible and infrared light, or radar energy. Unlike in situ methods, remote sensing can cover large areas almost instantaneously, especially when using satellites.

One established method involves monitoring the ocean surface using stereophotography (e.g., Alberello et al., 2022; Benetazzo, 2006; Toffoli et al., 2024). This technique reconstructs three-dimensional images of the ocean surface by synchronizing high-quality cameras to capture overlapping images of waves at regular intervals. Differences between the 3D images are used to calculate wave profiles (for technical details, see Benetazzo, 2006).

Conventional ship radar, typically used to detect solid obstacles, can also be adapted to detect reflections from softer surfaces such as waves (Derkani et al., 2021). These radar reflections are usually considered unwanted echoes. The radar's wavelength, often in the centimeter range, primarily reflects very short water waves like capillary waves, modulated by longer waves. This pattern appears on radar screens as images of longer waves from which information on individual waves as well as the wave spectrum can be inferred.

Airborne and satellite-mounted radars, known as synthetic aperture radar (SAR), compensate for altitude errors by emitting specific programmed signals. SAR images provide statistical characteristics of the surface rather than directly measuring elevation. SAR is limited to detecting waves longer than 150–200 m and so therefore is primarily used to observe swell (Aouf et al., 2021; Collard et al., 2009).

Laser altimetry uses visible or infrared light, employing a downward looking laser device to measure vertical distance to the sea surface. This technique is suitable for fixed platforms and airplanes but is less practical for satellites due to interference caused by weather. To address this issue, satellite altimetry (Ribal & Young, 2020) uses a radar beam to observe a footprint of the ocean surface several kilometers in diameter. This technique is not accurate enough to measure an individual wave, but it does detect the overall surface roughness of the ocean caused by waves. This roughness measurement correlates with the significant wave height.

3.5 Wind Waves: Generation, Dispersion, and Transformation

This section focuses on the types of ocean waves generated by wind, the ones that are sought after most by surfers. It details the basic principles of wind forcing, the mechanisms of wave generation, the theory explaining propagation and dispersion across the ocean, and the transformation of waves as they approach the coastline.

3.5.1 Wind Forcing and Generation of Waves

Wind is complex and varies randomly in space and time. To simplify calculations and make it more manageable for scientific study, wind is often reduced to one horizontal component and averaged over a set time (typically 10 minutes) at a fixed elevation (usually 10 m) above the mean sea surface. However, this fixed elevation is more a convention than a practical reality, as instruments are often installed at

3 Why Are There Waves?

different heights due to structural constraints. Thus, wind measurements at 10 m are often estimated from measurements at other levels using assumed vertical wind profiles.

The connection between wind and ocean waves is obvious to anyone sitting on a board in the sea. Interestingly the precise details of how energy is transferred from wind to waves are still not fully understood, although current models are highly reliable (e.g., Young, 1999).

One model for initial wave generation (starting from a flat water surface) involves the instability of the water surface layer where the wind generates a current (Lamb, 1945). Two fluids moving at different speeds (air and water) create instabilities at their interface if their densities and speeds differ sufficiently. This is evident when wind first blows over still water, creating small, short capillary waves that gradually grow longer and higher. Another model, proposed by Phillips (1957), suggests that waves are generated by resonance between wind-induced pressure waves (air pressure) and freely propagating water waves. Miles (1957) further developed this idea, showing that initial waves modify airflow and wind-induced pressure at the water surface to enhance their own growth.

Miles' theoretical model indicates that air pressure at the water surface reaches a maximum on the side of the wave crest facing into the wind and a minimum on the other (leeward) side. This means the wind exerts pressure on the water surface, pushing it down where the wave surface is rising (windward side of the crest) and lifting it up where the surface is descending (leeward side); see a schematic of this process in Fig. 3.6. As the wave grows, this energy transfer becomes more effective, allowing the wave to grow faster.

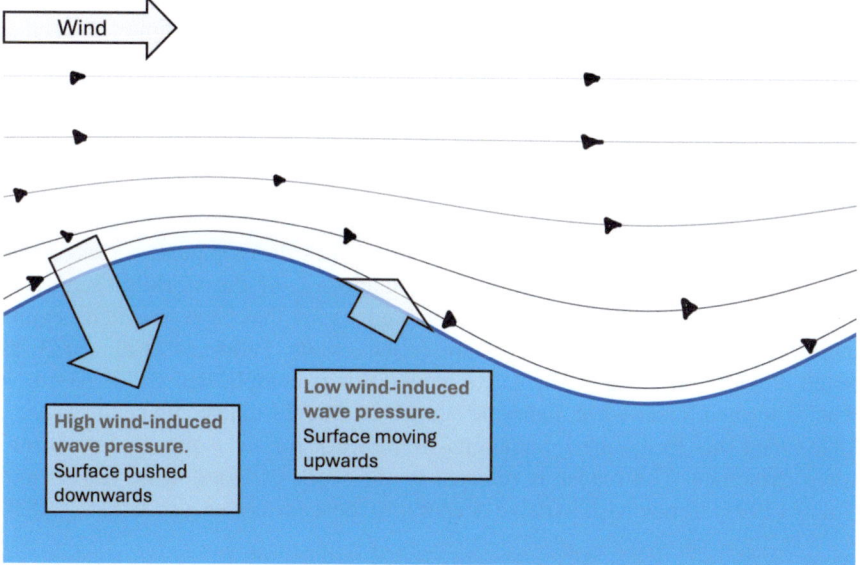

Fig. 3.6 The wave-induced wind-pressure variation over a propagating harmonic wave

It should be noted that measuring wave-induced variations in wind-induced air pressure at the wave surface to validate this theory is very difficult, leaving a significant degree of uncertainty. This uncertainty highlights the ongoing need for further research and the potential for alternative or supplementary theories.

3.5.2 Wave Height and Period as a Function of Wind and Fetch

Can one estimate wave parameters knowing wind speed? To answer this, it's important to recognize that wind speed alone may not be sufficient. Observations indicate that several factors influence wave growth besides wind speed (e.g. Toffoli et al., 2012). These include gustiness, wind direction, water depth, and the shape of the coast. One commonly examined idealized scenario is fetch-limited growth. Put simply, the fetch is the distance of water that the wind blows over: the longer the fetch, the larger the waves.

Observations in similar idealized fetch-limited conditions show that waves grow rapidly in height and period (and so also in length) at short fetches (see Bretschneider, 1952; Breugem & Holthuijsen, 2007; Kahma & Calkoen, 1992; Sverdrup & Munk, 1946; and Young & Verhagen, 1996, among many others). Under these circumstances, waves are considered young. However, this growth gradually slows and eventually stops because the wave speed increases as waves grow longer (see Sect. 3.5.3) and eventually matches or surpasses the wind speed. When this occurs, waves are fully developed, meaning the wind can no longer exert positive feedback on them (Young, 1999). Observations of fully developed wave fields in the North Atlantic Ocean (Pierson Jr & Moskowitz, 1964), where fetch is infinite, reveal that significant wave height and period depend solely on wind speed.

3.5.3 Wave Dispersion in the Open Ocean

After being generated by wind and growing to full development, waves spread out, or disperse, across the ocean. More than 75% of the world's oceans are believed to be dominated by storm-generated waves that have travelled beyond their generation area (these waves are called swells, Pathirana et al., 2023, as noted in Sect. 3.3). The mathematics behind this is complicated and is based on what is termed linear wave theory developed in the mid-nineteenth century (Airy, 1845), and this forms the foundational framework for subsequent research into wave dynamics. The reader is referred to the work of Holthuijsen (2010) and Young (1999) if they wish to delve deeper into these ideas (see also Eckart, 1952).

3 Why Are There Waves?

The important aspect is that gravity is a controlling factor on dispersal as is the density of the ocean. Left to their own devices, little energy is lost from waves in the deep ocean as they disperse from their storm areas toward the various surf breaks of the globe. When a group of waves with different periods propagates from their generation area, they travel at different speeds, causing them to separate based on their wavelengths. This dispersion process is pivotal in transforming irregular storm waves into the more regular swells observed during their travel across the ocean (e.g., Pathirana et al., 2023).

As waves disperse, they often encounter swells generated from storms in other parts of the world. Due to their different speeds, these two waves can reinforce each other at one moment when their crests coincide and cancel each other at another moment when one wave's crest coincides with the other's trough. Over time or space, this interference forms a wave group (see Fig. 3.7). The velocity of this group can be calculated as the ratio of the differences in frequencies and wavenumbers.

Additionally, it is important to note that the speed varies with water depth: in general, waves propagate faster in deeper water compared to shallower water. In deep water, as a wave passes, the water particles move in a circular pattern—these circular motions become smaller with depth. Interestingly, the wave induces a forward particle motion at the top of the wave crests and a backward motion at the bottom of the troughs. Therefore, a surfer waiting for the perfect wave would rhythmically move forward and then backward as waves pass by. Once a wave moves into shallow water, the orbital motion of the water interacts with the seabed and in doing so starts to transfer energy to the ocean bottom. This can be visualized as a squashing of the orbital motions. This means the wave transforms from a vehicle of energy transfer to one of erosive action (Fig. 3.8).

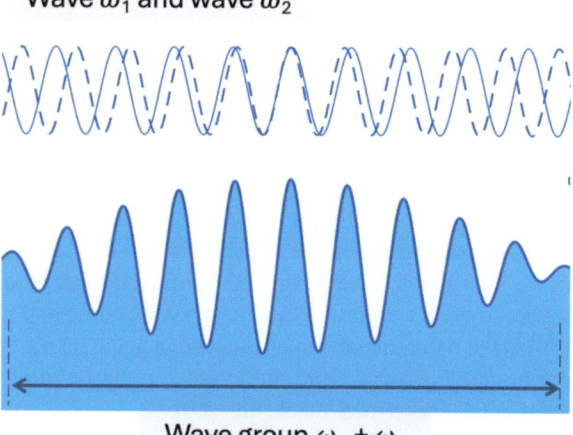

Fig. 3.7 Two sinusoidal waves with slightly different wavenumber (frequencies) forming wave groups

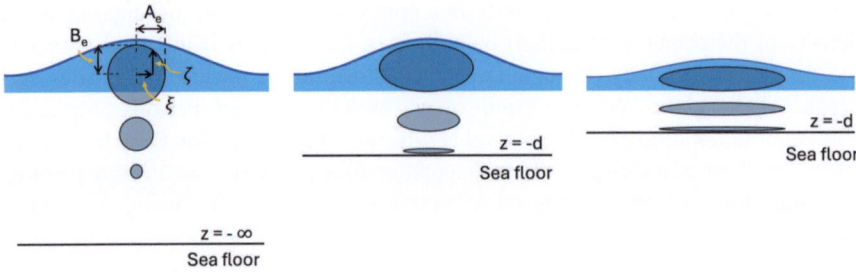

Fig. 3.8 The orbital motion in deep water, intermediate-depth water, and very shallow water

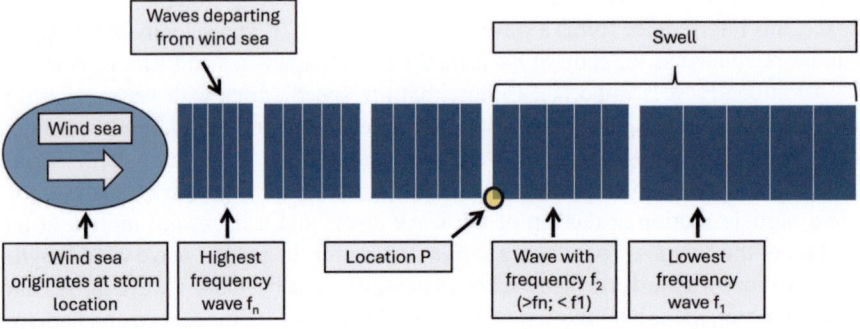

Fig. 3.9 A schematic of waves of various frequencies departing from a storm and transformation of wind sea in swell

3.5.4 Wind Sea and Swells

With a better understanding of wave generation and propagation, the concepts of wind sea and swell, previously introduced in Sect. 3.3, can be revisited. Under the effect of local wind (such as during a storm), waves begin to generate and gradually grow, until they reach full development. During this growth phase, the wave field is referred to as wind sea. According to linear theory applicable in deep water, the speed at which wave energy propagates depends on the frequency of waves. Thus, low-frequency waves (long waves) travel faster than high frequency waves (short waves). Consequently, the irregular wave field generated during a storm breaks down as it moves away from the storm. This breakdown results in fields of more regular waves, with low-frequency waves leading and high frequency waves trailing (this is the wave dispersion introduced in Sect. 3.5.3). Waves that disperse across the ocean, moving away from the initial wave generation region and the influence of local wind, are termed swell (Fig. 3.9).

A schematic of this process is depicted in Fig. 3.9. Initially, a storm forms somewhere in the ocean with a predominant wind direction, generating waves (wind sea). Outside the storm region, it is assumed there is no wind. The various components of the wave field, each traveling at the group velocity specific to their own frequency, depart the storm, with the fastest traveling waves (lowest frequency; f_1 in Fig. 3.9) leading the propagation. Consequently, the initial wave field breaks down into numerous individual wave fields due to variations in propagation speed.

Over time, the leading low-frequency wave component reaches a generic location *P* in the ocean. Subsequently, higher frequency components (e.g., f_2 in Fig. 3.9) follow, passing through the same location *P* but at different times. At any given moment at location *P*, only wave components within a certain frequency range (e.g., between f_1 and f_2) are present. The superposition of these few components forms the swell wave, which has separated from the wind sea. The swell is more regular and longer and carries less energy than the wind field from which it originated.

It should be noted that the wind has a three-dimensional nature, scattering waves in many directions. Consequently, it is reasonable to expect that waves would also disperse across multiple directions (Holthuijsen, 2010). Therefore, multiple swells can potentially originate from a single storm.

Surfers prefer to ride swells because their longer, more regular pattern is more predictable than a choppy wind sea and the break is smooth and uniform. This consistency allows surfers to better anticipate the wave's behavior, providing ideal conditions for catching waves, performing maneuvers, and enjoying a more satisfying and controlled surfing experience.

3.5.5 Transformations in the Nearshore

The following section discusses wave propagation over a sloping beach and the concurrent transformations due to changes in water depth. As waves enter shallow water, different parts of the waves find themselves in different water depths as the shoreline is reached (Fig. 3.10). The effect is the speed of the wave starts to differ along its length depending on water depth. The change in speed has important consequences on the wave profile and drives transformation processes nearshore. The three most obvious effects are wave shoaling, which is the change in shape, wave refraction, which is the change in direction, and wave breaking, which is the partial disintegration of the waves due to an excessive steepening of the wave form.

3.5.5.1 Wave Shoaling

As waves propagate into shallower water, both the phase speed and the group velocity decrease. Since the group velocity represents the speed at which energy propagates, variations in it lead to changes in local wave energy and, consequently, changes in wave height. For example, consider a wave propagating through shallow water toward a straight coastline (with parallel bottom contours) at normal incidence (perpendicular to the coastline; see the left panel in Fig. 3.10). In the absence of any generation or dissipation mechanism, the conservation of energy must be satisfied. This implies that an identical amount of wave energy enters and exits this

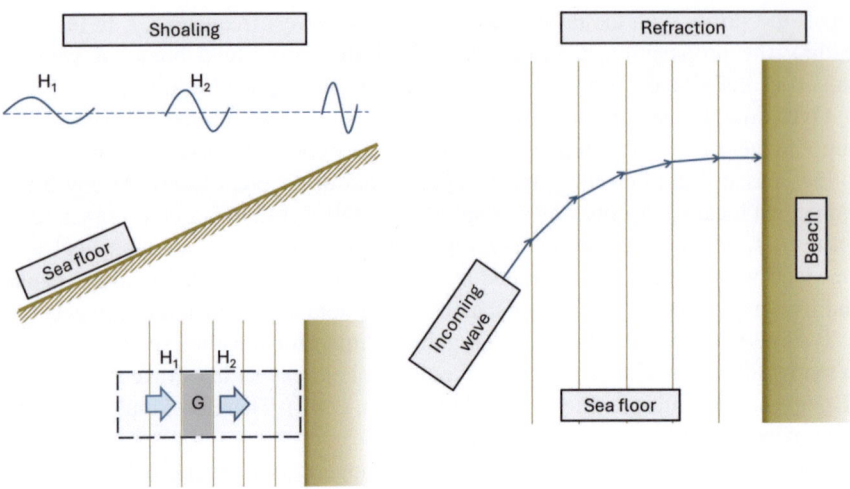

Fig. 3.10 Definition sketch showing the shoaling of water waves (left panel) and the refraction of waves over parallel depth contours (right panel)

volume through its up-wave and down-wave boundaries (assumed perpendicular to the wave direction).

The effect of shoaling is initially to decrease the wave height but then to increase it as the waves propagate further toward the shore. This effect may be referred to as energy bunching, which is the horizontal compacting of energy similar to a traffic jam when cars slow down. This is when a wave breaks—the wave form collapses and dissipates part of its energy as its profile becomes excessively steep, i.e., wave height increases and wavelengths decrease significantly.

3.5.5.2 Refraction

If waves approach the coast at an arbitrary angle, the crest line is not aligned with the seafloor contours (oblique incidence; right panel in Fig. 3.10). As a result, the part of the wave crest in deep water moves faster than the part in more shallow water, covering a greater distance over the same time interval. This causes the wave to turn toward the shallower water, i.e., toward the coast. This is why a person standing on the beach observes waves with their crest lines parallel to the beach.

3.5.6 Wave Breaking

As waves approach the beach, their heights increase due to refraction, and wavelength decreases due to reduction of speed, in response to the shallowing of the water. Consequently, the wave form becomes steeper and eventually

3 Why Are There Waves?

Fig. 3.11 Types of breaking waves

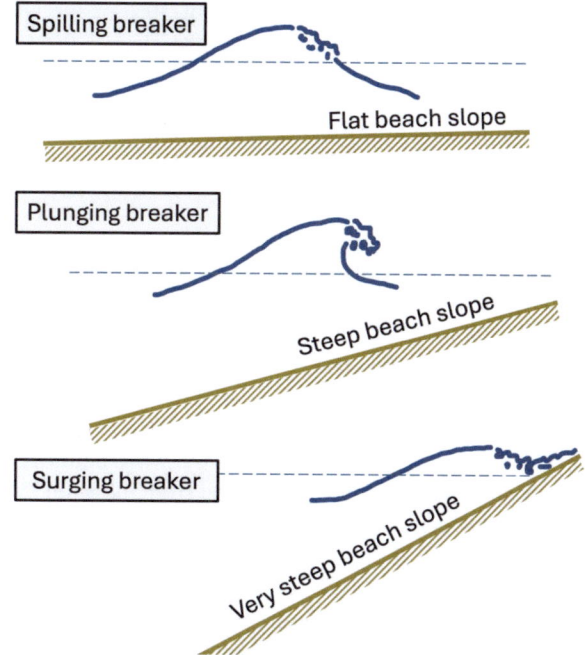

collapses (breaks) when the steepness reaches its maximum limit (Toffoli et al., 2010). This process remains poorly understood from a scientific viewpoint as steep waves begin to behave differently than linear theory predicts. Although there are numerical models capable of accurately modeling wave breaking by tracking individual water particles, depth-induced breaking conditions are typically identified using simplified empirical equations. The basic principle behind these empirical formulations is that breaking limits the wave height in shallow water. On average, waves break in a depth of 78% (range of 50–150%) of their height.

Wave breaking can take various forms depending on wave and beach characteristics (Fig. 3.11).

- Spilling: The crest of the wave gradually spills down the front face of the wave, creating a frothy, turbulent water surface. It occurs on gentle slopes where waves lose energy slowly.
- Plunging: The crest of the wave curls over and crashes down forcefully, creating a hollow, tubelike shape. It occurs on steeper slopes where waves lose energy rapidly.
- Surging: The wave does not spill or plunge but instead surges up the beach with a smooth, sliding motion. It occurs on very steep slopes, where the wave's energy is dissipated gradually as it moves up the shore.

The wave breaking event is followed by a dissipation of wave energy. Although quantifying the exact amount of energy lost during breaking is challenging (cf. Holthuijsen, 2010; Young, 1999), numerical simulations indicate approximate values: spilling waves dissipate about 30% of the energy, plunging waves about 60%, and surging waves more than 80% (Iafrati, 2011).

3.6 Wave Forecasting

Having understood the basics of wave generation and dispersion, as well as the concept of the wave spectrum, the focus now shifts to wave forecasting. This discipline involves predicting future conditions of the ocean surface, including parameters such as wave height, period, and direction. Accurate wave forecasts are critical for various maritime activities, including navigation, coastal protection, and recreational pursuits like surfing.

At the core of wave forecasting lies the energy balance equation, a fundamental tool that calculates the evolution of the wave field over space and time. This equation accounts for the energy received from the wind, energy transfer across various wave components through complex nonlinear processes, and energy loss due to wave breaking and bottom friction. While a detailed discussion of the balance equation exceeds the scope of this chapter, basic information is provided here. For more in-depth mathematical details, refer to Holthuijsen (2010) and Young (1994).

The general form of the balance equation indicates that variations in the wave energy spectrum over space and time must be balanced by external energy input, internal energy transfer, and dissipation. It can be expressed as follows:

$$\frac{\partial S}{\partial t} + \nabla \cdot \left(C_g \, E \right) = S_{in} + S_{nl} + S_{ds} \qquad (3.6)$$

where $\partial S/\partial t$ represents the rate of change of the wave spectrum S in time; $\nabla \cdot (C_g E)$ represents the propagation of energy in geographic space based on the group velocity C_g; S_{in} is the energy input term from wind; S_{nl} is the nonlinear interaction term, accounting for energy exchanges between waves; and S_{ds} represents the dissipation term, which includes energy losses due to wave breaking and friction.

As discussed in Sect. 3.5, wind is the primary source of energy for waves. As the wind blows across the ocean surface, it transfers energy to the water, generating waves. The efficiency of this energy transfer depends on several factors, including wind speed, the duration of the wind blowing over the water, and the distance over which it blows, known as the fetch. Stronger winds blowing for longer durations over a greater fetch result in larger waves.

The energy transfer refers to an interaction process between wave components. This regulates the transfer energy between them through complex nonlinear interactions. This includes the exchange of energy between waves of different frequencies and directions, leading to the growth of some waves at the expense of others. Another important process is wave dispersion, where waves of different wavelengths travel at different speeds, causing them to spread out and redistribute their energy over time.

The energy loss term accounts for various mechanisms. When waves reach shallow waters or encounter obstacles, they break, releasing their energy (see Sect. 3.5.6). In shallow water, waves can lose energy due to frictional interactions with the ocean floor. Additionally, in open waters (the deep ocean), waves can undergo a phenomenon called whitecapping, where they become so steep that they spontaneously break, dissipating energy in the process.

3.6.1 Numerical Wave Models

To solve the balance equation and generate accurate wave forecasts, scientists rely on numerical wave models. These models simulate wave behavior by integrating data such as wind fields and applying the principles encapsulated in the balance equation. Two widely used models are WAM (Wavewatch III Atmospheric Model) and WAVEWATCH III. These models are employed by meteorological centers to predict waves across entire oceans, focusing primarily on open ocean conditions due to their ability to simulate wave dynamics over large spatial scales (e.g. Liu et al. 2023).

In coastal regions, SWAN (Simulating WAves Nearshore) is a specialized model designed to predict wave conditions near the shore. SWAN takes into account local factors such as bathymetry and coastline shape, which significantly influence wave characteristics in nearshore environments. This model plays a crucial role in coastal engineering, navigation, and recreational planning by providing detailed forecasts that account for local variations in wave behavior.

Numerical wave models integrate data from atmospheric models to provide accurate and reliable forecasts over various timescales, ranging from a few hours to several days. These forecasts are indispensable for mariners, coastal managers, and recreational users who depend on timely and precise information to plan activities safely and efficiently.

Meteorological centers also use another tool to provide a comprehensive view of past wave conditions, known as hindcast. This tool consists of a reanalysis that integrates historical short-term weather forecasts with observational data using advanced data assimilation techniques. This process not only validates the accuracy of forecasts but also enhances our understanding of long-term wave climate trends, supporting climate research and efforts in hazard mitigation.

3.6.2 Accessing Wave Forecasts: Web Applications

With advancements in technology, wave forecasts have become accessible to everyone through various web applications. These platforms provide user-friendly interfaces to view real-time and forecasted wave conditions. One popular web app is Windy.com. Examples of wind and wave forecasts from Windy.com are shown in Fig. 3.12.

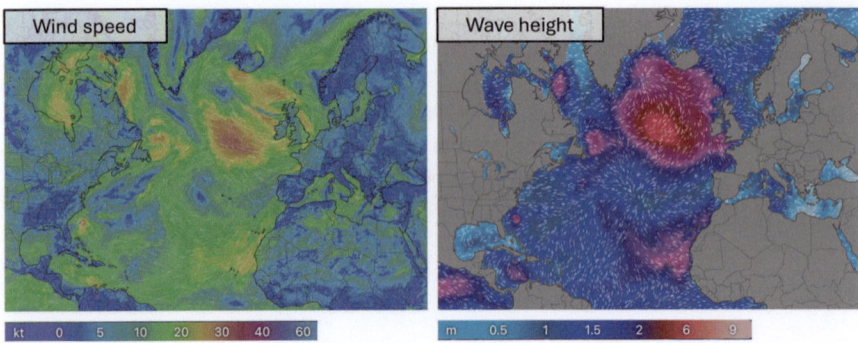

Fig. 3.12 Example of wind and wave forecasting products over the North Atlantic from the Windy.com app

This specific tool is an interactive platform that provides detailed weather and wave forecasts globally. The website features an intuitive map-based interface where users can visualize wind speed and direction as well as wave height, period, and direction among many other atmospheric and marine variables. Color-coded maps make it easy to interpret the data at a glance. Windy.com integrates data from multiple numerical wave models, including WAM, WAVEWATCH III, and SWAN, ensuring accurate and reliable forecasts. Users can customize the display to focus on specific regions or parameters, such as wind speed or sea surface temperature, making it particularly useful for sailors, surfers, and coastal managers who need tailored information. The platform provides both short-term and long-term forecasts, allowing users to plan their activities accordingly, with updates regularly reflecting the latest data.

3.7 Wave Climate

This section provides a brief overview of wave climate, which describes the long-term patterns and characteristics of significant wave height (H_s) across major ocean basins. Global distributions of significant wave height, including both mean values and extreme values represented by the 90th percentile (where only 10% of data points exceed this value), are illustrated in Fig. 3.13. Information presented here are based on the following sources: Cabral et al. (2022), Casas-Prat et al. (2024), Meucci et al. (2020), Young and Ribal (2019), and Young et al. (2020).

The Atlantic Ocean exhibits substantial variability in H_s, characterized by pronounced seasonal fluctuations and a distinct latitudinal gradient. Its values are lowest in the tropical Atlantic (2–3 m across the year) and reach their peak in the North Atlantic between Greenland and Europe in January (average of about 4 m and 90th percentile of approximately 7 m), influenced by intense storms and prevailing wind patterns. Historical trends in the northeast Atlantic show positive H_s until the 1990s

3 Why Are There Waves?

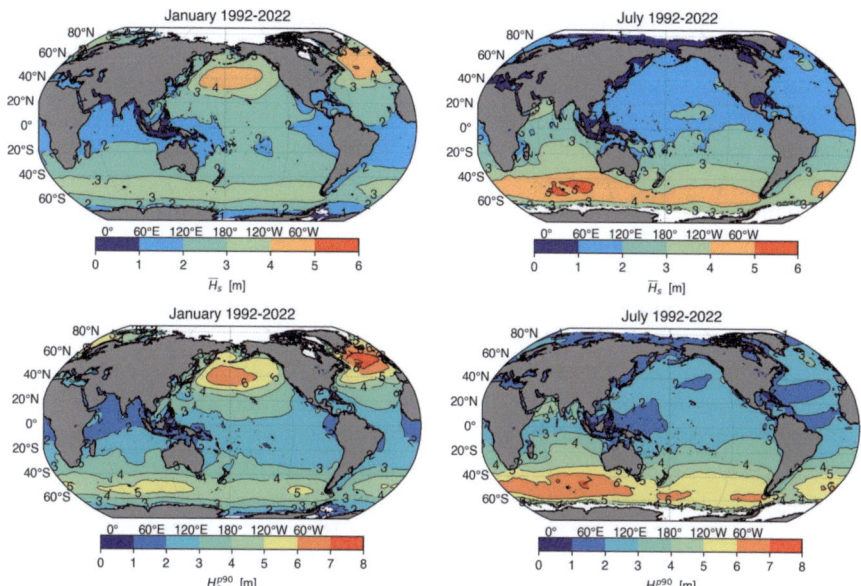

Fig. 3.13 Wave climate for the months of January and July based on a 30-year reanalysis data: average wave height (upper panels) and extreme values in the form of 90th percentile (lower panels)

with variability ranging from 0.5 to 3.4 cm, followed by no or slightly negative trends with variability of ±1 cm thereafter. Increasing trends in Hs have been observed in the tropical Atlantic since 1980, despite some evident data variability. The Southern Atlantic exhibits less consistent historical trends, except at higher latitudes where positive trends prevail.

Moving to the Pacific Ocean, its wave climate is diverse, shaped by geographical zones and atmospheric dynamics. Mid to high latitudes are dominated by extratropical storms. Wave energy peaks at the high latitudes of the North and South Pacific, with means of about 4 m and 90th percentile of about 6 m. Equatorial tropical regions display a complex wave climate influenced by Southern Ocean swells and trade winds, although means and extremes are low (2–3 m as in the Atlantic). Historical trends show varying patterns. The North Pacific has seen significant declines in annual mean H_s, although localized increases are noted in the northwest Pacific. Extreme wave events show varied trends across different regions, reflecting differences in both mean significant wave height (H_s) and the occurrence of extreme H_s values.

In the Indian Ocean, wave climatology is influenced by the monsoon winds and Southern Ocean swells. In the North Indian Ocean, annual monsoon wind reversals lead to seasonal fluctuations in mean and extreme H_s, whereas the South Indian Ocean experiences consistent westerly winds year-round, resulting in elevated mean and extreme H_s, reaching values of 5 and 7 m, respectively, in July. Southern Ocean swells propagate northward, influencing wave heights along the Indian Peninsula and surrounding coasts. Historical trends in wave characteristics show basin-wide increases in mean H_s, with stronger trends observed during specific seasons and varying regional impacts noted in areas like the Arabian Sea and Bay of Bengal.

The Arctic Ocean exhibits a wide range of wave climates across its regions. Semi-enclosed seas generally experience moderate wave heights, while the Norwegian and Greenland Seas see higher waves due to North Atlantic cyclones. Sea ice influences wave dynamics significantly, affecting wave growth and energy. Recent trends show increasing wave energy in response to declining sea ice, with mean wave heights rising by 1–3 cm per year. However, negative trends in wave heights have been observed in the Norwegian-Greenland Seas, potentially linked to reduced wind speeds associated with atmospheric changes.

Finally, the Southern Ocean, spanning from 40° to 60° south, is characterized by a robust wave climate driven by persistent westerly winds and long fetches. It boasts the highest annual mean Hs among ocean basins, with peak conditions observed between Africa and Australia. Recent data indicate a general increase in mean H_s of 1–3 cm per year from the 1980s to 2010s, influenced by strengthened westerlies and shifts in low-pressure systems. Extreme wave trends are less well understood due to data limitations but suggest heightened storm intensity and frequency, particularly in the Atlantic and Pacific sectors during the austral summer.

References

Airy, G. B. (1845). *Tides and waves*. B. Fellowes.

Alberello, A., Bennetts, L. G., Onorato, M., Vichi, M., MacHutchon, K., Eayrs, C., Ntamba, B. N., Benetazzo, A., Bergamasco, F., Nelli, F., et al. (2022). Three-dimensional imaging of waves and floes in the marginal ice zone during a cyclone. *Nature Communications, 13*(1), 4590.

Aouf, L., Hauser, D., Chapron, B., Toffoli, A., Tourain, C., & Peureux, C. (2021). New directional wave satellite observations: Towards improved wave forecasts and climate description in Southern Ocean. *Geophysical Research Letters, 48*(5), e2020GL091187.

Babanin, A. (2011). *Breaking and dissipation of ocean surface waves*. Cambridge University Press.

Benetazzo, A. (2006). Measurements of short water waves using stereo matched image sequences. *Coastal Engineering, 53*(12), 1013–1032.

Bitner-Gregersen, E. M., & Toffoli, A. (2014). Occurrence of rogue sea states and consequences for marine structures. *Ocean Dynamics, 64*, 1457–1468.

Bretschneider, C. L. (1952). The generation and decay of wind waves in deep water. *Eos, Transactions American Geophysical Union, 33*(3), 381–389.

Breugem, W. A., & Holthuijsen, L. H. (2007). Generalized shallow water wave growth from Lake George. *Journal of Waterway, Port, Coastal, and Ocean Engineering, 133*(3), 173–182.

Cabral, I. S., Young, I. R., & Toffoli, A. (2022). Long-term and seasonal variability of wind and wave extremes in the arctic ocean. *Frontiers in Marine Science, 9*.

Casas-Prat, M., Hemer, M. A., Dodet, G., Morim, J., Wang, X. L., Mori, N., Young, I., Erikson, L., Kamranzad, B., Kumar, P., Mene´ndez, M., & Feng, Y. (2024). Wind-wave climate changes and their impacts. *Nature Reviews Earth & Environment, 5*, 1–20.

Collard, F., Ardhuin, F., & Chapron, B. (2009). Monitoring and analysis of ocean swell fields from space: New methods for routine observations. *Journal of Geophysical Research: Oceans, 114*(C7).

Csanady, G. T. (2001). *Air-sea interaction: Laws and mechanisms*. Cambridge University Press.

Derkani, M. H., Alberello, A., Nelli, F., Bennetts, L. G., Hessner, K. G., MacHutchon, K., Reichert, K., Aouf, L., Khan, S. S., & Toffoli, A. (2021). Wind, waves, and surface currents in the Southern Ocean: Observations from the Antarctic circumnavigation expedition. *Earth System Science Data Discussions, 13*, 1189–1209.

Dudley, J. M., Genty, G., Mussot, A., Chabchoub, A., & Dias, F. (2019). Rogue waves and analogies in optics and oceanography. *Nature Reviews Physics, 1*(11), 675–689.

Eckart, C. (1952). 19. The propagation of gravity waves from deep to shallow water. In *Proceedings of NBS semicentennial symposium on gravity waves held at the NBS on June 18–20, 1951* (Vol. 521, p. 165). NBS.

Faltinsen, O. (1993). *Sea loads on ships and offshore structures* (Vol. 1). Cambridge University Press.

Finney, B., Howe, K., Irwin, G., Low, S., Neich, R., Salmond, A., Taonui, R., & Home, U. (2007). *Vaka moana. voyages of the ancestors: The discovery and the settlement of the pacific*. University of Hawaii Press.

Holthuijsen, L. H. (2010). *Waves in oceanic and coastal waters*. Cambridge University Press.

Iafrati, A. (2011). Energy dissipation mechanisms in wave breaking processes: Spilling and highly aerated plunging breaking events. *Journal of Geo- Physical Research: Oceans, 116*(C7).

Kahma, K. K., & Calkoen, C. J. (1992). Reconciling discrepancies in the Ob- served growth of wind-generated waves. *Journal of Physical Oceanog- Raphy, 22*(12), 1389–1405.

Lamb, H. (1945). *Hydrodynamics* (6th ed.). Dover Publications.

Liu, J., Meucci, A., Liu, Q., Babanin, A. V., Ierodiaconou, D., Xu, X., & Young, I. R. (2023). A high-resolution wave energy assessment of south-East Australia based on a 40-year hindcast. *Renewable Energy, 215*, 118943.

Massel, S. R. (1996). *Ocean surface waves: Their physics and prediction* (3rd ed.). World Scientific.

McAllister, M. L., Draycott, S., Adcock, T., Taylor, P., & Van Den Bremer, T. (2019). Laboratory recreation of the draupner wave and the role of breaking in crossing seas. *Journal of Fluid Mechanics, 860*, 767–786.

Meucci, A., Young, I. R., Hemer, M., Kirezci, E., & Ranasinghe, R. (2020). Projected 21st century changes in extreme wind-wave events. *Science Advances, 6*(24), eaaz7295.

Miles, J. W. (1957). On the generation of surface waves by shear flows. *Journal of Fluid Mechanics, 3*(2), 185–204.

Onorato, M., Residori, S., Bortolozzo, U., Montina, A., & Arecchi, F. (2013). Rogue waves and their generating mechanisms in different physical contexts. *Physics Reports, 528*(2), 47–89.

Pathirana, S., Young, I., & Meucci, A. (2023). Modelling swell propagation across the pacific. *Frontiers in Marine Science, 10*, 1187473.

Phillips, O. M. (1957). On the generation of waves by turbulent wind. *Journal of Fluid Mechanics, 2*(5), 417–445.

Pierson, W. J., Jr., & Moskowitz, L. (1964). A proposed spectral form for fully developed wind seas based on the similarity theory of S. A. Kitaigorodskii. *Journal of Geophysical Research, 69*(24), 5181–5190.

Reeve, D., Chadwick, A., & Fleming, C. (2018). *Coastal engineering: Processes, theory and design practice*. CRC Press.

Ribal, A., & Young, I. R. (2020). Calibration and cross validation of global ocean wind speed based on scatterometer observations. *Journal of Atmospheric and Oceanic Technology, 37*(2), 279–297.

Sverdrup, H. U., & Munk, W. H. (1946). Empirical and theoretical relations between wind, sea, and swell. *Eos, Transactions American Geophysical Union, 27*(6), 823–827.

Toffoli, A., & Bitner-Gregersen, E. M. (2017). Types of ocean surface waves, wave classification. In *Encyclopedia of maritime and offshore engineering* (pp. 1–8). (eds J. Carlton, P. Jukes and Y.S. Choo). John Wiley & Sons, Ltd. https://doi.org/10.1002/9781118476406.emoe077.

Toffoli, A., Babanin, A., Onorato, M., & Waseda, T. (2010). Maximum steepness of oceanic waves: Field and laboratory experiments. *Geophysical Research Letters, 37*(5).

Toffoli, A., Loffredo, L., Le Roy, P., Lefèvre, J.-M., & Babanin, A. (2012). On the variability of sea drag in finite water depth. *Journal of Geophysical Research: Oceans, 117*(C11).

Toffoli, A., Alberello, A., Clarke, H., Nelli, F., Benetazzo, A., Bergamasco, F., Ntamba, B. N., Vichi, M., & Onorato, M. (2024). Observations of rogue seas in the Southern Ocean. *Physical Review Letters, 132*, 154101.

Young, I. R. (1994). On the measurement of directional wave spectra. *Applied Ocean research, 16*(5), 283–294.

Young, I. R. (1999). *Wind generated ocean waves*. Elsevier.

Young, I. R., & Ribal, A. (2019). Multiplatform evaluation of global trends in wind speed and wave height. *Science, 364*(6440), 548–552.

Young, I. R., & Verhagen, L. (1996). The growth of fetch limited waves in water of finite depth. Part 1. Total energy and peak frequency. *Coastal Engineering, 29*(1-2), 47–78.

Young, I. R., Fontaine, E., Liu, Q., & Babanin, A. V. (2020). The wave climate of the Southern Ocean. *Journal of Physical Oceanography, 50*(5), 1417–1433.

Chapter 4
What Forms My Break?

Javier X. Leon and Tom D. Shand

4.1 Introduction

Surfing encompasses more than just the simple act of riding a wave (Chap. 2). The surfing experience is more nuanced and goes beyond just surfing a "perfect" wave, as evidenced by the mixed reception of the current artificial wave pools phenomenon (Brennan, 2024). Various surfing crafts, styles, or levels of performance are linked to particular types of waves or locations; hence, the actual meaning of surfing waves is tightly woven to the geographic concept of place. In other words, a "perfect" wave for one surfer might not be as good for another.

Broadly, the characterization of perfect waves is related to surf quality or amenity. Surf amenity includes the combination of water quality, wave dynamics, physical environment, surfer's safety, and interactions with other surfers (City of Gold Coast, 2015; Lazarow, 2007) (Chaps. 3 and 5). This chapter will only discuss surf amenity in the light of the interactions between wave dynamics and the physical environment. Furthermore, case studies, mostly geographically biased from the authors, will focus on wind-generated, long period swell waves (Chap. 3), as opposed to less common tidal bores, tsunami, glacier, ferry, or artificial pool waves, although most physical principles are shared across domains.

The scientific study of ocean waves has a long history (Chap. 3), but the detailed understanding of breaking waves through the surf zone became very significant during World War II. Knowledge of why, when, and where waves break was key for the

---

J. X. Leon (✉)
The University of the Sunshine Coast, Sunshine Coast, QLD, Australia
e-mail: jleon@usc.edu.au

T. D. Shand
The University of Auckland, Auckland, New Zealand
e-mail: t.shand@auckland.ac.nz

© The Author(s), under exclusive license to Springer Nature Switzerland AG 2025
D. M. Kennedy (ed.), *The Science and Culture of Surfing*,
https://doi.org/10.1007/978-3-031-80979-8_4

success of amphibious operations such as D-Day. The pioneering work of oceanographers Walter Munk and Harald Sverdrup was foundational for much of what we know nowadays about surf forecasting and breaking waves (Westwick, 2013).

After World War II, research continued bridging the gap between the accurate prediction of breaking waves and their interactions with the coastal zone. Programs such as the Waves Project of the University of California at Berkeley (Bascom & McCoy, 2020) and the Coastal Studies Institute at Louisiana State University (Roberts et al., 2014) focused on collecting coastal environmental data which helped advance our understanding on coastal morphodynamics, the interactions between coastal processes and morphology (Wright & Thom, 1977) . However, it was the seminal work by James "Kimo" Walker from the University of Hawaii in the early 1970s that specifically focused on answering the question: What makes surfable waves? His research on the components of natural reefs and surf parameters (Walker & Palmer, 1971; Walker, 1974; Walker & Kelly, 1973) set the scene for a new wave of surf science research trying to quantify the specific factors leading to high-quality surfing waves (Mead & Black, 2001b; Scarfe et al., 2003) and their conservation (Reineman et al., 2021; Touron-Gardic & Failler, 2022).

This chapter provides an overview of the interactions between waves and the seabed that result in the most common surfing break types. A more detailed analysis of the principal factors that make surfable waves "perfect," from preconditioning to wave breaking intensity and rate, is then presented. Finally, this chapter examines potential threats to surfing wave quality, including human interventions and climate change, and examples of actions to protect surf amenity and surfing ecosystems.

4.2 What Makes Good Surfing Waves?

Surfers are by nature scientists. Most people will look up from their books and beach balls long enough to notice that waves are big or small, tides are high or low. But surfers watch more intently, mentally calibrating their subconscious models of how subtle changes in wave dynamics and environmental factors affect the quality of waves at their local break, backing off one day, "sectiony" another, square and grinding or weak and gutless. But what happens to waves between their origin to that moment we turn and start the ride? Here, we describe wave characteristics and how they interact with environmental factors, particularly the shape of the seabed or bathymetry, to create good surfing waves.

4.2.1 Swell Characteristics

At global and regional scales, hotspots for good surfing waves share some common features including swell energy, direction, consistency, regional wind patterns, and temperature. Polynesian islands such as Hawaii, the cultural birthplace of modern surfing (Chap. 2), are a great example. Espejo et al. (2014) concluded that west-facing, low- to middle-latitude coasts, particularly those in the Southern Hemisphere, have higher probabilities of high-quality surf conditions due to the usual presence of long-crested extratropical swells and trade winds generally blowing offshore. Some of these swell wave systems with narrow spectrum and directional spread can travel uninterrupted across half the world, more than 20,000 km (Munk et al., 1963). Notable examples include the coasts of Western Africa, Indonesia, Western Australia, New Zealand, Chile, Peru, and Mexico, world-renowned surf destinations where surfing waves are big and consistent 290–320 days per year (Fig. 4.1). Tropical storm systems such as cyclones (also known as hurricanes or typhoons) and trade winds are also important swell contributors for coastal areas sheltered from westerly systems such as east-facing tropical and subtropical coasts and island archipelagos (Leon et al., 2024; Mortlock et al., 2023; Wandres et al., 2024).

Wave systems interact among each other, with most coasts having between two to four dominant swells (Mazzaretto & Menendez, 2024). Locations with dominant swell systems approaching from different directions can increase the chances of high-quality surfing waves throughout the seasons, effectively widening the swell window for a specific site. For example, the Baja California region experiences two significant swell wave systems, one generated in the South Pacific and one from the North Pacific, generally during their respective winter months.

Seasonality and interannual variability are important factors when describing the consistency of surfing waves. Globally, swells are more consistent during winter, although the Southern Ocean tends to be more consistent than the northern

Fig. 4.1 Long-crested swell waves reaching a west-faced coast, Chicama, Peru (Source: Tom Shand)

hemisphere during spring and summer as well due to the extensive fetch circling Antarctica. Interestingly, interannual climate variability (e.g., El Nino Southern Oscillation, North Atlantic Oscillation) is more important than seasonal variability for some regions, and preliminary analysis indicates a future increase in the consistency of surfing waves on west-facing coasts (Espejo et al., 2014). However, the exact impacts of increasing tropical and extratropical storminess in surfing quality, particularly in the context of climate change (Goodwin et al., 2016; Hemer et al., 2013; Wang et al., 2022), are yet to be resolved.

4.2.2 Tides, Currents, and Wind

Environmental factors such as tides, currents, and wind also play a role in surf quality. At a global scale, tides do not usually have a big influence on wave quality as areas with large tidal range (> 4 m) such as northeastern North America are not exposed to consistent, long-period swell (Espejo et al., 2014). However, at local scales, tides can impact surf quality through two mechanisms: changes in water elevation and interactions between waves and tide-induced currents.

As waves reach a depth about equivalent to their height, they break (Chap. 3). Changes in water elevation due to tidal fluctuations modulate the position and steepness of breaking waves. As water depth increases, the wave height to water depth ratio at breaking remains, and therefore position of breaking shifts closer to shore. On rapidly changing bathymetric profiles such as reefs, changes in depth can modulate wave steepness by modifying the wave height to water depth ratio at the point of breaking and thereby have significant impacts on surfing quality.

Tide-induced currents can be a dominant mechanism associated with longshore current variability under moderate wave conditions (Thornton & Kim, 1993). In areas where tidal currents are compressed, such as estuary mouths or reef passages, tide-induced currents interact with wave direction and height by inducing refraction, increasing steepness and focusing as they move into the opposing flow (Battjes, 1982). For example, wave heights at the mouth of inlets were found to increase up to 20% during ebb tides and decrease by up to 10% during flood tides (Dodet et al., 2013).

Nearshore tidal-, wind-, or wave-induced currents, such as longshore or rip currents, can be a surf hazard (Castelle et al., 2016) (Chap. 5). Longshore currents can flow at velocities between 0.8 and 2 m/s downdrift along a headland, making it very challenging for a surfer to paddle against them or keep position at the takeoff zone (Phillips et al., 2003). However, they can also work in favor of surfers trying to paddle back to the takeoff position by acting as a "conveyor belt" if flowing updrift or along topographically constrained rip channels (Chap. 5).

4 What Forms My Break?

Finally, surfing quality is often impacted by local winds. Onshore winds can cause waves to spill and break earlier in deeper water. Conversely, offshore winds tend to hold waves up, causing them to break later, in shallower waters and in a plunging form (Douglass, 1990). Hence, many surfers appreciate the presence of offshore winds or land breeze in the early morning, as opposed to the least preferred onshore or sea breeze, a common daily occurrence past midday in summer or warmer climates (Fig. 4.2).

Fig. 4.2 Contrasting wave conditions during (**a**) offshore (Photo: Agustina Montanaro on Pexels) and (**b**) onshore wind conditions (Photo: InfinitumProdux, iStock, with permission)

4.2.3 Bathymetry

Bathymetry is the most important factor determining how waves break. Bathymetry refers to the shape of the seabed which, similar to topography on land, may gently or steeply slope and contain irregular features such as ridges or channels formed by geomorphic processes such as deposition or erosion by past river channels, volcanism, and/or tectonics (Fig. 4.3). Bathymetric features within the shoreface (< ~200 m depth) are composed of sand, corals, or rock and combine in a multitude of ways to influence the way in which waves transform as they approach the shore. Waves slow down, shoal, bend, and focus on shallower bathymetry due to depth-induced refraction. Wave steepness increases as water depth becomes shallower, and eventually, waves become unstable and break (Chap. 3).

The configuration of different bathymetric features results in different patterns of wave breaking. The alongshore variation in wave height along the wave crest controls how waves break and have a considerable influence on surfing wave quality (Mead & Black, 2001a). Whether it's the consistency of reefs or the shifting dynamics of sandy bottoms, no two combinations are the same, and often different wave properties (height, period, and direction) will completely change how the bathymetry affects wave breaking; this is what makes forecasting surfing wave quality so challenging and local knowledge invaluable. These intricacies will be further discussed in Sect. 4.3.

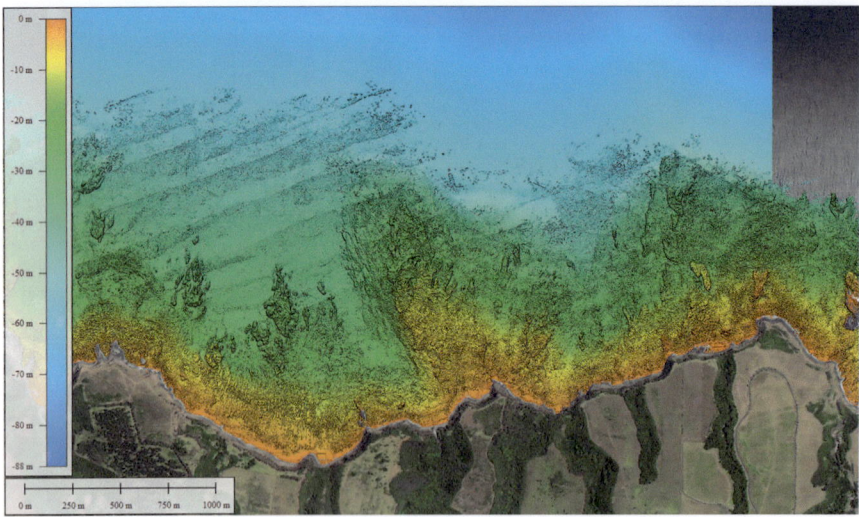

Fig. 4.3 Bathymetry of the seabed off Pe'ahi, Hawai'i. (Data: NOAA, 2020)

4.3 From Good to World Class: What Makes a "Perfect" Surfing Wave?

After debuting at the Tokyo 2020 Olympic Games, surfing made its second appearance in Paris 2024. However, instead of joining the other athletes in France, a country with good surfing waves, the surfers competed at the other side of the world in Tahiti. Why? Well, apart from the French coast resembling a lake during austral winter, the unique wave of Teahupo'o, which translates somewhat ominously as "place of skulls," is considered a world-class, "perfect" wave (Box 4.1). High-quality surfing waves, such as Teahupo'o, are the result of underlying modifications to wave shape prior to breaking. This preconditioning results in breaking surfing waves with highly sought-after characteristics including large breaking height, high face steepness, and long section length.

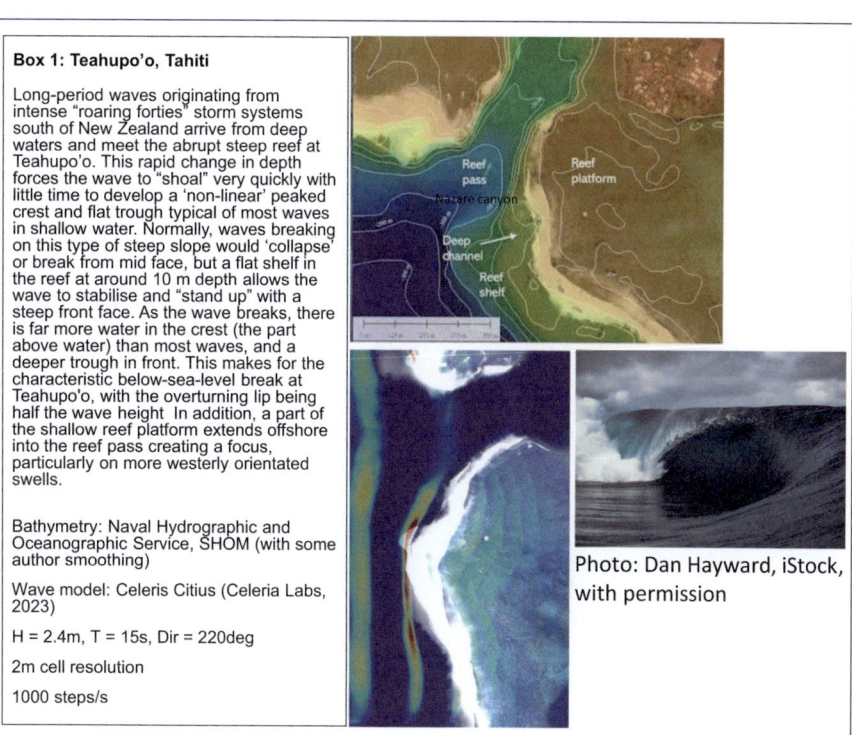

Box 1: Teahupo'o, Tahiti

Long-period waves originating from intense "roaring forties" storm systems south of New Zealand arrive from deep waters and meet the abrupt steep reef at Teahupo'o. This rapid change in depth forces the wave to "shoal" very quickly with little time to develop a 'non-linear' peaked crest and flat trough typical of most waves in shallow water. Normally, waves breaking on this type of steep slope would 'collapse' or break from mid face, but a flat shelf in the reef at around 10 m depth allows the wave to stabilise and "stand up" with a steep front face. As the wave breaks, there is far more water in the crest (the part above water) than most waves, and a deeper trough in front. This makes for the characteristic below-sea-level break at Teahupo'o, with the overturning lip being half the wave height In addition, a part of the shallow reef platform extends offshore into the reef pass creating a focus, particularly on more westerly orientated swells.

Bathymetry: Naval Hydrographic and Oceanographic Service, SHOM (with some author smoothing)

Wave model: Celeris Citius (Celeria Labs, 2023)

H = 2.4m, T = 15s, Dir = 220deg

2m cell resolution

1000 steps/s

Photo: Dan Hayward, iStock, with permission

Box 4.1 Teahupo'o, Tahiti. (Photo: Dan Hayward, iStock, with permission)

4.3.1 Preconditioning

Surfing waves, usually of long period (10–20 s), start interacting with the seabed at water depths of around 80–200 meters which are usually several kilometers offshore, except on steep-sloping islands (Montaggioni et al., 2019; Storlazzi et al., 2012). The region where surfing waves start transforming as they move onshore toward the breakpoint is referred to as the swell window or swell corridor (Atkin et al. 2019b). Throughout this area, waves interacting with bathymetric features at different spatial scales, or components (sensu Mead & Black, 2001b), are modified prior to breaking by a series of processes termed preconditioning.

Atkin et al. (2019a) defined two main types of preconditioning: disruptive and focus. Disruptive preconditioning occurs when wave crests are modified due to bifurcation and wave-wave interaction and result in multiple high-quality surfing peak waves (Fig. 4.4), such as those found along the beach breaks of Ocean Beach (California) or Martha Lavinia (King Island, Australia). A special case of this type of preconditioning occurs due to the constructive interference between two waves, a common occurrence where incoming waves refract or reflect off natural (e.g., headland) (Fig. 4.5) or artificial structures (e.g., groynes) and create wedge-type waves, such as Duranbah (Australia) or The Wedge (California) (Fig. 4.6).

Focus preconditioning occurs when wave rays converge around a bathymetric feature that initiates breaking in their lee and result in a focused peak such as Pe'ahi (Jaws) (Hawaii) (Fig. 4.3, 4.7). Focusing can also occur without initiating breaking.

Fig. 4.4 Disruptive preconditioning resulting in multiple peaks (Photo: Antonio Helio, with permission)

4 What Forms My Break?

Fig. 4.5 Constructive interference amplify wave heights as they reflect off the headland at Nazare, Portugal. (Photo: R.M Nunes, iStock, with permission)

Fig. 4.6 Waves reflect from the Newport Harbour Jetty and constructively interfere to form The Wedge, California. (Photo: Brandon James on Pexels)

Fig. 4.7 A focused peak resulting in big wave surfing at Pe'ahi, Hawai'i. (Photo: Peggy Johnson Philip Waikoloa on Pexels)

For example, islands and offshore structures such as shoals, sometimes called bomboras or bombies in Australia, can control the location and increase the consistency and quality of inshore surf breaks (Pitt, 2009).

Specific types of preconditioning rarely occur in isolation. More commonly, preconditioning modes combine and overlap at different spatial scales producing high-quality surfing waves. For example, offshore canyons result in wave rays diverging as waves propagate over them and subsequent wave rays converging and focusing on their shallow edges due to refraction. This type of channeling and focusing preconditioning is very common on the best beach breaks (Sect. 4.4.1) around the world, such as Puerto Escondido (Mexico), Hossegor (France), and Blacks Beach (California) (Shepard & Inman, 1950). The unique and extreme wave of Nazare (Portugal) occurs due to a combination of channeling, focusing, and constructive interference (do Carmo, 2022) (Box 4.2).

4.3.2 Breaking Intensity

One of the most important characteristics of high-quality surfing waves is the shape of the face when breaking. Most surfers seek steep and hollow plunging waves, as, arguably, surfing under the plunging lip of a wave, known as tube or barrel riding, is one of the pinnacles of the surfing experience (Fig. 4.2a).

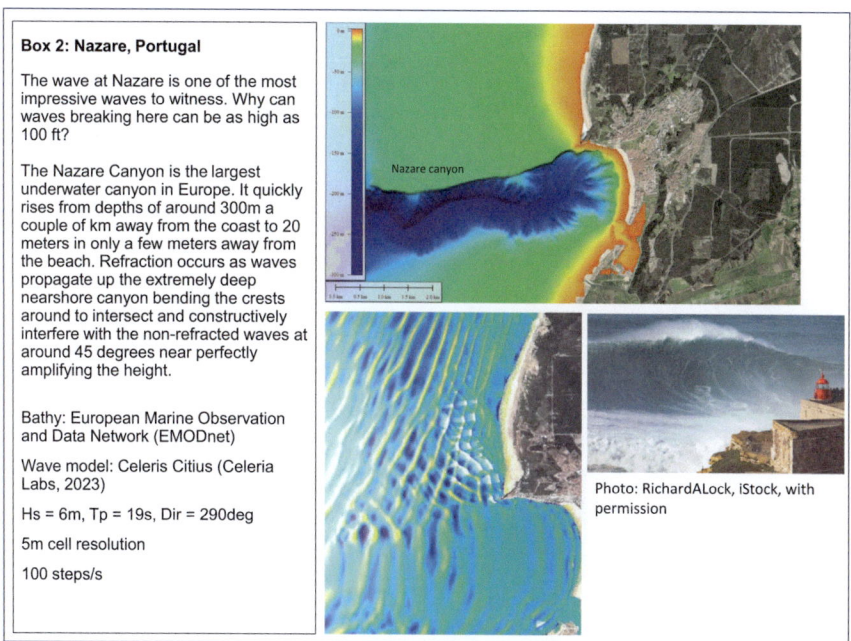

Box 2: Nazare, Portugal

The wave at Nazare is one of the most impressive waves to witness. Why can waves breaking here can be as high as 100 ft?

The Nazare Canyon is the largest underwater canyon in Europe. It quickly rises from depths of around 300m a couple of km away from the coast to 20 meters in only a few meters away from the beach. Refraction occurs as waves propagate up the extremely deep nearshore canyon bending the crests around to intersect and constructively interfere with the non-refracted waves at around 45 degrees near perfectly amplifying the height.

Bathy: European Marine Observation and Data Network (EMODnet)

Wave model: Celeris Citius (Celeria Labs, 2023)

Hs = 6m, Tp = 19s, Dir = 290deg

5m cell resolution

100 steps/s

Photo: RichardALock, iStock, with permission

Box 4.2 Nazare, Portugal. (Photo: RichardALock, iStock, with permission)

Wave breaking type depends on the ratio between wave steepness and the seabed slope, also known as the Iribarren number (Battjes, 1974). Steeper waves and flatter slopes result in gentle spilling waves, intermediate slopes result in energetic plunging waves, and lower waves and steeper slopes result in surging or collapsing waves (Galvin Jr., 1968). Generally, beginner surfers prefer gently sloping spilling waves and advanced surfers steep and hollow plunging waves (Hutt et al., 2001). However, preference for wave breaking types is as varied as the number of surfers and depends on factors such as style, level of performance, and/or selection of surfing crafts.

The degree of "hollowness" is, however, more nuanced than that captured by the Iribarren number and description of traditional breaking types. The shape of the plunging breaker, transitioning from gentle to extreme, is better characterized by the breaker intensity parameter (Eq. 4.1):

$$\text{BI} = w / h \qquad (4.1)$$

The breaker intensity (BI) is the ratio between the plunging barrel's width (w) to the barrel's height (h) (Sayce et al., 1999) (Fig. 4.8). For example, a value of 2 indicates that the barrel is twice as wide as it is high. Breaking waves with barrel ratios smaller than 1 are likely to collapse and greater than 3 are likely to gently plunge or spill. Mead and Black (2001c) described breaker intensity values ranging from 1.42

Fig. 4.8 How to parameterize a plunging wave in order to calculate its breaker intensity. (Photo: Emiliano Arano on Pexels)

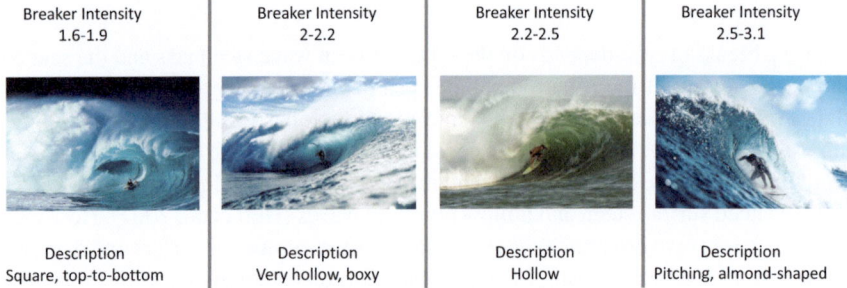

Fig. 4.9 Breaker intensity and description of the shape of the barrel (Modified from: Mead & Black, 2001c) (Photos: Kammeran Gonzalez-Keola on Pexels)

to 3.43 which could be empirically predicted by the gradient along the direction of wave propagation, known as the orthogonal seabed slope.

Breaker intensity values can be linked to common surfing terminology describing the shape of the barrel (Sayce et al., 1999). Low values (<2) are associated with very high intensity breaking waves, with the barrel shape commonly described as "square." Increasing intensity values (2–3) are associated with decreasing breaking intensity and more elongated barrel shapes (Fig. 4.9).

4 What Forms My Break?

4.3.3 Peel Angle and Rate

The rate at which waves break is another characteristic of high-quality surfing waves. This is complementary to wave height and breaking intensity, as the longer a big and/or barreling section of wave can be surfed, the higher the surfing quality (Scarfe et al., 2002).

Walker and Palmer (1971) first introduced the concept of peel angle as one of the most important parameters of recreational surfing waves. Peel angle is defined as the angle (0–90°) between the path of the broken wave and the unbroken crest (Fig. 4.10). The peel rate is, hence, defined as the velocity the wave breaks, or peels, along the wave crest.

Peel angle is a function of wave refraction. As waves undergo increasing refraction and closely align to the isobaths, peel angles decrease. As the peel angle approaches 0°, peel rate increases to the point the wave crest does not peel and breaks simultaneously, known as a closeout in surfing terms (Walker, 1974).

Peel angle is important because it determines the velocity required by surfers to stay just ahead of the breaking lip or curl. This can be approximated by adding the wave velocity perpendicular to the wave crest and the peel rate (Scarfe et al., 2009). Traditionally, surfers achieved speeds of about 12 m/s on larger waves (4–5 m) (Walker, 1974). However, with recent advances on equipment, surfers have reached velocities of up to 21 m/s on giant waves (20–30 m) (Pinto, 2024).

Higher (lower) surfing velocities are required for waves with small (high) peel angle. For example, peel angles <25° are deemed too challenging/fast for surfing, while peel angles >60° result in waves that break too slowly. Generally, peel angles between 30 and 60° result in the longest highest-quality surfing waves (Fig. 4.6) (Walker, 1974).

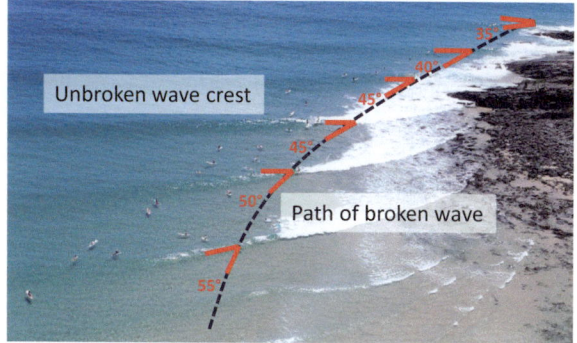

Fig. 4.10 Changing peel angle as waves propagate down the line up at Noosa, Australia. (Source: Javier Leon)

4.4 Common Surf Break Types

Surf break types can be grouped according to main geomorphic characteristics such as bathymetry and substrate composition. Previous classification schemes have proposed five or six categories based on geomorphology (headland/point, estuary, ledge) or substrate composition (e.g., sand, coral, rock) (Atkin et al., 2019a). However, as per most classification schemes, general types fall within a continuum spectrum and even overlap with each other (Scarfe et al., 2009). For example, some surf breaks might start as a rocky point break and finish as a sandy beach break. Here, we present four common surf break types and describe their main characteristics.

4.4.1 Beach Breaks

Beach breaks are one of the most common and more dynamic types of surf breaks. Open beaches have highly mobile sand bars and rip channels formed and shaped by waves and currents under different wave conditions (Castelle & Masselink, 2023) (Chap. 5). Pocket beaches tend to have more stable sand bars and fixed rip channels (Short, 2010). Under high energy, continuous bars are orientated parallel along the beach (Dissipative states, sensu Wright & Short, 1984), resulting in closeouts. However, under less energetic intermediate states, bars can rhythmically extend offshore and provide irregular seabed features for waves to peel along or to be focused over. As these bars and rips are typically aligned to the predominant swell direction, swells from nondominant direction result in much better surfing waves, for example, NW swells on the New Zealand west coast and NE swells on the Australian east coast.

World-class beach breaks can have strong preconditioning influences (Sect. 4.3.1). High-quality beach breaks commonly have multiple swells and waves constructively interfering, topographic controls such as headlands causing semi-permanent rips and bars or focusing offshore structures that substantially improve consistency and quality (Box 4.3: Manly, Australia).

4.4.2 Reef Breaks

Reef breaks are a catchall description of waves breaking on rock shelves, coral, or boulders. These breaks tend to be very robust due to the consolidated nature of their substrate, as opposed to the more dynamic beach breaks. Reef breaks vary widely, with some breaking on relatively flat bathymetry after refraction and focusing

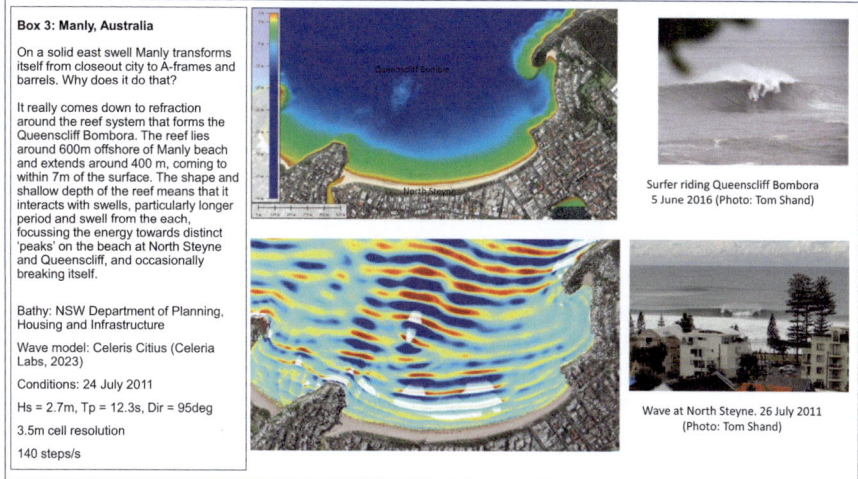

Box 4.3 Manly, Australia

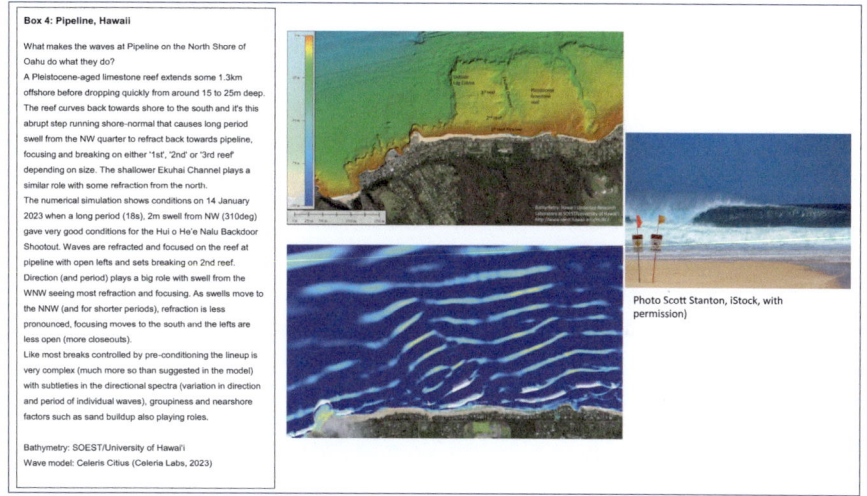

Box 4.4 Pipeline, Hawaii. (Photo Scott Stanton, iStock, with permission)

occurs further offshore (Box 4.4: Pipeline, Hawaii), while others break on abrupt bathymetry with little offshore wave conditioning, also known as slabs (Box 4.1: Teahupo'o, Tahiti). Both types are very consistent and produce high-quality waves given the right wave conditions, although the latter type is more sensitive to wave direction, period, height, and tidal elevation.

4.4.3 Point Breaks

Point breaks occur where a wave cannot bend or refract completely to align with seabed contours which causes the wave to peel for longer times. The more closely aligned the wave and bathymetric contours are, the faster the wave will peel (e.g., Ma'alaea, Hawaii), while waves that are more at right angles to the contours will peel more slowly (e.g., Malibu, California).

Point breaks, especially headland breaks, are less sensitive to offshore wave conditions than other types of surf break because waves wrap into the point, cleaning up and becoming aligned before breaking. In fact, as longer period swell refracts more efficiently than short-period wind waves, often only the ground-swell will wrap into the point, sometimes as close as 180°, while the short-period waves continue past (Box 4.5: Noosa, Australia). Moreover, headlands usually provide shelter from prevailing onshore winds, producing clean high-quality surfing waves.

Point breaks are generally sensitive to sediment transport on the seafloor. As opposed to the better known longshore and cross-shore sediment transport mechanisms affecting beach breaks, sediment headland bypassing is the least understood process of sediment transport in coastal geomorphology (Wishaw et al., 2021). The natural or human-driven intermittent flow of sand around points or headlands can have severe impacts in the quality of surfing waves, as evidenced by the positive, yet unexpected, formation of the Superbank from Snapper Rocks to Greenmount and the negative impacts to Kirra in the Gold Coast of Australia (Castelle et al., 2006; da Silva et al., 2020).

Box 5: Noosa, Australia

What makes waves at Noosa so long?

Noosa Headland, an exposed sandstone headland, provides protection from the prevailing SE winds and filters out shorter-period waves. Only the longest period SE waves and the more easterly and northerly oriented swells, common during summer months, refract and perfectly peel along the sand-bottom beaches. Clean waves peeling for up to 200m are not unusual but are sensitive to the availability and distribution of sand bars which tend to be very dynamic.

Bathymetry: Bathymetric LiDAR for Sunshine Coast/ QLD Government

Wave model: Celeris Citius (Celeria Labs, 2023)

Hs = 2.4m, Tp = 15s, Dir = 67.5deg

5m cell resolution

140 steps/s

Photo: Boyshots, iStock, with permission

Box 4.5 Noosa, Australia. (Aerial photo: Boyshots, iStock, with permission)

Fig. 4.11 Surf breaking across a bar at the entrance of a river in Costa Rica. (Photo: pilesasmiles, iStock, with permission)

4.4.4 Estuary/River Breaks

At the mouth of estuaries and rivers, sand accumulates as tidal flows pushing sand offshore become weaker and are balanced by waves pushing sediment onshore. These are known as ebb-tide deltas or bars and range in size and shape depending on the size of the estuary and outlet configuration (Harrison et al., 2017). Surfing waves can occur where the bar is shallow enough for waves to break on and is at an angle from the incoming swell to allow the waves peel (Fig. 4.11). This occurs most often when flow exits adjacent to a headland forming a triangular half-delta such as in Costa Rica or Mundaka (Spain) but can occur on non-headland-controlled open coasts with less consistency such as at Raglan (New Zealand).

4.5 Management of Surf Breaks

4.5.1 Threats

The quality of surf breaks is variable due to the interactions between waves and the physical environment. In some cases, surfing wave quality can improve, and in others, it can degrade due to natural or human-induced changes. These temporary or permanent changes operate over different spatial and temporal scales.

Negative impacts on surfing amenity due to natural changes are not uncommon but remain poorly understood and rarely quantified or mitigated (Bryan et al., 2019; Scarfe et al., 2009). A notable exception is the case of Mangamaunu, on the north east of New Zealand's South Island, where a Mw 7.8 earthquake occurred in November 2016 and caused parts of the coastline to be uplifted by up to 3 m, negatively impacting the surf break. As part of investigations for potential engineering works, comprehensive fieldwork was undertaken, which included the engagement of the local community including the collection of bathymetry, wave data, and breaking wave imagery and detailed modeling of pre- and post-earthquake conditions carried out (Shand et al., 2020).

Positive impacts on surfing amenity due to natural changes are also common but poorly documented. These are mostly related to cyclical changes in sediment supply, often occurring as "slugs" of sand along beach breaks or around headlands/infrastructure that can considerably improve surfing wave quality for months to years (Chapman, 1983; Phillips & Mead, 2009). These cycles are usually driven by climate variability (e.g., El Niño Southern Oscillation or the Interdecadal Pacific Oscillation) that influence the direction and frequency of waves (Adams et al. 2011; Wishaw et al., 2021).

The impacts of longer-term natural changes in surfing quality due to climate change remain uncertain (Sadrpour & Reineman, 2023). Increases in sea-level rise would, speculatively, make fixed seabeds such as deeper reefs, bad for low-tide only surfing waves such as Racetracks (Bali) but good for higher tide spots. Impacts on beaches will be highly variable, as some could "drown" while others adapt (Reineman et al., 2017). Moreover, changes in wave climate could be positive for some spots (Espejo et al., 2014) while negative for others (Goodwin et al., 2016).

Human influences on surf breaks are significant and can be negative or positive. These include direct modification of the seabed through construction of coastal structures or dredging/nourishing of sediment or indirect influences such as deforestation causing siltation of estuaries and rivers and damming of rivers reducing sediment supplies to the coast (Warrick et al., 2024).

Negative human impacts on surfing wave quality are usually caused by modifications within the swell corridor or at the breaking position. For example, the wave of Cabo Blanco (Peru) was shortened due to the upgrade of a fishing pier (Leon, 2012) and La Barre in Anglet (France) reportedly disappeared due to the construction of a dike to protect the port (Touron-Gardic & Failler, 2022). In other cases, oversights have been made in sediment sourcing resulting in dissimilar sediment being utilized for nourishment campaigns, changing the morphology-type of the beach and hence wave breaking processes affecting surfing amenity (Benedet et al., 2007). Coastal structures have also been beneficial for surfing when natural linear processes are interrupted forming cross-shore bars or reflecting waves (e.g., The Wedge, California; Sect. 3.1). Recently, an artificial surf reef at Palm Beach, Gold Coast (Australia), has resulted in waves which produce longer rides and more consistent breaking location compared with those on a nearby natural beach (Thompson et al., 2021).

Dredging or nourishment may have detrimental or beneficial effects depending on whether it adversely affects an existing wave such as occurred at Mundaka, Spain (Scheske et al., 2019) or augments or creates new waves such as the Superbank, Australia (Sect. 4.4). In general, human modifications around established waves are likely to be detrimental or at least pose a risk of degradation, while modifications around non-surfing waves can improve conditions by causing seabed irregularity and the potential for surf where there was previously none. These changes in surfing quality can also translate into negative or positive economic outcomes (Lazarow, 2007; McGregor & Wills, 2017) (Chap. 9).

4.5.2 Conservation

The conservation of surfing breaks aligns with initiatives to conserve marine and coastal environments (Touron-Gardic & Failler, 2022). Surf breaks are usually within or in close proximity to important carbon-dense coastal ecosystems such as coral reefs and mangroves, providing synergistic opportunities between biodiversity conservation and socioeconomic benefits for communities (Bukoski et al., 2024).

The management and conservation of surf breaks requires a holistic approach and understanding of the processes underlying surf amenity and should include the full swell corridor area (Atkin et al., 2019a; Atkin and Greer 2019). The future of surf break conservation will benefit from recent technological advancements on wave quality modeling (Tavakkol & Lynett, 2017), measuring (Vieira et al., 2025) and monitoring (Atkin et al., 2023; Murray et al., 2023; Thompson et al., 2021). Importantly, surfing break conservation frameworks need to be flexible and involve surfers and local communities in surf break co-management initiatives to develop community stewardship (Edwards & Stephenson, 2013; Shand et al., 2020).

Acknowledging the existence of traditional Polynesian *kapu* systems (Warren & Gibson, 2014), the first (modern) surfing reserve was declared in Bells Beach, Australia, in 1973 (Lazarow, 2010). In 2000, Peru was the first country to propose a law protecting surfing zones, which was implemented 13 years later effectively mandating the protection of surf breaks (Scheske et al., 2019). In 2010, New Zealand became the first country to established legal protection for surf breaks at a national scale (Orchard, 2020). Several national and regional legal frameworks and other protection mechanisms without legal status, such as the Australian National Surfing Reserves (Farmer & Short, 2007) or Save The Wave's World Surfing Reserves, and Surf Protected Area Networks, have resulted in notable benefits to surfing, including increased environmental awareness and biodiversity conservation (Arroyo et al., 2019; Reineman et al., 2021; Scheske et al., 2019). However, the lack of legal status has hindered their effectiveness as coastal conservation tools, although in some cases they have helped blocking coastal infrastructure and encouraged the monitoring of impacts due to development (Touron-Gardic & Failler, 2022).

References

Adams, P. N., Inman, D. L., & Lovering, J. L. (2011). Effects of climate change and wave direction on longshore sediment transport patterns in Southern California. *Climatic Change, 109*(1), 211–228.

Arroyo, M., Levine, A., & Espejel, I. (2019). A transdisciplinary framework proposal for surf break conservation and management: Bahía de Todos Santos World Surfing Reserve. *Ocean & Coastal Management, 168*, 197–211. https://doi.org/10.1016/j.ocecoaman.2018.10.022

Atkin, E. A., & Greer, D. (2019). A comparison of methods for defining a surf Break's swell corridor. *Journal of Coastal Research, 87*(sp1%J Journal of Coastal Research), 70–77, 78. https://doi.org/10.2112/SI87-007.1

Atkin, E. A., Bryan, K. R., Hume, T. M., Mead, S. T., & Waiti, J. (2019a). *Management guidelines for surfing resources*. Aotearoa New Zealand Association for Surfing Research. www.surfbreakresearch.org

Atkin, E. A., Mead, S. T., & Phillips, D. (2019b). Investigations of offshore wave preconditioning. *Journal of Coastal Research*, 78–90. https://www-jstor-org.ezproxy.usc.edu.au/stable/26851745

Atkin, E. A., Davies-Campbell, J., & McIntosh, R. (2023). Deep learning object detection application to surfing wave quality. *Coastal Engineering Proceedings, 37*, papers.25. https://doi.org/10.9753/icce.v37.papers.25

Bascom, W., & McCoy, K. (2020). *Waves and beaches: The powerful dynamics of sea and coast* (3rd ed.). Patagonia. https://books.google.com.au/books?id=2OF7zgEACAAJ

Battjes, J. A. (1974). Surf similarity. *Coastal Engineering*.

Battjes, J. A. (1982). A case study of wave height variations due to currents in a tidal entrance. *Coastal Engineering, 6*(1), 47–57. https://doi.org/10.1016/0378-3839(82)90014-X

Benedet, L., Pierro, T., & Henriquez, M. (2007). Impacts of coastal engineering projects on the surfability of sandy beaches. *Shore and Beach, 75*(4), 3.

Brennan, D. (2024). Wonder and the sublime in surfing and nature sports. *Journal of the Philosophy of Sport, 51*(2), 381–396. https://doi.org/10.1080/00948705.2024.2339887

Bryan, K. R., Davies-Campbell, J., Hume, T. M., & Gallop, S. L. (2019). The influence of sand Bar morphology on surfing amenity at New Zealand Beach breaks. *Journal of Coastal Research, 87*(sp1%J Journal of Coastal Research), 44–54, 11. https://doi.org/10.2112/SI87-005.1

Bukoski, J. J., Atkinson, S. R., Miller, M. A. S., Sancho-Gallegos, D. A., Arroyo, M., Koenig, K., Reineman, D. R., & Kittinger, J. N. (2024). Co-occurrence of surf breaks and carbon-dense ecosystems suggests opportunities for coastal conservation. *Conservation Science and Practice, n/a*(n/a), e13193. https://doi.org/10.1111/csp2.13193

Caldwell, P. C., Vitousek, S., & Aucan, J. P. (2009). Frequency and duration of coinciding high surf and tides along the North Shore of Oahu, Hawaii, 1981–2007. *Journal of Coastal Research, 25*(3 (253)), 734–743. https://doi.org/10.2112/08-1004.1

do Carmo, J. S. A. (2022). Dominant processes that amplify the swell towards the coast: The Nazaré Canyon and the giant waves. *Research, Society and Development, 11*(11), e578111133804. https://doi.org/10.33448/rsd-v11i11.33804

Castelle, B., & Masselink, G. (2023). Morphodynamics of wave-dominated beaches. *Cambridge Prisms: Coastal Futures, 1*, e1. https://doi.org/10.1017/cft.2022.2

Castelle, B., Lazarow, N., Marty, G., & Tomlinson, R. (2006). *Impact of beach nourishment on Coolangatta Bay morphology over the period 1995–2005*. Proceedings of the 15th NSW coastal conference.

Castelle, B., Scott, T., Brander, R. W., & McCarroll, R. J. (2016). Rip current types, circulation and hazard. *Earth-Science Reviews, 163*, 1–21. https://doi.org/10.1016/j.earscirev.2016.09.008

Chapman, D. M. (1983). *Sediment reworking on sandy beaches*. Sandy beaches as ecosystems: Based on the proceedings of the first international symposium on Sandy beaches, held in Port Elizabeth, South Africa, 17–21 January 1983.

City of Gold Coast. (2015). *Gold coast surf management plan*.

Dodet, G., Bertin, X., Bruneau, N., Fortunato, A. B., Nahon, A., & Roland, A. (2013). Wave-current interactions in a wave-dominated tidal inlet. *Journal of Geophysical Research: Oceans, 118*(3), 1587–1605. https://doi.org/10.1002/jgrc.20146

Douglass, S. L. (1990). Influence of wind on breaking waves. *Journal of Waterway, Port, Coastal, and Ocean Engineering, 116*(6), 651–663. https://doi.org/10.1061/(ASCE) 0733-950X(1990)116:6(651)

Edwards, A., & Stephenson, W. (2013). Assessing the potential for surf break co-management: Evidence from New Zealand. *Coastal Management, 41*(6), 537–560. https://doi.org/10.108 0/08920753.2013.842681

Espejo, A., Losada, I. J., & Méndez, F. J. (2014). Surfing wave climate variability. *Global and Planetary Change, 121*, 19–25. https://doi.org/10.1016/j.gloplacha.2014.06.006

Farmer, B., & Short, A. (2007). Australian National Surfing Reserves—Rationale and process for recognising iconic surfing locations. *Journal of Coastal Research, Special Issue, 50*, 99–103.

Galvin, C. J., Jr. (1968). Breaker type classification on three laboratory beaches. *Journal of Geophysical Research (1896–1977), 73*(12), 3651–3659. https://doi.org/10.1029/JB073i012p03651

Goodwin, I. D., Mortlock, T. R., & Browning, S. (2016). Tropical and extratropical-origin storm wave types and their influence on the East Australian longshore sand transport system under a changing climate. *Journal of Geophysical Research: Oceans.* https://doi.org/10.1002/2016JC011769

Harrison, S. R., Bryan, K. R., & Mullarney, J. C. (2017). Observations of morphological change at an ebb-tidal delta. *Marine Geology, 385*, 131–145. https://doi.org/10.1016/j.margeo.2016.12.010

Hemer, M. A., Fan, Y., Mori, N., Semedo, A., & Wang, X. L. (2013). Projected changes in wave climate from a multi-model ensemble [10.1038/nclimate1791]. *Nature Climatic Change.* Advance online publication. https://doi.org/http://www.nature.com/nclimate/journal/vaop/ncurrent/abs/nclimate1791.html#supplementary-information

Hutt, J. A., Black, K. P., & Mead, S. T. (2001). Classification of surf breaks in relation to surfing skill. *Journal of Coastal Research*, 66–81. http://www.jstor.org/stable/25736206

Lazarow, N. (2007). The value of coastal recreational resources: A case study approach to examine the value of recreational surfing to specific locales. *Journal of Coastal Research, 1*(1), 12–20.

Lazarow, N. S. (2010). *Managing and valuing coastal resources: An examination of the importance of local knowledge and surf breaks to coastal communities.* Australian National University.

Leon, J. (2012). A delicate balance: The Cabo Blanco and Panic Point (mis)management case, North coast, Peru. *Reef Journal, 2*, 36–45.

Leon, J. X., Manero, A., Lazarow, N., Spencer-Cotton, A., Wegener, T., Jarratt, P., & Pearce, T. (2024). Surfing at the Noosa world surfing reserve, Australia: Direct expenditure and travel cost analyses of recreational surfing. *Coastal Management, 52*, 449.

Mazzaretto, O. M., & Menendez, M. (2024). A worldwide coastal analysis of the climate wave systems [Original Research]. *Frontiers in Marine Science, 11.* https://doi.org/10.3389/fmars.2024.1385285

McGregor, T., & Wills, S. (2017). *Surfing a wave of economic growth* (CAMA working paper no. 31/2017, issue.

Mead, S., & Black, K. (2001a). Field studies leading to the bathymetric classification of world-class surfing breaks. *Journal of Coastal Research, 29*, 5–21.

Mead, S., & Black, K. (2001b). Functional component combinations controlling surfing wave quality at world-class surfing breaks. *Journal of Coastal Research*, 21–32. http://www.jstor.org.ezproxy.usc.edu.au:2048/stable/25736202

Mead, S., & Black, K. (2001c). Predicting the breaking intensity of surfing waves. *Journal of Coastal Research*, 51–65. http://www.jstor.org.ezproxy.usc.edu.au:2048/stable/25736205

Montaggioni, L. F., Collin, A., James, D., Salvat, B., Martin-Garin, B., Siu, G., Taiarui, M., & Chancerelle, Y. (2019). Morphology of fore-reef slopes and terraces, Takapoto atoll (Tuamotu archipelago, French Polynesia, Central Pacific): The tectonic, sea-level and coral-growth control. *Marine Geology, 417*, 106027. https://doi.org/10.1016/j.margeo.2019.106027

Mortlock, T. R., Nott, J., Crompton, R., & Koschatzky, V. (2023). A long-term view of tropical cyclone risk in Australia. *Natural Hazards.* https://doi.org/10.1007/s11069-023-06019-5

Munk, W. H., Miller, G. R., Snodgrass, F. E., Barber, N. F., & Deacon, G. E. R. (1963). Directional recording of swell from distant storms. *Philosophical Transactions of the Royal Society of London. Series A, Mathematical and Physical Sciences, 255*(1062), 505–584. https://doi.org/10.1098/rsta.1963.0011

Murray, T. P., Greaves, M. C., Vieira da Silva, G., Boyle, O. J., Wynne, K., Freeston, B., Ditria, L., Jardine, P., Ditria, E., & Strauss, D. (2023). *Utilising object detection from coastal surf cameras to assess surfer usage*. Australasian Coasts & Ports 2023 Conference.

National Oceanic and Atmospheric Administration (NOAA) (2020). Digital Coast Data Access Viewer. Custom processing of "2013 USACE NCMP Topobathy Lidar: Maui (HI) - LMSL". Charleston, SC: NOAA Office for Coastal Management. Accessed at https://coast.noaa.gov/dataviewer.

OCM Partners. (2024). *2013 USACE NCMP Topobathy Lidar: Maui (HI) from 2010-06-15 to 2010-08-15*. https://www.fisheries.noaa.gov/inport/item/49751

Orchard, S. (2020). Legal protection of New Zealand's surf breaks: Top-down and bottom-up aspects of a natural resource challenge. *Australasian Journal of Environmental Management, 27*(1), 6–21. https://doi.org/10.1080/14486563.2020.1719439

Phillips, D., & Mead, S. (2009). Investigation of a large sandbar at Raglan, New Zealand: Project overview and preliminary results. *The Reef Journal, 1*, 267–278.

Phillips, D., Mead, S. T., Black, K. P., & Healy, T. R. (2003). *Surf zone currents and influence on surfability*. Proceedings of the 3rd international surfing reef symposium.

Pinto, L. M. (2024). *How fast do surfers ride a wave?* https://www.surfertoday.com/surfing/how-fast-do-surfers-ride-a-wave

Pitt, A. (2009). *Surfing at bombora controlled beaches*. 5th Western Australian state coastal conference, Fremantle.

Reineman, D. R., Thomas, L. N., & Caldwell, M. R. (2017). Using local knowledge to project sea level rise impacts on wave resources in California. *Ocean & Coastal Management, 138*, 181–191. https://doi.org/10.1016/j.ocecoaman.2017.01.020

Reineman, D. R., Koenig, K., Strong-Cvetich, N., & Kittinger, J. N. (2021). Conservation opportunities Arise from the co-occurrence of surfing and key biodiversity areas [brief research report]. *Frontiers in Marine Science, 8*. https://doi.org/10.3389/fmars.2021.663460

Roberts, H. H., Coleman, J. M., & Walker, H. J. (2014). *Coastal studies institute: A history of science contributions for 60 years*.

Sadrpour, N., & Reineman, D. (2023). *The impacts of climate change on surfing resources*.

Sayce, A., Black, K., & Gorman, R. (1999). Breaking wave shape on surfing reefs. Coasts & Ports 1999: Challenges and directions for the new century; proceedings of the 14th Australasian coastal and ocean engineering conference and the 7th Australasian port and harbour conference.

Scarfe, B., de Lange, W., Chong, A., Black, K., & Mead, S. (2002). *The influence of surfing wave parameters on manoeuvre type from field investigations at Raglan, New Zealand*. Proc. for the 2nd Surfing Arts, Science and Issues Conf.(SASIC 2).

Scarfe, B. E., Elwany, M. H. S., Mead, S. T., & Black, K. P. (2003, June 22–25). *The science of surfing waves and surfing breaks – A review*. Proceedings of the 3rd international surfing reef symposium, Raglan, New Zealand.

Scarfe, B. E., Healy, T. R., & Rennie, H. G. (2009). Research-based surfing literature for coastal management and the science of surfing: A review. *Journal of Coastal Research, 25*(3), 539–665. http://www.jstor.org.ezproxy.usc.edu.au:2048/stable/27698350

Scheske, C., Arroyo Rodriguez, M., Buttazzoni, J. E., Strong-Cvetich, N., Gelcich, S., Monteferri, B., Rodríguez, L. F., & Ruiz, M. (2019). Surfing and marine conservation: Exploring surf-break protection as IUCN protected area categories and other effective area-based conservation measures. *Aquatic Conservation Marine and Freshwater Ecosystems, 29*(S2), 195–211. https://doi.org/10.1002/aqc.3054

Shand, T., Reinin-Hamill, R., Weppe, S., & Short, A. (2020). *Development of a framework for assessing effects of coastal engineering works on a surf break*. Australasian Coasts & Ports 2019 Conference, Hobart, Tasmania.

Shepard, F. P., & Inman, D. L. (1950). Nearshore water circulation related to bottom topography and wave refraction. *Eos, Transactions American Geophysical Union, 31*(2), 196–212. https://doi.org/10.1029/TR031i002p00196

Short, A. D. (2010). Sediment transport around Australia – Sources, mechanisms, rates, and barrier forms. *Journal of Coastal Research, 26*(3), 395–402. https://doi.org/10.2112/08-1120.1

da Silva, A. P., Woortmann, L. G., da Silva, G. V., Murray, T., Strauss, D., & Tomlinson, R. (2020). A 90-year morphodynamic analysis in Southern Queensland (Australia). *Journal of Coastal Research, 95*(SI), 438–442. https://doi.org/10.2112/si95-085.1

Storlazzi, C. D., Field, M. E., Dykes, J. D., Jokiel, P. L., & Brown, E. (2012). Wave control on reef morphology and coral distribution: Molokai, Hawaii. *Ocean Wave Measurement and Analysis, 2001*, 784–793. https://doi.org/10.1061/40604(273)80. (Proceedings).

Tavakkol, S., & Lynett, P. (2017). Celeris: A GPU-accelerated open source software with a Boussinesq-type wave solver for real-time interactive simulation and visualization. *Computer Physics Communications, 217*, 117–127. https://doi.org/10.1016/j.cpc.2017.03.002

Thompson, M., Zelich, I., Watterson, E., & Baldock, T. E. (2021). Wave Peel tracking: A new approach for assessing surf amenity and analysis of breaking waves. *Remote Sensing, 13*(17), 3372. https://www.mdpi.com/2072-4292/13/17/3372

Thornton, E. B., & Kim, C. S. (1993). Longshore current and wave height modulation at tidal frequency inside the surf zone. *Journal of Geophysical Research: Oceans, 98*(C9), 16509–16519. https://doi.org/10.1029/93JC01440

Touron-Gardic, G., & Failler, P. (2022). A bright future for wave reserves? *Trends in Ecology & Evolution, 37*(5), 385–388. https://doi.org/10.1016/j.tree.2022.02.006

Vieira, M., Guedes Soares, C., Guimarães, P. V., Bergamasco, F., & Campos, R. M. (2025). Nearshore space-time ocean wave observation using low-cost video cameras. *Coastal Engineering, 197*, 104694. https://doi.org/10.1016/j.coastaleng.2024.104694

Walker, J. R. (1974). Recreational surf parameters.

Walker, K., & Kelly, J. (1973). *Surf parameters: Final report.* University of Hawaii, James K.K. Look Laboratory of Ocean Engineering. https://books.google.com.au/books?id=G04TzwEACAAJ

Walker, J., & Palmer, R. (1971). *Surf parameters; a general surf site concept.* Department of Ocean Engineering, University of Hawaii.

Wandres, M., Espejo, A., Sovea, T., Tetoa, S., Malologa, F., Webb, A., Lewis, J., Lee, G., & Damlamian, H. (2024). A national-scale coastal flood hazard assessment for the atoll nation of Tuvalu. *Earth's Futures, 12*(4), e2023EF003924. https://doi.org/10.1029/2023EF003924

Wang, G., Wu, L., Mei, W., & Xie, S.-P. (2022). Ocean currents show global intensification of weak tropical cyclones. *Nature, 611*(7936), 496–500. https://doi.org/10.1038/s41586-022-05326-4

Warren, A., & Gibson, C. (2014). *Surfing places, surfboard makers: Craft, creativity, and cultural heritage in Hawaii, California, and Australia.* University of Hawai'i Press.

Warrick, J. A., Buscombe, D., Vos, K., Bryan, K. R., Castelle, B., Cooper, J. A. G., Harley, M. D., Jackson, D. W. T., Ludka, B. C., Masselink, G., Palmsten, M. L., Ruiz de Alegria-Arzaburu, A., Sénéchal, N., Sherwood, C. R., Short, A. D., Sogut, E., Splinter, K. D., Stephenson, W. J., Syvitski, J., & Young, A. P. (2024). Coastal shoreline change assessments at global scales. *Nature Communications, 15*(1), 2316. https://doi.org/10.1038/s41467-024-46608-x

Westwick, P. J. (2013). *World in the curl: An unconventional history of surfing.* New York Crown.

Wishaw, D., Leon, J., Fairweather, H., & Crampton, A. (2021). Influence of wave direction sequencing and regional climate drivers on sediment headland bypassing. *Geomorphology, 107708*. https://doi.org/10.1016/j.geomorph.2021.107708

Wright, L. D., & Short, A. D. (1984). Morphodynamic variability of surf zones and beaches: A synthesis. *Marine Geology, 56*(1–4), 93–118. https://doi.org/10.1016/0025-3227(84)90008-2

Wright, L. D., & Thom, B. G. (1977). Coastal depositional landforms: A morphodynamics approach. *Progress in Physical Geography, 1*, 412–459.

Further Reading

Celeria Labs. (2023). *Celeris Citius Software.* https://www.celerialabs.com

Chapter 5
Beach Safety and Surf Hazards

David M. Kennedy

5.1 Introduction

The surf zone is a dangerous environment (Chaps. 6 and 7). It is the area where energy is expended by waves as they crash onto the shore, thereby creating major stresses that the landform system must absorb. In fact, the coast can be considered the buffer, or shock absorber, between the energetic marine environment and land. The constant breaking of waves causes erosion of sediment and movement of sand and rock down to tens of meters water depth. In turn, this movement leads to the formation of coastal landforms such as beaches and their subcomponents such as bars and ridges. These features, created by waves, then affect where and when a wave will break (Chap. 4) and therefore the distribution of wave energy in the surf zone. This dynamic occurs every time a wave breaks, with the shoreline in a constant process of adjustment during both calm and stormy conditions. The interaction between waves, sediment movement, and morphology operates as a feedback loop (Wright & Thom, 1977) (Fig. 5.1). This feedback is the fundamental reason beaches, which are loose piles of sand, exist in an energy environment that eventually destroys human structures constructed within reach of waves (Fig. 5.1).

Surf zone bars, sand ridges which create lines of shallow water below the waves, are the primary morphology that dissipates wave energy on a beach. They may be either attached to the beach and be exposed at low tide or form many hundreds of meters offshore. Their size and position are directly related to the incoming wave energy, and they move in response to changes in marine weather (Davis & Fitzgerald, 2004). Under high wave energy conditions, bars move offshore, and during the intervening calmer periods, they progressively move back toward the beach.

D. M. Kennedy (✉)
The University of Melbourne, Melbourne, VIC, Australia
e-mail: davidmk@unimelb.edu.au

© The Author(s), under exclusive license to Springer Nature Switzerland AG 2025
D. M. Kennedy (ed.), *The Science and Culture of Surfing*,
https://doi.org/10.1007/978-3-031-80979-8_5

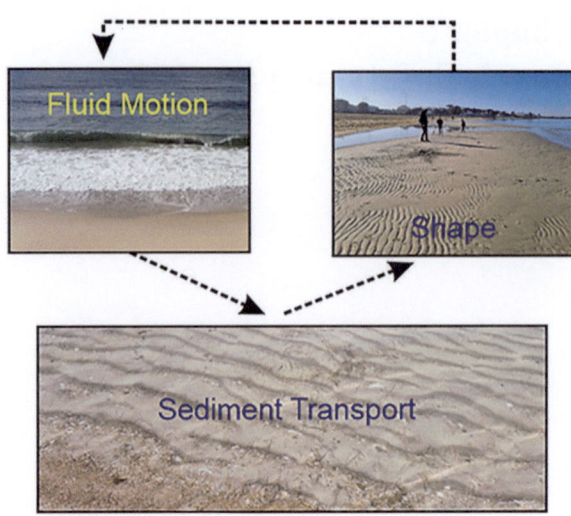

Fig. 5.1 The feedback loop of waves driving sediment transport which in turn creates features that affect where the waves will next break underpins all landform functioning on the coast. (Photos: David M. Kennedy)

Between and through the bars, fast flowing currents of water, called rips, are found. These currents direct water back offshore after a wave has broken and are also important in shaping the bars that they cross.

As noted in Chap. 3, before waves reach the coast, they are essentially just transferring energy through the ocean. Only once they break (Chap. 4) does significant water movement start to occur. It is the movement of this water, both toward and away from the shoreline, that causes a risk to people (Chap. 6). In this chapter, we will explore the dynamics of the surf zone and explore how fundamental principles of hydrodynamics and geomorphology create the hazards that are encountered by surfers and other beach users. We will then see how these basic principles are used to manage beach hazards in real time in order to save lives.

5.2 Rip Channels and Currents

Rips are defined as a narrow, concentrated, very strong current of water that flows through the surf zone into deepwater (Castelle et al., 2016; NOS, 2024; SLSNSW, 2024), and they account 85% of drownings on the coast in Australia at a rate of 26 deaths/year (SLSA, 2021). Understanding how, why, and when they form is therefore critical for ensuring public safety.

A breaking wave is an instantaneous event of great power. It is to experience this power that people surf. Once breaking has occurred, water starts to move and travels toward the shoreline as a bore. Each wave bore adds more water to the base of the beach. As water cannot endlessly accumulate at the shoreward side of the surf zone, it must escape back seaward to return to deep water. Rip currents are the main mechanism for this to occur.

5 Beach Safety and Surf Hazards

Table 5.1 A summary of the common terms used to define rips

Term	Component/ mechanism	Description
Rip current	Hydrodynamic (water flow)	A concentrated flow within the surf zone that transports water offshore
Rip channel	Morphology	A narrow trench or gully within the surf zone in which a rip current flows
Rip tide	Hydrodynamic (water flow)	An out-of-date term that sometimes is used to refer to rips that are found only at a certain elevation of the tide
Undertow	Hydrodynamic (water flow)	A defined sheetlike flow of water at the seabed. This is quite distinct from a rip and should not be used as an equivalent term

Like waves on the beach, and the breaks that are surfed, rips have a wide variety of characteristics. The typical rip forms as a result of the movement of water masses parallel to the edge of the beach (alongshore) which converge to produce a narrow, shore normal current. They may be fixed in one position for long periods of time or occur for only minutes rushing offshore in bursts. The types of rips and their position and strength can all be related to the boundary conditions of the coast, namely, the energy environment in which they are found (i.e., the waves) and the shape of the local environment (i.e., its geomorphology).

Before analyzing the formation and hydrodynamics of rips, it is worth considering some terminology. In the literature, rips are commonly referred to as rip currents, rip channels, rip tides, or undertow. Some of these terms are inaccurate and often they are incorrectly used. The common element however is they all involve water moving offshore. A general description of each term is provided below (Table 5.1).

5.2.1 Rip Current Flow

Traditionally rips are viewed as being part of a cell circulation system. In this cell, water moves toward the beach as a series of lines as each wave transforms into a bore after breaking. When the bore reaches the shore, the water then flows parallel along the base of the beach (Fig. 5.2). Once a threshold is reached, usually a function of the interaction between energy stresses and gravity-infragravity wave interactions (Chap. 3), the longshore currents abruptly turn offshore. The current then flows directly offshore as a very defined flow through the surf zone. Once outside the wave breaking zone, the water then disperses alongshore (Davis & Fitzgerald, 2004; Komar, 1998; Woodroffe, 2003) (Fig. 5.2).

Research using drifters, ranging from floating sensors to swimmers and surfers, has challenged the idea of whether a rip always fully exits the surf zone (MacMahan et al., 2010; McCarroll et al., 2014a, b) (Fig. 5.3). It has been found that when people float through a rip, they often start to move parallel to shore while still within

Fig. 5.2 (**a**) Rips and their related feeder currents (yellow) are often distinguished by areas of dark water in the surf zone as seen on Point Addis Beach, on the Great Ocean Road in Southern Australia. (Photo: David M. Kennedy). (**b**) A schematic diagrams of water flow across a typical surf zone and through a rip between the two bars

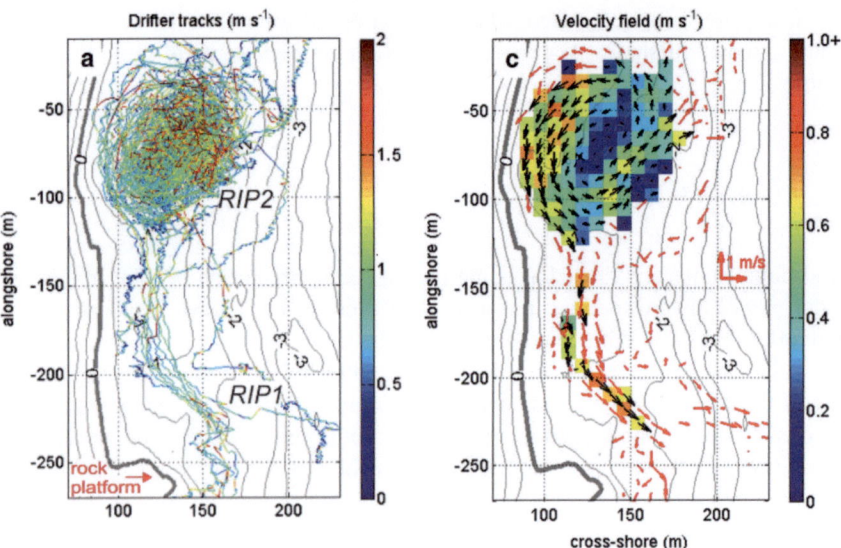

Fig. 5.3 (**a**) Individual drifter tracks and (**b**) a summary of the current velocity of the rip flows through two rip currents undertaken in an experiment in in Sydney Australia. (Reproduced with permission from (McCarroll et al., 2014a))

the breakers rather than exiting beyond the break. In an experiment at Whale Beach, Sydney, Australia, drifters were found to exit rips at a rate of 22–80% of the time depending on the position and type of rip. The higher rates of exiting occurred when the rip is located adjacent to a headland (McCarroll et al., 2014b). The importance of this research is that it shows that rips can also be confined within the surf zone as a series of semi-enclosed vortices (Castelle et al., 2016).

The precise nature of water movement through rips is extremely complex; however, for surfers, swimmers, and other beach users, the two broad mechanisms underpin much of the safety management strategies found on the worlds' coast (Fig. 5.4). That is, when caught in a rip, a person should either (1) swim diagonally across the rip or (2) tread water and drift with the current (NWS, 2024; SOS, 2024) (Fig. 5.4). There is no distinct strategy to escape from a rip as their energy conditions are highly variable as is an individual's swimming ability. The key is not to panic. For most surfers, rips do provide an easy, energy saving way to paddle through the break by riding the current rather than having to battle through the rolling wave bores to get offshore.

5.2.2 Types of Rips

Rips can be classified based on many different factors ranging from their size, causative factors, stability, and hydrodynamics (Castelle et al., 2016). For a hazard focus, Surf Life Saving Australia defines four types, namely, permanent, fixed, flash, and travelling (SLSA, 1987). This classification focuses on the dangers associated with drowning and incorporates many of the formative mechanisms of rips and their differing hydrodynamic characteristics.

Permanent rips are those which occur at a specific location at all times during the year (Fig. 5.5a). They are most often associated with hard structures in the surf zone (e.g., storm water pipes) or embayment-scale geomorphology (e.g., headlands). These rips are also termed boundary-controlled rips (Scott et al., 2011) and are characterized by water flow against a structure (Fig. 5.5a).

Fixed rips are ones that are also persistent through time (months to years), but not necessarily always located at the same location over several years (Fig. 5.5b). Their location can be determined as a function of the incoming wave dynamics, specifically the alongshore variability in wave breaking height and angle (also called bathymetric controlled (focused rips) by (Castelle et al., 2016)). In locations where it is purely energy that determines the position of a rip (i.e., no hard structures), the position of the rip is often related to the shape of infragravity frequency waves in the surf zone. The rip is positioned at a point where wave energy is most variable over period of up to 5 minutes (Komar, 1998).

Significant feedback exists between rips and the beach shape, as rips in addition to moving water through the surf zone also are significant transporters of sediment offshore (Loureiro et al., 2012a; Loureiro et al., 2012b) (Fig. 5.6). The result is rips can erode their own channels and in doing so create a feedback whereby the rip

Fig. 5.4 A community education poster used to educate people on the strategies to survive when being caught in a rip current. (Reproduced with permission from Surf Life Saving Australia)

Fig. 5.5 (**a**) A permanent rip adjacent to a rocky headland at Fingal Head; (**b**) tracer dye in a fixed rip at Tamarama Beach, Australia; (**c**) a channelized rip at Freshwater Beach, Australia; and (**d**) a series of channelized and fixed rips at Lighthouse Beach, Australia. (Photos **a**, **b**, **d** from Rob Brander, **c** from David M. Kennedy)

becomes fixed in its position during different size wave breaks (channelized or bathymetry controlled rips). In such cases, the rip is created during a high wave energy event, but as calm periods return, the current velocities are not strong enough to change the rips' position. It essentially becomes trapped within the channel formed during the high energy event (Fig. 5.5c, d). This feedback defines many different types of beaches and forms the basics of beach safety management schemes in Australia and elsewhere globally (see Sect. 5.4).

Flash rips are ones which suddenly appear or more commonly suddenly increase in intensity so as to be felt by swimmers and surfers, while traveling rips progressively move along the shoreline (SLSA, 1987). It is often hard to distinguish between the two when in the water, because when at a particular point whether the unnoticed rip increases in intensity or moves into the area, the result is still the same, namely, a sudden increase in the current force moving offshore. The hydrodynamics of these systems is complex and is related to the generation of eddies of water within the surf zone (Feddersen, 2014).

Fig. 5.6 Sand moving in rips (**a**) adjacent to a rock groyne and (**b**) as a defined flow beyond a low energy surf zone (end of rip marked by yellow arrow). (Photo (**a**) Rob Brander, (**b**) David M. Kennedy)

5.2.3 Rips, Sediment Movement, and Feedback

In swell-dominated environments, such as the east coast of Australia, usually there are one to two bars (Wright & Short, 1984). In sea-dominated environments, such as the North Sea coast of Europe and within embayments, over a dozen bars may be present on a given beach (Short & Aagaard, 1993) (Fig. 5.7). The position of bars offshore has been related to the interaction between gravity and infragravity frequency waves (Chap. 3). Gravity waves (the ones you surf) are the primary energy drivers of sediment movement, while infragravity waves (waves with a period of up to 5 minutes) modulate where this energy in concentrated through time. As sea- and swell-dominated environments have different wave energies regimes, we therefore observe stark variations in the number of sand bars in the surf zone (Fig. 5.7).

As the interaction of gravity and infragravity waves determines the number and offshore position of bars, it also affects the movement of water within the surf zone and therefore the number and positioning of rips. In areas dominated by short period waves (e.g., strong wind embayments or moderate energy seas), there are many small rips, up to 12 per 1 km of shoreline (Fig. 5.8). As energy increases within the surf zone, the number of rips starts to decrease but their size significantly increases. On high energy west coast swell environments, there are often only two rips per kilometer of coast, and their width is double to triple that of the strong wind bays. At the highest energy end, strong longshore currents are persistent and only one rip may be present, such as found at Muriwai Beach in New Zealand (Short & Brander, 1999). This pattern of change does not always occur on all beaches because it

Fig. 5.7 (**a**) A series of multiple bars found at a sea-dominated fetch-limited beach at St Leonards, Port Phillip Bay, Australia. (Photo: Ruth Reef). (**b**) A single shore-attached bar dissected by a rip on the swell-dominated coast of Philip Island, Australia. (Photo: David M. Kennedy). (**c**) Two shore parallel bars at Raumati Beach, New Zealand, where the beach is exposed to high energy sea and swell. (Photo: Nick Boyens)

assumes that the rips are formed solely through hydrodynamic factors, that is, purely on the basis of the distribution of energy within the surf zone. In many, or even most, cases, currents in the surf zone will encounter a solid structure and be diverted offshore, forming a topographically controlled, permanent or fixed type rip.

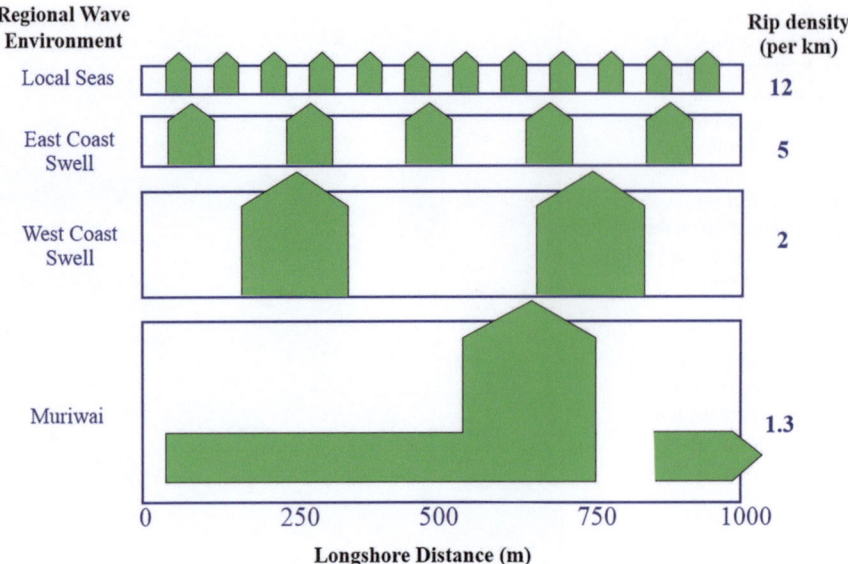

Fig. 5.8 The number of rips and their size can be directly related to the wave energy found at a particular coastal location. (Modified from Short & Brander, 1999)

5.3 Beach Morphodynamic Models

The feedback that occurs within the surf zone between water moving toward the beach first as a steepening wave and then as a bore and water returning offshore through rip currents results in large amounts of sediment movement (Fig. 5.6). This causes a continual adjustment in beach form as bars move offshore during storms and back toward the beach during calmer periods. As the bars and rips move, it affects the pattern of wave breaking. In high energy settings, wave energy is dissipated before it reaches the beach, but in low energy settings, it can reach the beach directly and surges onto the shore. These two scenarios define the end members of a morphological continuum of beach types, namely, reflective and dissipate beaches (Bryant, 1982). This continuum has been conceptualized in the beach bar models of Wright and Short (1984) and Lippmann and Holman (1990). The following section summarizes these beach models.

The lowest energy beach state is a reflective beach. There are no bars on this type, and waves can surge up the beach or plunge directly on its toe. One of the characteristic morphologies found in this setting is a cusp. This is shaped like a mini bay with a trough (or embayment) occurring between two horns (Fig. 5.9a). While shore breaks are rarely used for surfing, they are important in a hazard context as people are often caught unaware by the wave uprush due to the lack of an offshore surf zone. When the beach is composed of gravel-size material, subtidal bars do not generally form due to the energy required to move the larger size sediment (Jennings

Fig. 5.9 (**a**) Beach cusps found in gravelly sediment at Whirinaki Beach, New Zealand. (**b**) A low-tide terrace at Manly Beach, Australia. (**c**) North Cronulla Beach, Australia, showing a transition between low energy reflective forms through a low-tide terrace (LTT) to a higher energy transverse bar rip (TBR) morphology as wave height increases along the beach. (Photos: David M. Kennedy)

& Shulmeister, 2002). Gravel beaches are very common in New Zealand, and the shore breaks can be very energetic easily being meters in height (Fig. 5.9a).

As wave energy increases, a subtidal, or low intertidal elevation, terrace is formed that is near horizontal in slope (Fig. 5.9b). Termed a low-tide terrace (LTT), wave energy is quite variable during different stages of the tide. At high tide, the beach face is often reflective, but at low tide, waves become spilling and dissipate most of their energy across the terrace. Water movement across the intertidal profile can increase in energy to the extent that rips start to be incised into the terrace, but often they are only active during mid tidal elevations.

At the next stage, with greater wave energy, there are larger volumes of water moving through the surf zone, and in turn the strength of the rip currents increase. The rips are able to rework the subtidal terrace, and the feeder currents can also start to move sand. The result is the rip channels become deeper, and the bars rather than being a continuous terrace feature become elongated and point offshore. For these beaches, rips are very common, and like the bars, the rip channels form at angles to the beach. The morphology therefore becomes a series shallow bars and troughs orientated transverse to the shoreline in an alongshore direction (transverse bar rip

(TBR) beach) (Fig. 5.9c). The bars are generally attached to the beach at their edge. The resulting surf break is broken with waves preferentially breaking on the end of each bar.

Further increases in wave height lead to the bars to become detached from the beach. Large troughs develop between the bars, and this can lead to meters' difference in water depth over short distances in the surf zone. Strong rip circulation still occurs on these beaches, and this is manifest in a rhythmic, cuspate-like shape to the bar systems (called a rhythmic-bar beach). As energy increases, the bars become more continuous and less rhythmic (longshore bar trough) (Fig. 5.7c) until a fully dissipative beach state is reached. In this final stage, there are no rips, and water moves through the surf one as a sheetlike flow along the seabed, termed undertow.

The morphodynamic state of the beach, that is, the form and position of the bars and rips in the surf zone, directly determines the type of break that is available for surfers. Reflective beaches only have shore breaks. These are dangerous locations as the wave energy is expended directly on the exposed seabed, leaving no room for error when riding the wave. As wave energy is transformed as it moves through the surf zone, shore breaks can also occur when waves reform after breaking on offshore bars. Such reflective environments at the toe of the beach are therefore not unusual, except when the beach is fully dissipative.

Intermediate-type beaches, those where the bars form offshore and are shaped by rips, have the most diverse form of surf breaks. Left and right handers can occur over each of the bars and around the rip heads all due to the complex three-dimensional topography of the seafloor in the surf zone (Chap. 4).

Fully dissipative beaches on the other hand are characterized by lines of breakers that progressively become smaller the closer to the shoreline. Spilling breaks are the characteristic form, and while such beaches are found in very high energy locations, accessing the open water is very difficult as the beach can be considered "closed out" for most of the time.

5.4 Australian Beach Safety Management Program (ABSAMP)

The interconnected relationship between beach shape, waves, rips, and the resulting surf breaks has been the focus of much attention for hazard managers. Big wave events will break further offshore and create sand bars separated from the shore by deep troughs, while low energy waves break closer to the shore and create shore breaks and surge up and down the beach. The question for hazard managers is whether a more nuanced and predictive relationship can be found—one that can be accessible to all beach users regardless of their general and local knowledge about a location.

An underlying component of the principle of morphodynamics is sediment, that is, the size of the sand and gravel of which the beach is composed. In general, higher energy beaches especially those which are, or are close to being, dissipative in nature are finer grained than those which are reflective (Davis and Fitzgerald, 2004). For hazard management, it has been found that there is a close and predictable relation between the form of the beach, its composition, and the waves that break upon its shore. This relationship is termed the dimensionless fall velocity (Ω) (Gourlay, 1968):

$$\Omega = \frac{H_b}{T w_s}$$

where H_b is the breaking wave height, T is the wave period, and w_s is the fall velocity of sand. Fall velocity is the speed at which sand falls through water, and it is directly related to the size of the particle. Other relationships also exist such as the surf scaling parameter which calculates the degree of reflection and dissipation of waves in the surf zone (Guza & Inman, 1975), but the dimensionless fall velocity has proven itself to be the best predictor of the different types of intermediate-type beaches (Table 5.2).

Using this fundamental relationship between sand and wave size, Surf Life Saving Australia developed an integrated hazard framework called the Australian Beach Safety Management Programme (ABSAMP) and applied it to every beach on the Australian coast (Short & Hogan, 1994; Short et al., 1993). It has subsequently been applied in the UK, the USA, and New Zealand (Short, 1996). ABSAMP includes other factors such as ease of access and distance to emergency services to rate beaches on a scale of 1–10. As the dimensionless fall velocity is calculated on wave parameters of height and period, the ABSAMP framework is also dynamic in that it can be updated in accordance with the wave conditions at any given time. The end product is the BeachSafe app, where wave forecasts are directly attributed to static variables such as sediment size to provide a real-time assessment of beach hazard (Fig. 5.10). For surfers, this provides a powerful information tool, which, when combined with their own knowledge of a particular break, allows an assessment of the surf conditions to be made without ever leaving the house.

Table 5.2 The dimensionless fall velocity is an excellent predictor of beach type

Beach type	Dimensionless fall velocity
Dissipative	>6
Longshore bar trough	~5
Rhythmic-bar beach	~4
Transverse bar rip	~3
Low-tide terrace	~2
Reflective	<1

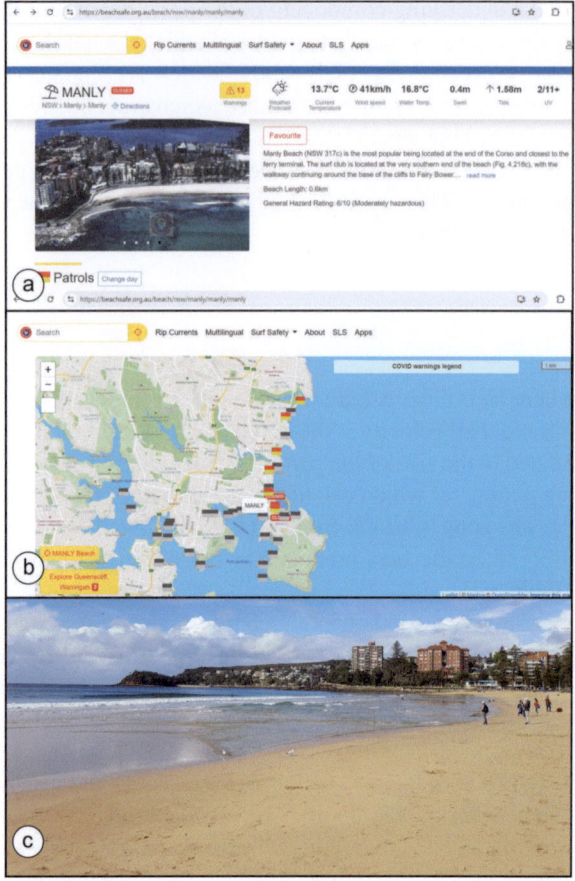

Fig. 5.10 (**a**) The opening page of the BeachSafe program for Manly Beach in Sydney, Australia. The page shows the weather conditions at the time of viewing and the general hazard rating (6/10). It also indicates that on this day, the beach is closed due to unsafe water conditions. (**b**) A map view of beaches around Manly, including others in the Northern Beaches of Sydney and Sydney Harbour. (**c**) A view of Manly Beach on a calm day. (Photo: David M. Kennedy)

5.5 Headland Breaks and Rocky Coast Safety

The success of the ABSAMP framework in preventing fatal drownings is undeniable (SLSA, 2014b), but it is not transferrable to other landform systems such as rocky coasts. The surf breaks here are developed on rocky formations such as reefs or submarine projections of headlands that allow waves to wrap around and create point breaks (Chap. 4). As there is little to no sand, an index based on sand size is therefore irrelevant. The challenge is therefore to identify the fundamental variables of the landscape which can be used to predict hazard.

This is important because while drownings on beaches have reduced, particularly relative to population growth, they have increased on rocky coasts and now account for 19% of coastal fatalities (SLSA, 2014a, b). An important element of assessing the risk for these locations is that, unlike beaches, most users of shore platforms such as rock fishers and walkers do not plan on entering the water and therefore are neither equipped in an emotional or physical sense for sudden immersion. A classic example is a mass rescue of over 100 people at the Fig. 8 Pools south of Sydney when day trippers to a series of rock pools were caught unawares by a larger breaking wave surging across the shore platform (Mail, 2016). This is unlike surfers who often use these areas as a jump-off spot for accessing harder to reach breaks.

Rock coasts are landforms that form vertical cliffs that plunge into the sea, or when geological conditions are right, ledges termed shore platform may develop (Stephenson et al., 2013). Shore platforms are the result of the retreat of a landward cliff that leaves a level ledge of rock at intertidal elevations. Their exact form is the result of a complex interaction of erosive processes both on land and from the sea, and they may take hundreds of thousands of years to form (Kennedy et al., 2014). In general, the harder the rock in which they are cut, the higher elevation of the shore platform. Their shape is semi-horizontal in the intertidal zone but can have a slope of several degrees in areas with a high tide range (Trenhaile, 1987). In microtidal settings, the edge of the platform is often characterized by a drop-off (termed a seaward cliff), though it may also be a ramp or a ledge at several meters water depth (Fig. 5.11). As waves break as a function of the water depth they encounter

Fig. 5.11 (**a**) A shore platform near Curl Curl Beach in Sydney, Australia, with a steep drop-off on its seaward edge where people are fishing. (**b**) A platform which slopes gradually offshore at Cape Paterson, Victoria, Australia. (Photos: David M. Kennedy)

close to shore—how quickly a shore platform deepens will determine where waves break. This in turn affects how much waves can inundate the intertidal ledges and therefore catch people unawares.

For hazard management, using the water depth immediately offshore of the intertidal edge of the platform has been shown to be a useful metric for analyzing danger (Kennedy, 2016; Kennedy et al., 2012, 2017). In this framework, the water depth within 100 m of the intertidal edge of the platform provides a basis for calculating hazard as it can account for the range of seaward morphologies that can occur in nature. This morphological exposure approach (Kennedy et al., 2013) has been successfully tested on the southern rocky shores of the Australian coast (Kennedy et al., 2017) and holds great potential for increasing safety for rock coasts worldwide.

5.6 Conclusion

Reducing the risk of drowning on the coast has been a challenge since people first started to use the water. Only in the past century has this become a specific focus of coastal managers, especially as the beach has become more popular as more people started surfing. Stationing paid and volunteer life guards at locations of high risk is one of the most effective strategies of saving lives. However, this is resource intensive in terms of both equipment, infrastructure and personnel and therefore cannot be undertaken at every location that people visit. Even with the benefit of bystander rescues (Chap. 6), of which surfers are highly adept, it is still impossible to be everywhere at every time.

Understanding the fundamentals of the principal dangers of the surf zone, namely, breaking waves and rips, allows planners to target their limited resources to priority locations. The critical natural features for planning are the position and strength of rips, as well as the broad pattern of bars, all of which are relatively predicable based on the boundary conditions of each location. Areas of larger waves produce a limited number of large rips, with bars being found offshore. In stormy environments, multiple bars are found and in moderate conditions the bars and rips form intricate shapes along the beach. The precise shape is determined by the wave height and period as well as the sediment size. This is encapsulated in the dimensionless fall velocity parameter. As the sediment found on a beach generally remains the same size through time, the shape of the surf zone can then be quickly predicted using just the wave parameters. As a result, beach safety programs such as ABSAMP can provide real-time danger ratings for beaches through linking with wave forecasting apps. Similar geomorphologically derived frameworks can be also applied to rocky coasts where the depth immediately offshore of a shore platform is also a proxy for the danger of a particular location. Understanding the basic principles of earth science and wave physics therefore has a direct impact on safety on the coast and reducing the number of fatal and near-fatal drownings.

References

Bryant, E. (1982). Behavior of grain size characteristics on reflective and dissipative foreshores, Broken Bay, Australia. *Journal of Sedimentary Petrology, 52*, 431–450.

Castelle, B., Scott, T., Brander, R. W., & McCarroll, R. J. (2016). Rip current types, circulation and hazard. *Earth-Science Reviews, 163*, 1–21.

Davis, R. A., & Fitzgerald, D. M. (2004). *Beaches and coasts*. Blackwells.

Feddersen, F. (2014). The generation of surfzone eddies in a strong alongshore current. *Journal of Physical Oceanography, 44*(2), 600–617.

Gourlay, M.R., 1968. Beach and dune erosion tests.

Guza, R. T., & Inman, D. L. (1975). Edge waves and beach cusps. *Journal of Geophysical Research, 80*, 2997–3012.

Jennings, R., & Shulmeister, J. (2002). A field based classification scheme for gravel beaches. *Marine Geology, 186*, 211–228.

Kennedy, D. M. (2016). The subtidal morphology of microtidal shore platforms and its implication for wave dynamics on rocky coasts. *Geomorphology, 268*, 146–158.

Kennedy, D. M., Brighton, B., Woodroffe, C. D., Weir, A., Sherker, S. (2012). *Meeting the challenge of preventing drowning deaths on the rocky coast*, Australian Water Safety Council Conference, Sydney.

Kennedy, D. M., Sherker, S., Brighton, B., Weir, A., & Woodroffe, C. D. (2013). Rocky coast hazards and public safety: Moving beyond the beach in coastal risk management. *Ocean and Coastal Management, 82*, 85–94.

Kennedy, D. M., Stephenson, W. J., & Naylor, L. A. (2014). *Rock coast geomorphology: A global synthesis*. Geological Society of London.

Kennedy, D. M., Ierodiaconou, D., Weir, A., & Brighton, B. (2017). Wave hazards on microtidal shore platforms: Testing the relationship between morphology and exposure. *Natural Hazards, 86*, 741–755.

Komar, P. D. (1998). *Beach processes and sedimentation* (2nd ed.). Prentice Hall.

Lippmann, T. C., & Holman, R. A. (1990). The spatial and temporal variability of sand bar morphology. *Journal of Geophysical Research, 95*, 11575–11590.

Loureiro, C., Ferreira, Ó., & Cooper, J. A. G. (2012a). Extreme erosion on high-energy embayed beaches: Influence of megarips and storm grouping. *Geomorphology, 139*, 155–171.

Loureiro, C., Ferreira, Ó., & Cooper, J. A. G. (2012b). Geologically constrained morphological variability and boundary effects on embayed beaches. *Marine Geology, 329–331*, 1–15.

MacMahan, J., Brown, J., Brown, J., Thornton, E., Reniers, A., Stanton, T., Henriquez, M., Gallagher, E., Morrison, J., & Austin, M. J. (2010). Mean Lagrangian flow behavior on an open coast rip-channeled beach: A new perspective. *Marine Geology, 268*(1-4), 1–15.

Mail, D., 2016. Tourists obsessed with getting the perfect Instagram picture are 'risking their lives' at the figure 8 rock pool – Where a freak wave smashed hundreds of swimmers.

McCarroll, R. J., Brander, R. W., MacMahan, J. H., Turner, I. L., Reniers, A. J. H. M., Brown, J. A., Bradstreet, A., & Sherker, S. (2014a). Evaluation of swimmer-based rip current escape strategies. *Natural Hazards, 71*(3), 1821–1846.

McCarroll, R. J., Brander, R. W., Turner, I. L., Power, H. E., & Mortlock, T. R. (2014b). Lagrangian observations of circulation on an embayed beach with headland rip currents. *Marine Geology, 355*, 173–188.

NOS. (2024). *What is a rip current?* National Ocean Service, National Oceanic and Atmospheric Administration.

NWS. (2024). *How to avoid getting caught in a rip*. National Weather Service.

Scott, T. M., Russell, P., Masselink, G., Austin, M., Wills, S., & Wooler, A. (2011). 14 rip current hazards on large-tidal beaches in the United Kingdom. In *Rip currents: Beach safety, physical oceanography, and wave modeling* (pp. 225–243). CRC Press.

Short, A. (1996). *Beaches of the Victorian Coast and Port Phillip Bay. Australian Beach Safety and Management Programme*. University of Sydney Press.

Short, A. D., & Aagaard, T. (1993). Single and multi-bar beach change models. *Journal of Coastal Research, 15*, 141–157.

Short, A. D., & Brander, R. W. (1999). Regional variation in rip density. *Journal of Coastal Research, 15*, 813–822.

Short, A., & Hogan, C. L. (1994). Rip currents and beach hazards: Their impact on public safety and implications for coastal management. *Journal of Coastal Research, SI12*, 197–209.

Short, A., Williamson, B., Hogan, C. L. (1993). *The Australian beach safety and management programme - surf life saving Australia's approach to beach safety and coastal planning, 11th Australasian conference on coastal and ocean engineering* (pp. 113–118). Institution of Engineers, Australia, Barton, ACT.

SLSA. (1987). *Surf life saving training manual*. Surf Lifesaving Association of Australia.

SLSA. (2014a). *Annual report 2013/14*. Surf Life Saving Australia.

SLSA. (2014b). *National coastal safety report 2014*. Surf Life Saving Australia.

SLSA, 2021. Coastal safety brief – rip currents 2021, .

SLSNSW. (2024). *What is a rip?* Surf Life Saving New South Wales.

SOS. (2024). *Rip current safety*. Dr Rips Science of the Surf.

Stephenson, W. J., Dickson, M. E., & Trenhaile, A. S. (2013). Rock coasts. In D. Sherman (Ed.), *Coastal geomorphology*. Academic Press.

Trenhaile, A. S. (1987). *The geomorphology of rock coasts*. Clarendon Press.

Woodroffe, C. D. (2003). *Coasts: Form, process and evolution*. Cambridge University Press.

Wright, L. D., & Short, A. D. (1984). Morphodynamic variability of surf zones and beaches: A synthesis. *Marine Geology, 56*, 93–118.

Wright, L. D., & Thom, B. G. (1977). Coastal depositional landforms: A morphodynamic approach. *Progress in Physical Geography, 1*, 412–459.

Chapter 6
Lifeguards in the Lineup: Surfers as Rescuers

Robert W. Brander, William A. Koon, and Amy E. Peden

6.1 Introduction

Beach safety is a topic of global significance, and drowning at beach locations is a recognized global health and management issue associated with significant emotional, societal, and economic costs (Houser et al., 2021; Sherker et al., 2008). Tragically, each year many people drown while swimming, wading, or surfing at beaches around the world (Koon et al., 2021a; Lawes et al., 2023), and while the exact number of people who drown each year along beaches is unknown, largely due to challenges associated with accurate and consistent data reporting (Franklin et al., 2020; Koon et al., 2021b), it is likely in the order of thousands, if not tens of thousands.

In Australia, which is known for its beach and surf culture (Ellison & Brien, 2020; Jaggard, 2007), approximately 70–80 people unintentionally drown each year on beaches (Surf Life Saving Australia, 2023), and according to US Lifesaving Association (USLA) statistics, at least 110 people drown on beaches in the USA annually. Elsewhere, analysis of Costa Rican drowning data found that an average of 49 deaths per year occurred between 2001 and 2019 on the country's beaches (Segura et al., 2022). In Brazil, open water locations, including beaches, accounted for 41% of the 7005 unintentional drowning deaths per year (Szpilman et al., 2020). Undoubtedly, due to their appeal, beaches unfortunately represent a common location for drowning.

The number of beach drowning fatalities would be even higher if not for the presence of lifeguards (Branche & Stewart, 2001). Globally, lifeguards make tens of thousands of rescues at beaches each year (Leatherman et al., 2024; Szpilman et al., 2020). However, the reality is that not all beaches are patrolled by lifeguards and

R. W. Brander (✉) · W. A. Koon · A. E. Peden
UNSW Sydney, Sydney, NSW, Australia
e-mail: rbrander@unsw.edu.au; w.koon@unsw.edu.au; a.peden@unsw.edu.au

© The Author(s), under exclusive license to Springer Nature
Switzerland AG 2025
D. M. Kennedy (ed.), *The Science and Culture of Surfing*,
https://doi.org/10.1007/978-3-031-80979-8_6

even on beaches that are, lifeguards are not present on the beach at all times, or in all locations. It is therefore not surprising that the vast majority of beach drownings occur on unpatrolled beaches, significant distances away from lifeguard services, or at times outside of lifeguard patrol hours (Koon et al., 2023a, b; Surf Life Saving Australia, 2023; Venkateswarlu et al., 2023).

In the absence of lifeguards in these situations, it is not uncommon for bystanders to make a rescue in beach and coastal environments (Brander et al., 2019; Franklin & Pearn, 2011; Franklin et al., 2019; Lawes et al., 2020; Moran et al., 2017). The response and action of bystanders helps save lives by both rescuing someone from the water and providing assistance to that person after the rescue, such as the application of cardiopulmonary resuscitation (CPR). However, the majority of bystander rescuers are not trained or experienced in water-based rescue and basic first aid and CPR (Franklin et al., 2019), and as a result, these situations frequently place both the bystander rescuer and the person they are attempting to rescue at great risk (Moran et al., 2017). Tragically, it is not uncommon for bystander rescuers to drown while attempting to save the life of another (Franklin & Pearn, 2011; Lawes et al., 2020).

Given the heightened risk of bystander rescue situations, one group particularly well placed to act as bystander rescuers at beaches is recreational surfers. Surfers of all types (shortboard, longboard, bodyboard, kite, stand-up) are generally physically fit and competent swimmers. They are experienced in wave and surf conditions and are particularly familiar with their local beaches and surfing locations. They generally have a flotation device in the form of their board to assist in conducting a rescue (Berg et al., 2021). Previous research has identified the importance of flotation devices in assisting bystander rescues (Barcala-Furelos et al., 2021; Beale-Tawfeeq, 2019) and that a key risk factor in bystander rescuer fatalities is attempting to conduct the rescue without a floatation device (Lawes et al., 2020).

While most beaches are only lifeguarded during warmer months, particularly summer, surfing is a year-round activity, and surfers also frequently surf at unpatrolled beaches that do not have any lifeguard services. On patrolled beaches, surfers are also often in the water both before and after lifeguard services are operational, such as early in the morning or late in the evening, or at locations away from the lifeguard patrols and crowds of bathers during patrol hours (Attard et al., 2015; Mead et al., 2024). Surfers are therefore present in many locations and at times when lifeguards are not.

Surfers also commonly use offshore flowing rip currents, as an easier pathway to paddle out through areas of intense wave breaking and also as a quick way to navigate through heavy and potentially dangerous shore break when present (Berg et al., 2021) (Chap. 5). Surfers also often frequently surf near headlands which are typically characterized by the presence of boundary rip currents (Castelle et al., 2016) and near channelized rip currents along open stretches of beach, which often promote better surfing breaks. Given that rip currents are the primary hazard to swimmers and bathers on surf beaches (SLSA, 2023), the fact that surfers both use and surf near them means that they are well placed to assist people who are experiencing trouble in rip currents (Fig. 6.1).

Fig. 6.1 Surfer Zeb Boeskool (center) successfully rescues a teenage girl from the surf at Grand Haven State Park Beach, Michigan, in Lake Michigan in the Great Lakes. The dramatic rescue is described in the documentary Rip Current Rescue, which was shown on the Public Broadcasting Service in the USA. (Image courtesy of Mike Dixon)

With an estimated global surfer population of between 17 and 35 million (CBI, 2018; Nathanson, 2020; Sports Medicine Australia, 2021; Surfer Today, 2018), there are always a lot of surfers in the water, all over the world. It is therefore not surprising that several studies have shown that surfers of all types make a considerable number of rescues (Berg et al., 2021; De Oliveira et al., 2023; Dehez et al., 2024; Mead et al., 2024). However, this valuable societal role in drowning prevention by surfers often goes unnoticed and unrecognized.

This chapter explores rescues conducted by surfers and the essential role surfers have in beach safety. In doing so, it will provide valuable information on the extent, characteristics, and context of rescues conducted by surfers, their opinions and attitudes toward making rescues, as well as important information about the surfing population itself. It will also explore the value and benefits of programs that train surfers in basic rescue, first aid, and CPR, discussing existing programs that are available to surfers globally in this regard. The overall aim of the chapter is to provide a greater understanding of the role that surfers play as unofficial, but crucial, guardians of the surf and to identify future opportunities to better recognize, and strengthen understanding of, this vital lifesaving contribution.

6.2 Surfers Conducting Rescues

Surfers, including surfing pioneer Duke Kahanamoku himself (Tomizawa, 2016) (Chap. 2), have undoubtedly been saving lives in the ocean for as long as surfing has been a popular recreational activity. However, while the surfing community and those who have been helped by surfers are largely aware of the important role that surfers play as informal lifeguards, the general public are not. While some high-profile stories of surfers making rescues have received media attention (e.g., Bradley, 2023; Meacham, 2021; Payne, 2016; Rachwani, 2021), the vast majority of surfer rescues go unnoticed and undocumented. Only recently have surfer rescues begun to gain formal attention, primarily through a number of research studies involving online surveys of surfers that have been conducted globally (Table 6.1). Collectively, these studies have yielded important information about the extent and scope of surfer rescues, the circumstances contributing to these rescues, the person(s) being rescued, the value of rescue training for surfers, and also some of the inherent challenges associated with documenting the occurrence of surfer rescues.

6.2.1 Characteristics of Australia and New Zealand Surfer Rescues

The first attempt at documenting surfer rescues began in 2013 when a group of researchers from the University of New South Wales in Sydney, Australia, developed an online survey designed to attempt to quantify the number of rescues performed by Australian surfers (of all types). The survey inquired about the geographic location and environmental conditions associated with these rescues and how the surfers perceived and reacted to their rescue experiences, including what they considered to be their most serious rescue (Attard et al., 2015). The survey was primarily promoted through the social media network of Swellnet Australia, a popular surf forecasting organization in Australia, and the almost immediate response of over 500 surveys was indicative of the high interest and engagement that surfers had on this topic.

Results of the survey and the eventual published study by Attard et al. (2015) showed that, even using the most conservative estimates and assumptions, surfers were making an equal number of rescues on Australian beaches each year as those conducted by professional beach lifeguards and volunteer surf lifesavers. Approximately 80% of the surfers had performed at least two rescues over their surfing career with almost a quarter indicating they had performed more than five (Table 6.1). Surfers who had previously had formal water safety training, which in Australia is primarily through the large and popular volunteer surf lifesaving movement (Cornwall, 2007), were more likely to have performed a higher number of rescues. Surfing experience and self-rated surfing ability were also important with more experienced and skilled surfers performing a higher proportion of rescues.

Table 6.1 Summary of key characteristics of surfer rescues based on existing research studies utilizing online surveys of surfers

Study	Attard et al. (2015)	Berg et al. (2021)	De Oliveira et al. (2023)	Mead et al. (2024)	Dehez et al. (2024)	GSS
Region Characteristics	Australia	Europe	Portugal/Spain	New Zealand	France	NSW Australia
Year surveyed	2013	2015	2020	2022	2021–2022	2021–2023
# responses	545	1705	1190/858	418	569	773
% male-female	95-5	76-24	91-9	87-13	81-19	75-25
Surfer rescuers						
Performed rescue	80%	39%	79%	62%	56%	68%
% male-female performed rescue	n/a	45-22	n/a	n/a	n/a	n/a
# rescues lifetime	n/a	n/a	n/a	3	n/a	3.9
Focus rescue Type: Most	Serious	–	Serious	Recent	Recent	Recent
Rescuee swimming	63%	–	–	57%	55%	64%
Rescuee surfing	25%	–	–	29%	40%	35%
Rescuee male	n/a	–	n/a	71%	73%	72%
Most common rescue age	18–29 (48%)	–	–	15–25 (36%)	20–30 (37%)	20–30 (37%)
Rip current related	75%	–	86%	75–100%	66%	80%
Patrolled locations	45%	–	n/a	27%	20%	32%
Unpatrolled locations[a]	53%	–	51%	73%	74%	61%
Surfer training[b]						
Interested in training/course	–	n/a	–	–	65%	57%
Already taken course/training	–	53%	64%	25%	–	29%
Agree surfers should do course	–	35%	–	–	82%	76%

GSS = Global Surfer Survey; NSW = New South Wales. "n/a" means that the information is available, but is not reported in the publication. Full publication details are provided in the References
[a]Includes both unpatrolled locations and outside of lifeguard patrol times
[b]Training refers to lifeguard, lifesaving, or cardiopulmonary resuscitation (CPR) training or courses

While most (78%) of the surfers who conducted rescues said they were "happy to help," approximately a quarter also expressed feelings of annoyance or inconvenience, generally toward the rescues for being in dangerous locations or toward the lifeguards/lifesavers for not making the rescue first or being aware that someone was in trouble. In terms of the reported "most serious rescue," 68% of surfers felt that their intervention had saved a life (Attard et al., 2015). This latter finding, in particular, illustrates the vital role that surfers play in reducing the coastal drowning toll in Australia.

In terms of the circumstances of the rescues themselves, Australian surfers performed their most serious rescue on beaches that were both patrolled by lifeguards/lifesavers (45%) and beaches that were unpatrolled (53%; Table 6.1). Most of the rescue events occurred during conditions of moderate wave heights and energy conditions and fine, sunny weather with rip currents being identified by surfers as the primary cause of their rescues in general (75%; Table 6.1). While most (63%) of the people rescued by surfers were swimmers, a quarter of the rescues were of other surfers, either due to rip currents (55%) or impact injuries (30%). The largest demographic of the people rescued by surfers were males aged between 18 and 29 years old (Attard et al., 2015).

More recently, the UNSW Beach Safety Research Group, the International Drowning Researchers Alliance (IDRA), and researchers in France (who are surfers) from the Université de Bordeaux, National Centre for Scientific Research (CNRS), and the National Research Institute for Agriculture, Food, and Environment (INRAE) developed the online Global Surfer Survey (GSS), which contains a section on surfer rescues and was launched in November 2021 at https://www.beachsafetyresearch.com/gss. In collaboration with Surfing New South Wales (SNSW), the leading authority responsible for promoting competitive and recreational surfing in New South Wales (NSW), Australia's most populated state, the GSS was promoted through SNSW social media pages and a media release and generated 773 responses from surfers in NSW (Table 6.1; Peden et al., 2023). Approximately two-thirds (68%) of the surfers had rescued someone while surfing with an average of approximately four people rescued during their surfing career (Table 6.1). When combined with estimates of the number of surfers in NSW (SportAus, 2019), the results of the GSS study suggest that an average of almost 12,000 rescues are made by NSW surfers and approximately 2000 lives are saved per year (Peden et al., 2023).

In 2022, an online survey of surfers in New Zealand found that surfers conducted an average of three rescues during their surfing career, primarily at their local surfing location with 75% being done in the absence of lifeguards (Table 6.1; Mead et al., 2024). Similar to the Australian study (Attard et al., 2015), more experienced and skilled surfers in New Zealand were more likely to make rescues, and most rescues were related to rip currents (Mead et al., 2024). The New Zealand study asked surfers for more details about their most recent rescue with 75% of the surfers reporting that they had some, or a lot of, difficulty conducting the rescue and almost half feeling that they had saved a life. Importantly, Mead et al. (2024) identified the need for governments, policymakers, and beach and water safety practitioners working in drowning prevention to be aware of, acknowledge, and work closely with surfer communities on ways to reduce drowning at surf beaches.

6.2.2 Characteristics of European Surfer Rescues

In 2015, members of Surfing Medicine International (SMI) adapted the Australian-based survey of Attard et al. (2015) and promoted an online survey geared primarily to European surfers through Magicseaweed, a popular European surf website, the World Surf League (WSL), O'Neill, and various other surfing associations and surf magazines (Berg et al., 2021). A total of 1705 surveys were used for analysis representing surfers from 23 different countries with most respondents being from the Netherlands (24%), the UK (19%), and Germany (12%). Results of the survey found that 39% of all respondents had previously rescued someone in the surf, which is approximately half of what was reported by Australian surfers in the Attard et al. (2015) study (Table 6.1). Similar to the Australian study, however, surfers with more years of surfing experience and surfers of higher self-reported ability were more likely to have conducted a rescue (Berg et al., 2021). Importantly, there was a noted geographic pattern among the occurrence of European surfer rescues with 97% of surfers who had reported making a rescue having done so on beaches along the Atlantic Ocean coast, which is characterized by larger waves and better surfing conditions than beaches along the North Sea and Mediterranean Sea (Berg et al., 2021).

The study on European surfers by Surfing Medicine International also took a public health approach in their survey by asking questions about injury, first aid training, and health and safety perceptions and precautions of the surfers (Berg et al., 2021).

The survey results showed that surfers who had previous lifeguard and/or CPR training were more likely to have made a rescue while surfing. Surfers who had conducted a rescue were also more likely to believe that some kind of basic lifesaving and CPR course, similar to those described in Sect. 6.3 of this chapter, should be obligatory for all surfers. Approximately two-thirds (68%) of the surfers indicated they were interested in receiving information about how to stay healthy and safe while surfing (Berg et al., 2021).

More recently, surveys have been conducted which have focused on Portuguese and Spanish surfers (De Oliveira et al., 2023) and French surfers (Dehez et al., 2024) and their experiences conducting rescues on beaches in their own country. The Portuguese/Spanish study distributed an online survey in 2020 via common social media platforms of popular surfing communities in Portugal and Spain and newsletters of the Portuguese and Spanish surfing federations (FPS and FES, respectively). The French study utilized the Global Surfer Survey (GSS) that was promoted in France with the support of the French National Surfing Federation (FSS), online French surfing magazines, and French surfing social networks. A total of 2048 Portuguese/Spanish and 713 French surfer surveys were obtained (De Oliveira et al., 2023; Dehez et al., 2024) with 79% and 56% of the Portuguese, Spanish, and French surfers reporting to have conducted at least one rescue during their surfing career, respectively (Table 6.1). In the Portuguese/Spanish study, 14% of the respondents reported having made ten or more rescues while surfing. Similar to the studies described above, the number of rescues performed by surfers increased with years of surfing experience and self-reported surfing ability.

The Portuguese/Spanish surfer survey was very similar to the original Australian survey (Attard et al., 2015) in question design and focus on the surfers' most serious rescue. In general, most rescues occurred on sunny days (68%) with rough seas (53%) although almost half (47%) were conducted during low to moderate (0.5–1 m) wave heights. Most rescues were conducted when not many people or surfers were around (34%). Like the Australian study, the majority of surfers most serious rescues were related to rescuing someone who was caught in a rip current (96%). Approximately half occurred on either unpatrolled beaches (29%) or outside of patrol hours on lifeguard patrolled beaches (22%; Table 6.1). After conducting their most serious rescue, most surfers (75%) were happy to have helped, and 60% felt they had saved the life of the person they rescued. The majority of surveyed surfers did not have essential knowledge about rescue and resuscitation (De Oliveira et al., 2023). This study also used conservative hypothetical estimates based on their results to suggest that if 80% of Portuguese and Spanish surfers made at least one rescue each over a 25-year period, that would equate to 16,000 rescues per year, and by association, many lives saved (De Oliveira et al., 2023).

The French surfer survey (Dehez et al., 2024) found similar results to previous studies in that most surfer rescues took place during sunny and fine days (64%), took place where and when no lifeguards were present (74%), and were primarily related to rip currents (66%; Table 6.1). They also asked specific questions relating to surfers' opinions about the role of surfers in making rescues as well as their willingness to attend rescue training programs available to surfers. In terms of the former, almost 90% of the surveyed surfers either somewhat agreed (27%), agreed (22%), or strongly agreed (40%) with the statement that surfers have a responsibility to look after the safety of others while surfing. A majority of surfers also somewhat agreed (35%), agreed (17%), or strongly agreed (30%) with the idea that that all surfers should complete a basic lifesaving course, and 65% indicated that they were willing to take part in a lifesaving and rescue training course designed for surfers if it was free.

6.2.3 Other Characteristics of Surfer Rescues

The surveys of surfer rescues described above had some commonalities in terms of the information acquired and reported (Table 6.1), but all of them varied to some degree in terms of their particular focus and types of questions asked, some of which shed additional light on the characteristics of rescues made by surfers. For example, in terms of how their most serious or recent rescues were initiated, Attard et al. (2015) and Peden et al. (2023) found that approximately 80% of the Australian surfers surveyed saw the person in trouble and decided to act themselves. Attard et al. (2015) also reported that 55% of the surfer rescuers paddled their rescuee to shore while 25% paddled them to a safer and shallower location where they could stand up.

It's also important to note that surfers do not always rescue people when they are actually surfing—many happen when surfers are simply on the beach or coast. Experienced and skilled surfers in particular are very comfortable in engaging with potentially hazardous surf conditions and, similar to the publicized incident in 2021 involving Australian professional surfer Mikey Wright on Oahu's North Shore (Rachwani, 2021), can also identify someone in trouble from the beach and enter the water to conduct a rescue, either with their board or without (Fig. 6.1). Results from the Global Surfer Survey of NSW surfers (Peden et al., 2023) showed that surfers make a considerable amount of rescues while not surfing, but from the shore while they are just visiting the beach or coast, sometimes in excess of the number of people they have rescued while surfing.

Surfers also do not always make rescues on their own, or only of one person. Peden et al. (2023) found that approximately a quarter of the surveyed NSW surfers' most recent rescue involved the help of other surfers while 20% involved rescuing two people or more and in their surfing career, 42% of surfers indicated that they had been involved in a mass rescue involving a group of people in trouble. De Oliveira et al. (2023) reported that a quarter of Portuguese and Spanish surfers had been involved in a rescue involving five or more people, and Mead et al. (2024) also found that New Zealand surfers described regularly being involved in mass rescues.

6.2.4 Surfer Injuries and Fatalities

While surfers mostly rescue beachgoers who have found themselves in trouble in the water, the studies described above have also shown that surfers are also rescuing other surfers. Surfing can be a hazardous activity (Chap. 7), and there are numerous ways in which a surfer may find themselves in need of rescue by another surfer. Although experienced surfers are generally comfortable with dynamic wave, current, and tidal conditions, no surfer is immune to finding themselves in situations and conditions outside of their abilities where they need assistance, especially in the case of those novice surfers or those who are still learning. Surfers may also experience unexpected contact with either their own surfboard or someone else's, or a hard surface of sand, rock, or coral, all of which can lead to impact injuries ranging from minor to major skin lacerations and contusions, sprains and strains, dislocations, and fractures (Klick et al., 2016).

A systematic review of published information on surfer injuries (McArthur et al., 2020) found that while severe injuries are rare, most were from being struck by their own board (39%) followed by hurting themselves while paddling or performing a maneuver while surfing (20%) and impact with the seafloor (18%). The most common body region injured in surfer injuries has been shown to be the face, head, and neck and lower limbs (Klick et al., 2016; McArthur et al., 2020; Minasian & Hope, 2022). While it is not possible to quantify the number of surfer injuries, Dehez et al. (2024) found that 82% of surveyed French surfers had previously sustained an injury while surfing with 36% requiring medical attention or a visit to a doctor.

Surfers, particularly older surfers, may also suffer cardiac events. A study of surfer and bodyboarder deaths in Australia between 2004 and 2020 found that 33% of the 155 deaths were related to cardiac conditions (Lawes et al., 2023). This study also found that drowning was the most common cause of death (58%), with more fatalities in general among older surfers aged 55 + years. Regardless of the type or cause, surfer injuries and fatalities are not uncommon suggesting that surfers should also be aware of the importance of not only rescuing fellow surfers but the importance of having first aid and CPR training to assist those who may require lifesaving treatment.

6.2.5 A Synthesis of Surfer Rescues

Studies on surfer rescues reveal consistent patterns regardless of geographic location and have highlighted that surfers perform a significant number of rescues, potentially more than those made by beach lifeguard services. Importantly, they are saving thousands of lives each year. These rescues are predominantly carried out by experienced and skilled surfers, and the minority of those with previous lifeguard, lifesaving, or first aid/CPR training are more likely to make rescues compared to those without this training. Surfers often exhibit a willingness to assist individuals in distress, providing critical lifesaving support on both guarded and unguarded beaches. Their contributions underscore the importance of bystander rescues in enhancing beach safety.

Most surfer rescues occur during conditions of low to moderate wave heights and fine, sunny weather, which supports findings of other research that has explored the relationships between environmental conditions and the occurrence of rip current rescues, fatalities, and surf zone injuries (Castelle et al., 2019; Scott et al., 2014). Rip currents are the main hazard resulting in a surfer rescue which is again consistent with global research and drowning reports that have established rip currents as the main drowning and rescue hazard to swimmers on surf beaches in general (Arozarena et al., 2015; Brander & Scott, 2018; Brewster et al., 2019; Brighton et al., 2013; Ishikawa et al., 2014; Koon et al., 2023a, b; Surf Life Saving Australia, 2023) (Chap. 5). In Australia and the USA, rip currents are estimated to result in an average of 26 and 100 drowning fatalities each year, respectively (Brewster et al., 2019; Cooney et al., 2020), while in Costa Rica, approximately 54 people fatally drown in rip currents on average each year (Arozarena et al., 2015). No doubt the drowning toll due to rip currents would be even higher without the efforts of surfers.

Most individuals rescued by surfers are swimmers and most are young males, which supports the findings of many coastal drowning studies (Anary et al., 2010; Bessereau et al., 2016; Koon et al., 2023a, b; Lawes et al., 2021; Suresh Kumar Shetty & Shetty, 2007). However, it should be noted that surfers are also rescuing fellow surfers, either due to rip currents or impact injuries, which supports findings of research on the occurrence of surf zone injuries (Castelle et al., 2018; Thom et al., 2022; Venkateswarlu et al., 2023). While many surfers have not received any rescue or lifesaving training, the studies have shown that overall, there is support within the surfing community to take part in available training programs for surfers.

6.3 Surfer Safety Training

The findings of the surfer rescue studies described above have clearly shown that surfers who have had some sort of lifesaving or CPR training are more likely to make a rescue and most surfers in general agree that surfers should complete a basic lifesaving course (Table 6.1). Given the important role that surfers play in saving lives in the surf, ensuring they have the skills to act in an emergency is clearly an important facet of injury and drowning prevention at the beach and at surf breaks. In fact, so important is the role of surfers, particularly in preventing drowning—the World Health Organization (WHO) has identified surfers as a key beneficiary group for safe rescue and resuscitation training (World Health Organization, 2017). In this regard, there are a range of programs and training courses around the world which aim to provide safety training to surfers, including recreational surfers, competitive surfers, and big wave surfers. Training commonly comprises theoretical and practical skills on topics such as recognizing someone in distress and how to conduct safe rescues with a surf board, CPR, and first aid skills. This section describes some of the different kinds of programs available around the world, including the theory and practical content taught, as well as the documented impact of such programs.

6.3.1 Training Programs

One of the first rescue training programs for recreational surfers was the Surfers Rescue 24/7 (SR24/7) program developed in 2012 by Surfing NSW, the leading authority responsible for promoting competitive and recreational surfing throughout the Australian state of New South Wales (Surfing New South Wales, n.d.-a). The SR24/7 program receives state government funding and is free for participants. Since its inception, an estimated 15,000 surfers have participated in the Surfers Rescue 24/7 program (Fig. 6.2), and the growth of the program is evident from the fact that 11,000 of those participants completed the course between 2021 and 2024 (Laweson, 2024). More recently, the program has expanded to three other Australian states (Victoria, Queensland, and Western Australia) and to New Zealand (Koon et al., 2023a).

The Surfers Rescue 24/7 program typically consists of two 90-min courses that comprise both theoretical and practical components. The theoretical component is delivered either indoors or on the beach, and the practical components are conducted both on the beach (Fig. 6.3) and in the water under the supervision of trained instructors (Koon et al., 2023a). To be eligible to participate in the program, surfers must be competent with handling, paddling, and surfing on a surfboard, bodyboard,

Fig. 6.2 Happy surfers having completed the practical component of a Surfers Rescue 24/7 program in NSW, Australia. (Image courtesy of Surfing NSW)

Fig. 6.3 Paddling with a patient training on the beach during a Surfers Rescue 24/7 program in NSW, Australia. (Image courtesy Surfing NSW)

or stand-up paddleboard (SUP), as well as being a competent swimmer, able to swim 100 m in open water unassisted (Surfing New South Wales, n.d.-a). The program is advertised online and via boardrider clubs.

Outside of Australia and New Zealand, there are a range of other programs teaching basic rescue techniques and CPR to surfers in the USA, Brazil (Sociedade Brasileira de Salvamento Aquatico - SOBRASA), and across parts of Europe (e.g., the Association of Surf Schools in Portugal). In Europe, the Surfing Medicine International (SMI) provides courses to surfers including an Advanced Surf Life Support course, which covers topics such as drowning and hypothermia; rescue techniques; theoretical content about waves, tides, and rip currents; site safety; and patient assessment (Surfing Medicine International, 2023). The course is taught by facilitators with a range of expertise including doctors (who are surfers), wilderness medics, drowning prevention specialists, and professional surf lifeguarding instructors.

Another program is offered by the Academy of Surfing Instructors (ASI), a leading training and professional membership organization and international governing body for instructors, coaches, and schools for surfing, stand-up paddle, and body boarding. ASI provides a half day course providing basic rescue skills and knowledge to rescue both conscious and unconscious people using rescue boards or similar boards, in the ocean in small surf conditions (Academy of Surfing Instructors, 2024). This course is provided for a fee and covers topics such as the identification and response to emergency situations, application of surf rescue techniques, demonstration of simulated in-water rescues on conscious and unconscious patients, identification of the signs of potential spinal injuries and how to respond and treat accordingly, and emergency signals and incident reporting.

Given the compounded risk for big wave surfers, there are also a range of courses offered by organizations such as the Big Wave Assessment Group (Hawaii, USA), the Irish Tow Surf Rescue Club (Ireland), Step Up To Bigger Waves (South Africa), K38 Rescue (Global), and online by The Ocean Warrior. These courses are designed for big wave surfers, providing skills training specific to the big wave context, including wave forecasting, breath hold techniques, and safety information toward big wave surf spots (Big Wave Surfing, n.d.).

6.3.2 Rescue Techniques and Cardiopulmonary Resuscitation (CPR)

Most of the existing surfer rescue programs teach surfers important skills around how to recognize someone in distress in the water, how to approach the patient to avoid the surfer getting into trouble themselves (i.e., approaching someone in trouble with the surfboard between the surfer and the patient), and how to get a patient onto the surfboard and paddle back to shore. Techniques taught differ for a conscious or an unconscious patient (Surfing New South Wales, n.d.-c). For a conscious patient, they may be placed along the board as is traditional, or the board may be used horizontally (Fig. 6.4). If the patient is able, they can assist in paddling back to shore. For an unconscious patient, strategies differ as they will be heavy, slippery, and unbalanced making them extremely difficult to manage in still water, let alone in the surf zone. Some in-water rescue techniques for surfers include the sideways paddle and the leg hook (Fig. 6.4). The steps to take for performing these rescues are described in Table 6.2.

Many courses also teach participants the steps of CPR as noted in Table 6.2. CPR is vital in cases of drowning and other emergencies where the airway is impaired and breathing is impacted (Szpilman & Morgan, 2021). Research focused on lifeguards has shown that in-water resuscitation on a large rescue board is feasible both at the time of rescue and also when towing the patient (Barcala-Furelos et al., 2024). Commencing resuscitation in-water shortens the reoxygenation time for the patient but delays arrival to shore. The study of Barcala-Furelos et al. (2024) found that conducting in-water resuscitation using lifeguard rescue boards did not contribute to greater perceived levels of fatigue of the trained lifeguard rescuer, but it should be acknowledged that conducting such practice for a typical surfer on their recreational boards would be extremely challenging. In fact, existing training courses for surfers do not teach in-water resuscitation for this reason.

Beyond any specific training a surfer may receive, it has been shown that other factors, such as the level of experience of the surfer, familiarity with the surf environment, and time in the water (such as surfing frequency), contribute to the likelihood of a successful rescue (Barcala-Furelos et al., 2021). Similarly, the more

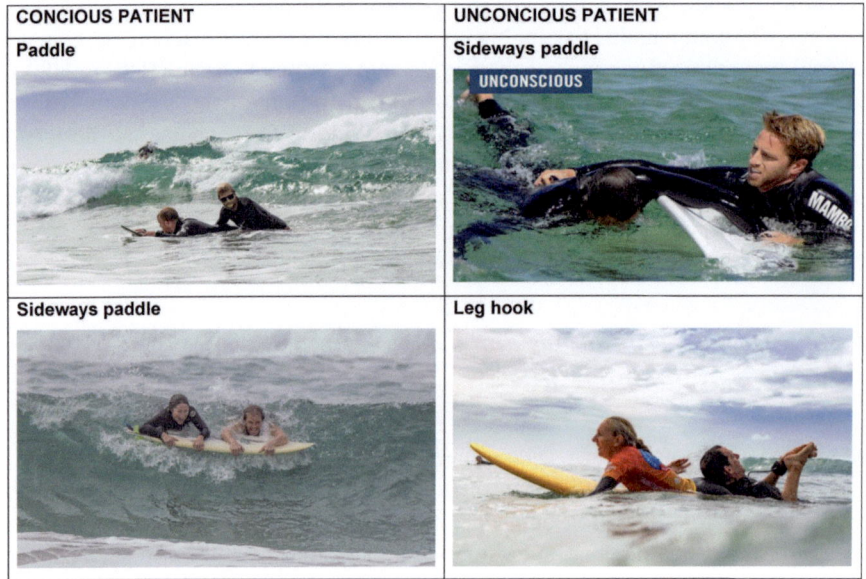

Fig. 6.4 Surfer rescue techniques for conscious and unconscious patients. (Information and images courtesy of Surfing NSW)

Table 6.2 Steps taken for performing sideways paddle or leg hook in-water rescues of unconscious patients (Surfing New South Wales, n.d.-c)

Sideways paddle	Leg hook
Place hand up on the front rail of a fins-up board	Roll the patient onto their back; ensure their head is out of the water
Secure hand	Maneuver your body to hook your legs under the patient's armpits (Fig. 6.4)
Roll board toward rescuer, dragging patient onto deck of the board	Slide board underneath you so that the patient lies head first up on your buttock
Secure patient	
Lay on top of patient toward the tail of the board	Hook your legs up under their armpits at a 90° angle
Slide up the patient and attempt to paddle	Cross your ankles over the patient's chest to lock their body in a secure position
Signal for assistance	Paddle the patient away from the danger zone and into shore if safe
	Remain mindful of your patients position and of keeping their head out of the water
	Administer CPR and call for an ambulance if required

Information provided courtesy of Surfing NSW

experience a surfer has gained via frequent and regular participation in surfing, the higher likelihood they have of reporting having conducted a rescue (Attard et al., 2015; Berg et al., 2021; Dehez et al., 2024).

6.3.3 First Aid

In addition to CPR and rescue skills, which are vital for successfully responding to drowning emergencies, surfers often witness, and may also experience, and attend to other types of injuries in the surf (Sect. 6.2.4). These include cervical spine injuries, traumatic injuries due to fin chop (lacerations caused by surfboard fins) and shark attack, marine envenomation and contusions, sprains, strains, and fractures (Nathanson et al., 2002, 2007; Chang et al., 2006; Dimmick et al., 2013; Doelp et al., 2018) (Chap. 7). However, it must be noted that, when adjusted for exposure (i.e., time in water performing the activity), surfers have a lower mortality rate than other in-water activities (Lawes et al., 2023).

Therefore, beyond board rescue and CPR skills and knowledge, first aid skills are also incredibly valuable for surfers to know (Fig. 6.5). Courses such as the Advanced Surf Life Support course provided by Surfing Medicine International cover information regarding diagnosis and treatment of wounds, marine envenomation, and musculoskeletal injuries (Surfing Medicine International, 2023).

Fig. 6.5 Surfers applying first aid to attend to fin chop suffered in the water. (Image courtesy Surfing NSW)

6.3.4 Impact of Safety Training Programs for Surfers

There has been limited independent evaluation of rescue and safety training programs for surfers. One review of the Surfers Rescue 24/7 program in Australia and New Zealand employed mixed methods to capture insights from past participants of the course via surveys and in-depth interviews while also capturing changes in surfers' assessments of their confidence in the skills needed to perform rescues before and after course participation (Koon et al., 2023a).

The evaluation found that course participants had high satisfaction regarding their experience of the course (Koon et al., 2023a). In particular, respondents indicated improved observation and awareness of safety concerns while surfing, as well as improved understanding of ocean conditions and hazards. Participants of Surfers Rescue 24/7 programs indicated they had learned new rescue techniques and skills, and information with respect to their own safety in the ocean, resulting in improved self-reported confidence in responding to an emergency situation in the surf. Some participants noted they had conducted rescues in real life since their course completion, and many participants indicated they would recommend that other surfers attend the course (Koon et al., 2023a, b).

Overwhelmingly, the study found that surfer rescue and safety training courses were successful in instilling an ethos of personal safety among participants, empowering them to perform rescues without increasing their own risk. This study also shed light on the less obvious role surfers play as guardians of the surf, namely, the preventive actions performed that negate the need for a rescue to be conducted (Koon et al., 2023a). These include checking in on swimmers in the water before they get into trouble and providing safety information or warnings to beachgoers before they enter the water. In addition to expanding investment in and availability of surfer rescue courses to supplement the limited coverage of lifeguards and lifesavers, it was acknowledged the course content could be further improved to provide practical skills in how to provide warnings to beachgoers, for those not comfortable in doing so (Koon et al., 2023a).

The Koon et al. (2023a) study also yielded important observations and recommendations based on gender considerations related to the Surfers Rescue 24/7 program. Half of the in-depth interview conducted in their study were with female surfers, and there was some acknowledgement from participants that the course training techniques could be adapted for bodies of smaller stature and physical strength and that all-female courses might prove beneficial in this regard.

6.4 Knowledge Gaps, Challenges, and Recommendations

As awareness of surfers' role in performing rescues and saving lives along the coast grows, there remain a number of important knowledge gaps and challenges related to gauging the true contribution that surfers make to coastal safety and increasing the availability, awareness, uptake, and effectiveness of surfer rescue training programs.

First, there is a need to better quantify the number of rescues conducted, including lives saved, by surfers. None of the studies described in Sect. 6.2 were able to successfully achieve this and relied upon conservative estimates based on a select group of surfers who responded to surveys. To capture better data, there is a need to promote improved documentation of surfer rescues in quasi-real time. Several organizations already encourage surfers to report the details of their rescues, such as Surfing NSWs "Report your Rescue" online portal (Surfing New South Wales, n.d.-b), and these organizations should work toward consistent data reporting and sharing as well as sharing and encouraging other organizations to use their surfer rescue data capture tool. The data capture tool should also be easily accessible and attractive to surfers to use. However, this system will not work without considerable buy-in from surfers regarding the importance of reporting their rescues, and it may take considerable time before this becomes an intrinsic component of surfing culture. Creating viable and effective ways to promote the value and ease of reporting rescues to the surfing community will be a major challenge to overcome.

Existing knowledge on surfer rescues is also limited in extent, both geographically and in terms of gender. As evident in Table 6.1, despite surfing being a global recreational activity, there is almost no published information about surfer rescues outside of Australia/New Zealand and Europe. Knowing more about surfer rescues, and surfers' experiences and opinions about conducting rescues, in the Americas, South Africa, the South Pacific, and elsewhere would provide additional information regarding the relative importance and contribution of surfers, particularly in areas where lifeguard services may still be developing. There may also be both coastal geomorphological and sociocultural differences which factor into the occurrence of surfer rescues and the people who surfers are rescuing. As shown in Table 6.1, existing surveys have also been dominated by responses obtained from male surfers. While it is difficult to find global estimates of the proportion of surfers who are female, Manero et al. (2024) reported that 33% of Australian surfers were female. In contrast, the proportion of survey responses from female surfers in studies related to surfer rescues ranges from 5 to 25% (Table 6.1).

A potential solution to both of these sampling issues may be in further promotion of the Global Surfer Survey as described in Sect. 6.2.1. The GSS is ongoing surfer survey, and while success has been achieved in promoting it to surfers in the state of New South Wales in Australia (Peden et al., 2023) and France (Dehez et al., 2024), it has not yet received significant promotion or traction in other regions. One of the limitations with the GSS is the survey length, which can take up to 30 minutes to complete as it asks surfers of all types of questions about their demographics, surfing background and experience, rescues, and opinions on a variety of environmental and coastal management issues. However, it provides a rich dataset about surfers and with proper promotion has the potential to reach a considerable global population of surfers and yield extremely valuable information for the surfing industry. A concerted effort by Surfing NSW to promote the GSS to surfers in New South Wales also resulted in a high proportion of female surfer respondents (Table 6.1).

Research has shown that in addition to rescues, surfers also conduct a range of preventive activities, such as checking in on swimmers in the water or verbally warning people on the beach about ocean hazards, before a rescue is required to be performed (Koon et al., 2023a). In addition to capturing the true extent of the contribution of surfers in performing rescues, further research is required to better understand, and attempt to capture, the full range of preventive actions surfers are undertaking. This information may prove to be a valuable addition to existing surfer rescue and safety training programs. Further research is also required to evaluate the effectiveness of safety training programs for surfers outside of the Australian context (Koon et al., 2023a), including in low and middle income contexts where the vast majority of drowning deaths occur (World Health Organization, 2014). Such research will provide the evidence to support increasing investment into, and expansion of, effective surfer safety and rescue training programs.

It is clear that many surfers are carrying out rescues while surfing without having any prior training (De Oliveira et al., 2023) and the vast majority of these rescues are going unreported. While there is clearly a societal need to expand surfer rescue training programs and to improve documentation of surfer rescues, there is also a strong need to attempt to quantify the economic value of lives saved through the interventions of recreational surfers. The direct impact of adult surfers into the Australian economy alone based on retail and travel expenditure has been estimated to be A$2.71 billion per year (Manero et al., 2024). This value would be even higher if the cost of human lives saved by surfers was considered. The study of New Zealand surfers by Mead et al. (2024) estimated that if only 1% of New Zealand surfers conducted one rescue that saved a life per year, it would prevent an economic cost of NZD $6.4 billion. Calculating the economic value of surfer rescues in other countries and globally would not only be of benefit to beach and water safety providers and the surfing industry itself, but it would illustrate to governments and potential funding bodies the strong return on any investment into surfer rescue training programs.

However, a fundamental limitation to addressing these knowledge gaps in relation to both surfer rescues and surfer rescue training programs is the difficulty in promoting the importance of both topics to surfers themselves and getting their buy-in. Researchers involved in studies involving online surveys of surfers have acknowledged both formally (Manero et al., 2024) and anecdotally the difficulties in reaching out and engaging with the surfing community and industry about the topic of surfers conducting rescues, including assistance in recruitment for research. Through many conversations with surfers themselves, the authors of this chapter have found that surfers are generally content simply with the act of surfing and while they generally accept that they serve an important role in helping others, they do not necessarily see the value in documenting these acts, or being celebrated for it (Peden et al., 2023). Finding ways to overcome some of these barriers is a necessary step to help ensure better quality information on the role of surfers as rescuers in the future.

6.5 Conclusion

Surfers regularly perform rescues in the coastal environment and save many lives, be it through physically performing rescues, providing CPR, or performing first aid, but also in undertaking preventative actions such as checking in on swimmers in the water and educating people on the beach. These actions generate significant societal and economic benefit; however, there remains significant challenges in quantifying the true impact of surfers in this regard. What is clear is that surfers are performing important functions in ensuring public safety both on patrolled and, more commonly, on unpatrolled beaches, representing a valuable layer of prevention and protection that supplements a beach safety system which relies heavily on lifeguards. In order to further strengthen the unofficial role of surfers as guardians of the surf, there is a need to expand delivery of free or highly subsidized surfer rescue training programs and also more widely promote those which are already in existence, particularly those proven to be effective.

References

Academy of Surfing Instructors. (2024). *Course overview – ASI surf water safety rescue award*. Academy of Surfing Instructors. Retrieved 19-06-2024 from https://www.academyofsurfing.com/courses/surf-safety-water-rescue-award

Anary, S. H. S., Sheikhazadi, A., & Ghadyani, M. H. (2010). Epidemiology of drowning in Mazandaran Province, North of Iran. *American Journal of Forensic Medicine and Pathology, 31*(3), 236–242. https://doi.org/10.1097/PAF.0b013e3181e804de

Arozarena, I., Houser, C., Echeverria, A. G., & Brannstrom, C. (2015). The rip current hazard in Costa Rica. *Natural Hazards, 77*, 753–768.

Attard, A., Brander, R., & Shaw, W. S. (2015). Rescues conducted by surfers at Australian beaches. *Accident Analysis and Prevention, 82*, 70–78.

Barcala-Furelos, R., Graham, D., Abelairas-Gomez, C., & Rodriguez-Nunez, A. (2021). Lay-rescuers in drowning incidents: A scoping review. *The American Journal of Emergency Medicine, 44*, 38–44.

Barcala-Furelos, R., De Oliveira, J., Duro-Pichel, P., Colón-Leira, S., Sanmartín-Montes, M., & Aranda-García, S. (2024). In-water resuscitation during a surf rescue: time lost or breaths gained? A pilot study. *The American Journal of Emergency Medicine, 79*, 48–51.

Beale-Tawfeeq, A. K. (2019). Triennial scientific review: assisting drowning victims: effective water rescue equipment for Lay-responders. *International Journal of Aquatic Research and Education, 10*(4), 8.

Berg, I., Haveman, B., Markovic, O., van de Schoot, D., Dikken, J., Goettinger, M., & Peden, A. E. (2021). Characteristics of surfers as bystander rescuers in Europe. *The American Journal of Emergency Medicine, 49*, 209–215. https://doi.org/10.1016/j.ajem.2021.06.018

Bessereau, J., Fournier, N., Mokhtari, T., Brun, P., Desplantes, A., Grassineau, D., et al. (2016). Epidemiology of unintentional drowning in a metropolis of the French Mediterranean coast: a retrospective analysis (2000–2011). *International Journal of Injury Control and Safety Promotion, 23*(3), 317–322. https://doi.org/10.1080/17457300.2015.1047862

Big Wave Surfing. (n.d.). *Top 5 big wave rescue courses in the world,*. Big Wave Surfing. Retrieved 19-06-2024 from https://www.bigwavesurfing.com/top-big-wave-rescue-courses-

Bradley, Z. (2023). *'Amazing and freeing': Aussie dad helps woman he rescued from surf get back in the water'*. Retrieved 1-07-2024 from https://9now.nine.com.au/a-current-affair/john-gordon-save-olivia-titor-dangerous-rip-ocean-new-south-wales/34654a6a-40fc-4bb8-ac3b-38dae39aa52a

Branche, C. M., & Stewart, S. (2001). *Lifeguard effectiveness: A report of the working group*. Centers for Disease Control and Prevention, National Center for Injury Prevention and Control.

Brander, R., & Scott, T. (2018). Science of the rip current hazard. In *The science of beach lifeguarding* (pp. 67–85). CRC Press.

Brander, R. W., Warton, N., Franklin, R. C., Shaw, W. S., Rijksen, E. J. T., & Daw, S. (2019). Characteristics of aquatic rescues undertaken by bystanders in Australia. *PLoS One, 14*(2), e0212349. https://doi.org/10.1371/journal.pone.0212349

Brewster, B. C., Gould, R. E., & Brander, R. W. (2019). Estimations of rip current rescues and drowning in the United States. *Natural Hazards and Earth System Sciences, 19*(2), 389–397.

Brighton, B., Sherker, S., Brander, R., Thompson, M., & Bradstreet, A. (2013). Rip current related drowning deaths and rescues in Australia 2004-2011. *Natural Hazards and Earth System Sciences, 13*(4), 1069. https://doi.org/10.5194/nhess-13-1069-2013

Castelle, B., Scott, T., Brander, R.W., McCarroll, R.J. (2016). Rip current types, circulation and hazard. *Earth-Science Reviews, 163*:1–21.

Castelle, B., Brander, R.W., Tellier, E., Simonnet, B., Scott, T., McCarroll, J., Campagne, J-M., Cavailhes, T., Lechevrel, P. (2018). Surf zone hazards and injuries on beaches in SW France. *Natural Hazards, 93*, 1317–1335.

Castelle, B., Scott, T., Brander, R., McCarroll, J., Robinet, A., Tellier, E., De Korte, E., Simonnet, B., & Salmi, L.-R. (2019). Environmental controls on surf zone injuries on high-energy beaches. *Natural Hazards and Earth System Sciences, 19*(10), 2183–2205.

CBI – Centre for the Promotion of Imports from Developing Countries. (2018). *Surf tourism from Europe*. Retrieved 8-11-2020 from https://www.cbi.eu/market-information/tourism/surfing-tourism/europe

Chang, S. K. Y., Tominaga, G. T., Wong, J. H., Weldon, E. J., & Kaan, K. T. (2006). Risk factors for water sports–Related cervical spine injuries. *Journal of Trauma and Acute Care Surgery, 60*(5). https://journals.lww.com/jtrauma/Fulltext/2006/05000/Risk_Factors_for_Water_Sports_Related_Cervical.18.aspx

Cooney, N., Daw, S., Brander, R., Ellis, A., & Lawes, J. (2020). *Coastal safety brief: Rip currents*. S. L. S.

Cornwall, J. (2007). Between the Flags: One Hundred Summers of Australian Surf Lifesaving edited by Ed Jaggard. *Public History Review, 14*.

De Oliveira, J., Lorenzo-Martínez, M., Barcala-Furelos, R., Queiroga, A. C., & Alonso-Calvete, A. (2023). Surfers as aquatics rescuers in Portugal and Spain: Characteristics of rescues and resuscitation knowledge. *Heliyon, 9*(5), e16032.

Dehez, J., Castelle, B., Carayon, D., Peden, A. E., & Brander, R. W. (2024). The role of surfers in beach safety management: Insights from French respondents to a global surfer survey. *Ocean & Coastal Management, 248*, 106973.

Dimmick, S., Brazier, D., Wilson, P., & Anderson, S. E. (2013). Injuries of the spine sustained whilst surfboard riding. *Emergency Radiology, 20*, 25–31.

Doelp, M. B., Puleo, J. A., Cowan, P., & Arford-Granholm, M. (2018). Characterizing surf zone injuries from the five most populated beaches on the Atlantic-fronting Delaware coast. *The American Journal of Emergency Medicine, 36*(8), 1372–1379.

Ellison, E., & Brien, D. L. (2020). *Writing the Australian beach: local site, global idea*. Springer.

Franklin, R. C., & Pearn, J. H. (2011). Drowning for love: the aquatic victim-instead-of-rescuer syndrome: drowning fatalities involving those attempting to rescue a child. *Journal of Paediatrics and Child Health, 47*(1), 44–47.

Franklin, R. C., Peden, A. E., Brander, R. W., & Leggat, P. A. (2019). Who rescues who? Understanding aquatic rescues in Australia using coronial data and a survey. *Australian and New Zealand Journal of Public Health, 43*(5), 477–483. https://doi.org/10.1111/1753-6405.12900

Franklin, R. C., Peden, A. E., Hamilton, E. B., Bisignano, C., Castle, C. D., Dingels, Z. V., et al. (2020). The burden of unintentional drowning: global, regional and national estimates of mortality from the Global Burden of Disease 2017 Study. *Injury Prevention, 26*(Supp 1), i83–i95. https://doi.org/10.1136/injuryprev-2019-043484

Houser, C., Arbex, M., & Trudeau, C. (2021). Economic impact of drowning in the Great Lakes Region of North America. *Ocean & Coastal Management, 212*, 105847.

Ishikawa, T., Komine, T., Aoki, S. I., & Okabe, T. (2014). Characteristics of rip current drowning on the shores of Japan. *Journal of Coastal Research, 72*, 44–49.

Jaggard, E. (2007). Bodysurfers and Australian beach culture. *Journal of Australian Studies, 31*(90), 89–98.

Klick, J., Jones, C. M. C., & Adler, D. (2016). Surfing USA: an epidemiological study of surfing injuries presenting to US EDs 2002 to 2013. *The American Journal of Emergency Medicine, 34*(8), 1491–1496. https://doi.org/10.1016/j.ajem.2016.05.008

Koon, W., Peden, A., Lawes, J. C., & Brander, R. W. (2021a). Coastal drowning: A scoping review of burden, risk factors, and prevention strategies. *PLoS One, 16*(2), e0246034. https://doi.org/10.1371/journal.pone.0246034

Koon, W., Schmidt, A., Queiroga, A. C., Sempsrott, J., Szpilman, D., Webber, J., & Brander, R. (2021b). Need for consistent beach lifeguard data collection: Results from an international survey. *Injury Prevention, 27*(4), 308–315.

Koon, W., Peden, A. E., & Brander, R. W. (2023a). Impact of a surfer rescue training program in Australia and New Zealand: a mixed methods evaluation. *BMC Public Health, 23*(1), 2193.

Koon, W. A., Peden, A. E., Lawes, J. C., & Brander, R. W. (2023b). Mortality trends and the impact of exposure on Australian coastal drowning deaths, 2004–2021. *Australian and New Zealand Journal of Public Health, 47*(2), 100034. https://doi.org/10.1016/j.anzjph.2023.100034

Lawes, J. C., Rijksen, E. J. T., Brander, R. W., Franklin, R. C., & Daw, S. (2020, September). Dying to help: Fatal bystander rescues in Australian coastal environments. *PLoS One, 15*(9), e0238317. https://doi.org/10.1371/journal.pone.0238317

Lawes, J. C., Ellis, A., Daw, S., & Strasiotto, L. (2021). Risky business: a 15-year analysis of fatal coastal drowning of young male adults in Australia. *Injury Prevention, 27*(5), 442. https://doi.org/10.1136/injuryprev-2020-043969

Lawes, J. C., Koon, W., Berg, I., van de Schoot, D., & Peden, A. E. (2023). The epidemiology, risk factors and impact of exposure on unintentional surfer and bodyboarder deaths. *PLoS One, 18*(5), e0285928.

Laweson M. *Estimated participation numbers for SR24/7 program.* Personal communication received 1-07-2024.

Leatherman, S. P., Leatherman, S. B., & Rangel-Buitrago, N. (2024). Integrated strategies for management and mitigation of beach accidents. *Ocean & Coastal Management, 253*, 107173.

Manero, A., Yusoff, A., Lane, M., & Verreydt, K. (2024). A national assessment of the economic and wellbeing impacts of recreational surfing in Australia. *Marine Policy, 167*, 106267.

McArthur, K., Jorgensen, D., Climstein, M., & Furness, J. (2020). Epidemiology of acute injuries in surfing: Type, location, mechanism, severity, and incidence: A systematic review. *Sports, 8*(2), 25. https://www.mdpi.com/2075-4663/8/2/25

Meacham, S. (2021). *Surfer hailed a hero after seeing boat capsize on NSW South Coast.* Retrieved 1-07-2024 from https://www.9news.com.au/national/surfer-hero-after-boat-capsized-bulli-new-south-wales-south-coast/d1c2451e-0524-45f8-86f3-e34f3d54f2e9

Mead, J., Le Dé, L., & Moylan, M. (2024). The unexplored role of surfers in drowning prevention: Aotearoa, New Zealand as a case study. *Environmental Hazards, 23*(2), 150–166.

Minasian, B., & Hope, N. (2022). Surfing on the world stage: a narrative review of acute and overuse injuries and preventative measures for the competitive and recreational surfer. *British Journal of Sports Medicine, 56*(1), 51–60.

Moran, K., Webber, J., & Stanley, T. (2017). The 4Rs of aquatic rescue: educating the public about safety and risks of bystander rescue. *International Journal of Injury Control and Safety Promotion, 24*(3), 396–405.

Nathanson, A. (2020). Injury prevention in the sport of surfing: An update. *Muscles, Ligaments & Tendons Journal (MLTJ), 10*(2).

Nathanson, A., Haynes, P., & Galanis, D. (2002). Surfing injuries. *The American Journal of Emergency Medicine, 20*(3), 155–160. https://doi.org/10.1053/ajem.2002.32650

Nathanson, A., Bird, S., Dao, L., & Tam-Sing, K. (2007). Competitive surfing injuries: A prospective study of surfing-related injuries among contest surfers. *American Journal of Sports Medicine, 35*(1), 113–117. https://doi.org/10.1177/0363546506293702

Payne, N. (2016). *Kelly Slater helps rescue toddler and mother from giant wave.* Retrieved 1-07-2024 from https://www.outsideonline.com/outdoor-adventure/water-activities/kelly-slater-helps-rescue-toddler-and-mother-giant-wave/

Peden, A. E., Koon, W., Brander, R. W. *Analysis of NSW responses to the Global Surfer Survey* (Report No. 2304). UNSW Beach Safety Research Group.

Rachwani, M. (2021). *'Hold my beer': Australian surfer Mikey Wright charges into Hawaii surf to rescue struggling swimmer'.* The Guardian. Retrieved 1-07-2024 from https://www.theguardian.com/sport/2021/jan/02/hold-my-beer-australian-surfer-mikey-wright-charges-into-hawaii-surf-to-rescue-struggling-swimmer

Scott, T., Masselink, G., Austin, M. J., & Russell, P. (2014). Controls on macrotidal rip current circulation and hazard. *Geomorphology, 214*, 198–215.

Segura, L., Arozarena, I., Koon, W., & Gutiérrez, A. (2022). Coastal drowning in Costa Rica: incident analysis and comparisons between Costa Rican nationals and foreigners. *Natural Hazards, 110*(2), 1083–1095.

Sherker, S., Brander, R., Finch, C., & Hatfield, J. (2008). Why Australia needs an effective national campaign to reduce coastal drowning. *Journal of Science & Medicine in Sport, 11*, 81–83. https://doi.org/10.1016/j.jsams.2006.08.007

SportAus. (2019). *Surfing state of play report*. AusPlay. Retrieved 1-07-2024 from https://www.clearinghouseforsport.gov.au/__data/assets/pdf_file/0005/843062/State_of_Play_Report_Surfing.pdf

Sports Medicine Australia. (2021). *Surfing fact sheet*. Retrieved 26-05-2021 from https://sma.org.au/resources-advice/surfing/

Suresh Kumar Shetty, B., & Shetty, M. (2007). Epidemiology of drowning in Mangalore, a coastal Taluk of South India. *Journal of Forensic and Legal Medicine, 14*, 410–415.

Surf Life Saving Australia. (2023). *National coastal safety report 2023*. Surf Life Saving Australia.

Surfer Today. (2018). *How many surfers are there in the world?* Retrieved 8-11-2020 from https://www.surfertoday.com/surfing/how-many-surfers-are-there-in-the-world

Surfing Medicine International. (2023). *Advanced surf life support course*. Surfing Medicine International. Retrieved 19-06-2024 from https://www.surfingmed.com/asls-course-advanced-surf-life-support/

Surfing New South Wales. (n.d.-a). *About surfers rescue 24/7*. Surfing New South Wales. Retrieved 19-06-2024 from https://www.surfersrescue247.com/about

Surfing New South Wales. (n.d.-b). *Report your rescue*. Surfing New South Wales. https://www.surfersrescue247.com/report-your-rescue

Surfing New South Wales. (n.d.-c). *Surfers Rescue 24/7*. Surfing New South Wales. Retrieved 19-06-2024 from https://sandshoesbr.com.au/wp-content/uploads/2019/06/surf-rescue-classes.pdf

Szpilman, D., & Morgan, P. J. (2021). Management for the drowning patient. *Chest, 159*(4), 1473–1483.

Szpilman, D., Mello, D. B., Queiroga, A. C., Emygdio, R. F. E., & R. F. (2020). Association of drowning mortality with preventive interventions: A quarter of a million deaths evaluation in Brazil. *International Journal of Aquatic Research and Education, 12*(2), 3.

Thom, O., Roberts, K., Leggat, P.A., Devine, S., Peden, A.E., Franklin, R.C. (2022). Cervical spine injuries occuring at the beach: epidemiology, mechanism of injury and risk factors. *BMC Public Health*, 22:1404.

Tomizawa, R. (2016). *Duke Kahanamoku Part 2: The day he saved eight souls*. Retrieved 1-07-2024 from https://theolympians.co/2016/07/20/duke-kahanamoku-part-2-the-day-he-saved-eight-souls/

Venkateswarlu, C., Surisetty, V. A. K., Somani, A., Gireesh, B., & Naidu, C. (2023). Surf zone-related drownings and injuries based on lifeguard records in Goa beaches (2008–2020). *Natural Hazards, 117*(1), 313–337.

World Health Organization. (2014). *Global report on drowning: Preventing a leading killer*. World Health Organization.

World Health Organization. (2017). *Preventing drowning: an implementation guide*. World Health Organization.

Chapter 7
Surf Medicine and Health

James Furness

7.1 Introduction

Surfing is an invigorating, intense, and dangerous yet exhilarating sport. Paddling and riding are physical activities that require a high level of skill and fitness, combining many different aspects of human physiology. This exertion puts stress upon the body, thereby improving physical and mental well-being. The power of the waves and the equipment used to ride them result in accidents and injury. In this chapter, we explore the influence of surfing on key human body systems, physiological processes, and injuries. Each of the relevant body systems is explored from an anatomical level outlining their function and how surfing influences each of them from both a benificial and adverse viewpoint. Practical applications for readers and recreational surfers are also provided on how to physically prepare the body for surfing and the management of injuries.

Please note that any advice provided is advisory only. All medical and health decisions should be made in consultation with a qualified medical or health practitioner.

To undertake this analysis, a systematic literature search was conducted, using a broad strategy around the concept of "surfing" and associated key words in academic databases including PubMed, SportsDiscus, and Google Scholar. In addition, a "snowballing" technique was further undertaken whereby hand searching of key articles and associated reference lists was undertaken. Articles were included if they were related to the following key topics: cardiovascular system and associated physiological processes, musculoskeletal system including bones, and integumentary system (skin). Articles were excluded if they were related to other sports such as

J. Furness (✉)
Bond University, Robina, Gold Coast, QLD, Australia
e-mail: jfurness@bond.edu.au

surf lifesaving, mental health and surfing, the experiences and the history of surfing, studies specific to hormone levels, and surfing performance. Furthermore, newspaper and magazine articles were excluded.

All identified studies were then imported to an automation tool known as systematic review accelerator (Clark et al., 2020) to remove duplicate articles and then screen the remaining articles for relevance. A total of 322 were classified as relevant and are therefore used within this chapter. The article selection process for this chapter can be seen in Fig. 7.1.

7.2 Activity Requirements of Surfing

Surfing has been performed for hundreds and possibly thousands of years (Chap. 2), and during this time, boards and surfing styles have changed dramatically. Surfboards were originally carefully selected carved pieces of wood, but today they are most commonly made from carbon fiber, polyurethane foam, or expanded polystyrene and covered in a fiber glass cloth with either a polyester or epoxy resin. As a result, the weight, size, and maneuverability have dramatically changed. In addition, modern boards often have refined rails and pointed noses, all of which changes the potential and nature for injury.

The act of surfing has also changed through time, both driving technological change and being a result of it. From simply paddling beyond the whitewash and then catching a wave back toward the shore to performing turning maneuvers or aerials (where the surfer projects themselves and board into the air to perform an acrobatic maneuver) or even using foils, the stresses imparted on the human body has greatly evolved. Despite these changes, recreational surfing can still be broadly broken down into three key phases: (1) paddling, (2) sitting and waiting for a wave, and (3) riding the wave. These account for approximately 50%, 45%, and 5% of the

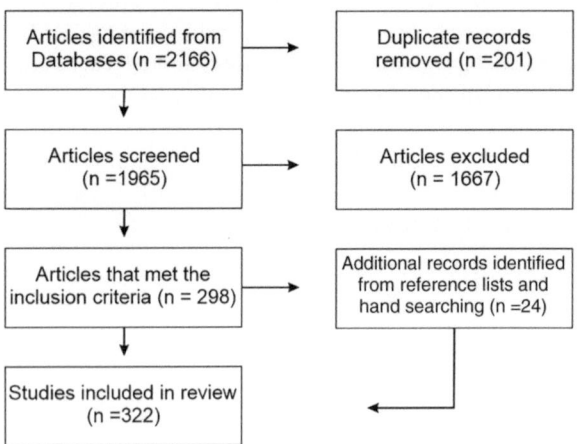

Fig. 7.1 Identification of articles via databases and hand searching

Fig. 7.2 Time spent during different activities while undertaking a surfing session. Riding the wave, while the most exhilarating part of surfing, is also by far the smallest time fraction of an entire session

time, respectively (see Fig. 7.2) (Barlow et al., 2014; Meir et al., 2015; O'Neill et al., 2021). Typically a surfer can spend anywhere between 20 minutes and 4–5 hours of surfing (Borgonovo-Santos et al., 2021). Therefore, a 1–2-hour surfing session would result in approximately 30–60 minutes being spent paddling and only 180–360 seconds spent riding waves (Fig. 7.2).

Paddling primarily uses the upper body. It can be broken down into intermittent high intensity sprints (catching a wave) followed by short rest periods (sitting waiting for waves) and repeated low-intensity endurance paddling in a prone position (paddling into position to catch a wave) with intermittent breath holds (duck diving under waves). Video analysis work completed by Secomb et al. (2015d) has shown that during a 2-hour surf session, a surfer may complete 20–40 paddling bouts lasting for approximately a minute (e.g., returning to the lineup after catching a wave). Furthermore, a surfer may sprint paddle for 5 seconds, for a wave 30–50 times.

What needs to be highlighted is that sitting and waiting is interspersed between paddling. This creates an activity that is intermittent, rather than purely continuous, such as endurance running or cycling. This has a unique influence on the cardiovascular system and uses two different energy systems (see Sect. 7.3). It is well known that as the demands of exercise increase, so does our heart rate in order to increase the delivery of oxygenated blood to our muscles. To quantify the intensity of an exercise, a common method is to determine what percentage of your heart rate maximum (HRmax) you are working at during the exercise. This is calculated by subtracting your age from 220. The World Health Organization guidelines recommend approximately 150–300 minutes of moderate intensity exercise per week, with American College of Sports Medicine (2022) determining moderate intensity being between 64% and 76% (HRmax). Using the example below, a 40-year-old recreational surfer's HRmax would be 180 bpm (beats per minute), and a moderate intensity would be between 115 and 137 bpm.

Let's firstly look at how the heart responds during surfing. Over a session, the surfer has an average heart rate around 60% of their HRmax; however, given the intermittent nature of surfing and the act of sprint paddling for a wave, the surfer will experience fluctuations in their heart rate with heart rate peaks around 80–90% HRmax. Individual heart rate data is presented below for a 40-year-old male recreational surfer from the study by O'Neill et al. (2021). During a 2-hour surf session (Fig. 7.3), we see that the average rate ranges between 50% and 80% of the surfer's heart rate maximum. So, it is fair to say that a surf session challenges the heart!

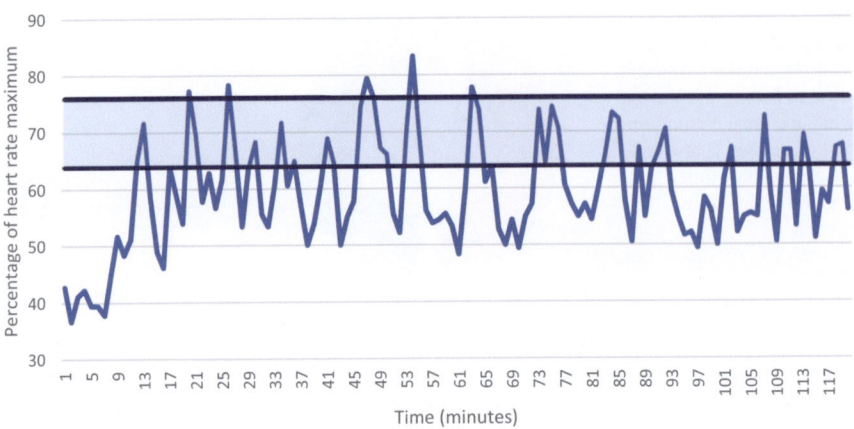

Fig. 7.3 Profile of the heart rate maximum percentages for a 40-year-old male recreational surfer in a beach break over a 2-hour period. Note: The shaded area is the moderate intensity zone

In the largest study to date, LaLanne et al. (2017) investigated the effect of surfing on the cardiovascular system in 228 surfers, ranging in age from 18 to 75 years old who spent around 8.5 hours a week in the water. It was found that there was no difference in predicted heart rate maximum when surfing intensity varied between moderate and vigorous (50–85% of age-predicted maximum HR). Furthermore, older surfers spent the same amount of time paddling compared to their younger counterparts, and the average amount of time surfing was approximately 1 hour for all ages. This level of exercise has been shown to have many health benefits across many activities (Sallis, 2015), and based on the LaLanne et al. (2017) study, it can be assumed that surfing may have protective benefits in older individuals.

7.3 Energy Systems

These physiological requirements of paddling require upper-body muscular strength, power, and endurance while requiring aerobic and anaerobic energy systems (Redd & Fukuda, 2016). Broadly speaking, there are three primary energy systems the body uses during exercise. In short, the phosphagen system relies on a chemical compound known as adenosine triphosphate (ATP) and phosphocreatine (PCr) stored in the muscles. The breakdown of these chemicals releases energy that powers muscle contractions for very short duration of high intensity which last around 10 seconds. The anaerobic energy system converts glycogen into glycose to produce ATP and power muscle contractions. This system is for moderate to high intensity activities lasting around 10–120 seconds. Both the ATP-PCr and anaerobic systems don't require oxygen and are used more during maximally exercise efforts of short durations (<120 seconds). The third system is the aerobic energy system, which breaks down glucose using oxygen and is more dominant during continuous low to moderate activities (>120 seconds).

To determine an athlete's aerobic fitness, the aerobic energy system has been commonly assessed using what is called a maximal oxygen consumption test (VO2-max or peak test). Essentially this test assesses the maximal volume of oxygen your body can use when exerted. A significant number of studies have been completed in competitive and recreational surfers to understand aerobic fitness (Farley et al., 2012; Furnesset al., 2018a, b; Loveless & Minahan, 2010). It is found that recreational and competitive surfers have VO2 peak scores approximately 20–30% higher than untrained individuals (Mendez-Villanueva & Bishop, 2005). When comparing surfers with trained individuals such as college swimmers, VO2 peak scores are similar; however, surfers have revealed scores 10–20% higher (Mendez-Villanueva & Bishop, 2005). Research has also illustrated differences in max scores, with competitive surfers having approximately 25% higher VO2 peak scores than recreational surfers, indicating this is one of key attributes that competitive surfers need to possess.

To determine peak anaerobic power, the phosphagen system (ATP-PC) has been commonly assessed using sprint paddle test. This test requires the surfer to paddle as fast as possible for 10 seconds. Like VO2 peak scores, surfers possess high peak power output scores in both competitive and recreational cohorts. Furthermore, competitive surfers have significantly higher power output scores when compared to recreational surfers (Furness et al., 2018a). One study revealed a correlation between competitive ranking and peak power output, meaning higher ranked surfers displayed higher power outputs during sprint paddling (Farley et al., 2012).

Given the need for all energy systems within surfing, surfers need to ensure that these systems are adequately trained. This may be either through surfing itself or additional conditioning exercises especially during periods when not in the water due to weather or injury. Strength and conditioning coaches advocate prioritizing training of the shorter energy systems, and this is commonly performed through high intensity interval training. Using the work to rest ratios developed through the time motion analysis studies, a 3:1 ratio may be appropriate (e.g., paddle for 45 seconds; rest for 15 seconds). An example of what this type of training may look like can be seen in training considerations for surfers within Tables 7.1 and 7.2.

Table 7.1 Training considerations for a recreational surfer wanting to improve paddling fitness

Focus	Upper-body exercise	Considerations	Progressions
Minimal surfing exposure and want to improve paddling fitness	High intensity interval training	Consider short high intensity exercises with short rest periods. To simulate surfing, 5–10 seconds of sprint paddling in pool or swimming with pool buoy or upper-body exercise such as rowing machine. Followed by ten squats body weight or burpees. Followed by 30-second rest. Complete five to ten times	Add speed to squat, jump squat/reduce rest period or increase repetitions

Disclaimer: Please ensure you consult with the appropriate health professional before undertaking any exercise program

Table 7.2 Training considerations for recreational surfer to improve paddling speed

Focus	Upper-body exercise	Considerations	Progressions
Improving paddling speed	Pronated pull-up	8–12 repetitions at 4 sets	Single arm assisted pull-up

Disclaimer: Please ensure you consult with the appropriate health professional before undertaking any exercise program

Fig. 7.4 A surfer in a paddling position as they prepare to catch a wave. (Photo: James Furness)

7.3.1 Upper-Body Strength: Paddling

Another requirement of paddling is upper-body strength and power. Research has shown a connection between a surfer's sprint paddling times at 5, 10, and 15 m and strength scores, with faster times being correlated with higher strength scores (Sheppard et al., 2012). In fact, when recreational and competitive surfers complete upper-body strength training, their sprint and endurance paddling times significantly improve (Coyne et al., 2017) (Fig. 7.4).

So which muscles would be crucial for paddling? One way to test muscle use during any activity is to record electrical activity which occurs during muscle contraction. This is assessed through electromyography (EMG), and studies reveal the latissimus dorsi and the deltoid muscles are heavily used during the propulsive component of paddling. These muscles were actively recruited earlier in the stroke when the surfer moved from an endurance paddle to a sprint paddle (Nessler et al., 2019). The middle and upper trapezius were activated more during the beginning (arm entry) and the end of the stroke cycle (arm exit). With this understanding, both the latissimus and the deltoid muscles are paramount in the propulsion phase of surfing and should be considered for land-based programs (Fig. 7.5).

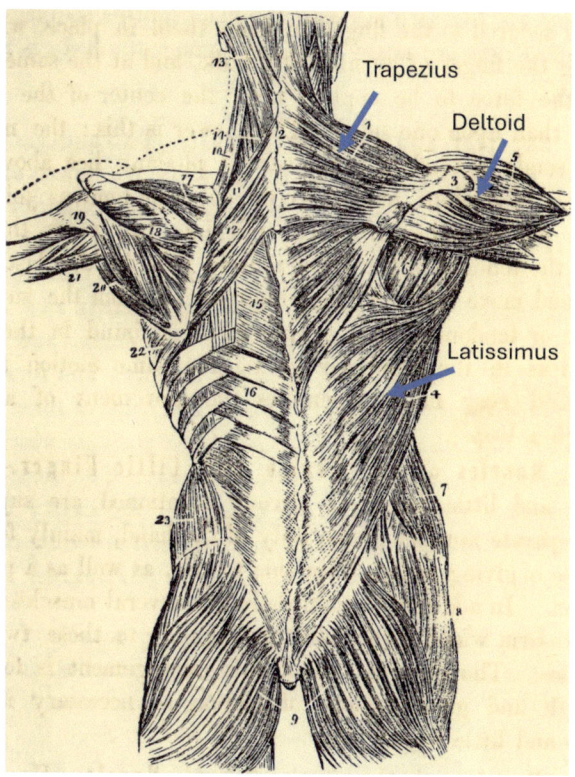

Fig. 7.5 Key musculature involved in paddling (blue lines). (Source: rawpixel.com, reproduced under a CC-0 licence)

Fig. 7.6 A surfer in the beginning of the pop-up stage of wave riding. (Photo: James Furness)

Fig. 7.7 A surfer at the end of the pop-up stage of wave riding. (Photo: James Furness)

Table 7.3 Training considerations for a recreational surfer to improve pop-up phase

Focus	Upper-body exercise	Considerations	Progressions
Improving pop-up speed	Push-up with body weight	8–12 repetitions, at 4 sets	Complete a push-up on an unstable surface (Swiss ball, foam etc.) Add speed to the push-up, 4 reps × 5 sets

Disclaimer: Please ensure you consult with the appropriate health professional before undertaking any exercise program

7.3.2 Upper-Body Strength: Pop-Up Phase

The pop-up is the quick transition from prone to standing on the surfboard and appears to occur seamlessly between paddling and standing on the board (Figs. 7.6 and 7.7). A surfer will symmetrically push down on the board and rapidly stand up, with approximately 95% of the surfer's body weight being exerted through the board (Eurich et al., 2010). Given this movement involves speed and force, exercises that target the pectoralis major muscle and involve a fast movement would be appropriate to consider for a land-based exercise (see Table 7.3).

7.4 Balance and Proprioception

Even though actual wave riding only takes up approximately 5% of the total time surfing, it is the reason why people catch waves. The activity requires coordination, good reaction time, flexibility, dynamic balance, and force development. While wave riding can be unique with various types of turning maneuvers and aerials, it can also be simplified into a bottom turn (Fig. 7.8), followed by a top turn (Fig. 7.9).

Fig. 7.8 A surfer conducting a bottom turn. (Photo: James Furness)

Fig. 7.9 A surfer conducting a top turn. (Photo: James Furness)

The bottom turn involves a sustained ankle, knee, hip, and trunk flexion, and the top turn is followed by explosive extension and rotation movement.

As the surfer stands up, the primary interface between the board and the surfer is the ankle joint. The ability of the surfer to sense and integrate the sensory signals (e.g., pressure) and determine body position and movement in space involves what is known as proprioception. In conjunction with proprioception, the surfer is needing to adapt to what the wave is doing (e.g., wave height, peel angle, strength, and length of the wave) to position themselves correctly.

Breaking waves are a highly unstable environment, and therefore the surfer requires excellent dynamic balance, which is the ability to control their posture and body position while riding the wave and performing maneuvers. Balance is influenced by what the surfer can see, their proprioceptive input, key centers in the brain, and the balance system in the ear known as the vestibular system. It has been proposed that these

Table 7.4 Training considerations for a recreational surfer to improve balance and proprioception

Focus	Lower body exercise	Considerations	Progressions
Balance/proprioception	Single leg standing	30 seconds at four sets	Add soft surface/eyes closed Practice a surf stance while standing on a soft surface

Disclaimer: Please ensure you consult with the appropriate health professional before undertaking any exercise program

activity requirements develop the balance system with a potential long-term influence. Frank et al. (2009) explored this long-term effect on a group of male surfers in their 50 and 60s who had surfed for 40–50 years and compared them to age match controls (those that cycle, swim, or walk). Interestingly, the surfers had significantly better postural control when balance was challenged (standing on a soft surface) and when their eyes were closed, relying more on proprioception.

Numerous parts of our anatomy generally peak between our 20s and 30, for example, muscle mass and bone mass, and then they gradually decline as we age (Locquet et al., 2019; McLeod et al., 2016). The same can be said regarding our balance systems and ability. It may be inferred from the previously mentioned study that surfing provides an adequate stimulus to challenge our balance systems and therefore has a protective benefit on the decline of this system. Table 7.4 provides some training considerations to target balance and proprioception.

7.5 Flexibility

Wave riding requires joint flexibility. While surfing, these joints (ankle, hip, and spine) are often at the end of their range of motion. As the surfer compresses their body to lower their center of gravity during the bottom turn, the ankle joint will move into dorsiflexion (the foot moving toward the head). The hip will move into a flexed and internally rotated position (where your knee is positioned towards the board and primarily involves the rear leg), and the spine will move from a flexed position to an extended and rotated position. This requires the surfer to have mobility within these joints to adequately complete the required maneuvers which means restriction in some areas may result in greater stress placed on other joints and potentially injury (Sect. 7.7).

The range of motion of joints does appear to differ between competitive and recreational surfers. Competitive surfers display greater ankle dorsiflexion, hip internal rotation, and spinal (thoracic spine) rotation (Furness, 2016). It is likely these increases in range of motion (ROM) are required within these cohorts; what is unknown is whether these are adaptations from surfing itself or physical attributes that competitive surfers already have. Regardless, in most cases, joint mobility and muscle flexibility are adaptable and can be changed (Table 7.5).

Table 7.5 Training considerations for a recreational surfer to improve flexibility at key joints

Focus	Lower body exercise	Considerations
Lower body mobility	Knee to wall	15–30-second hold
	Hip internal rotation	
Spinal mobility	Thoracic and lumbar rotation	
	Thoracic and lumbar extension	

Disclaimer: Please ensure you consult with the appropriate health professional before undertaking any exercise program

7.6 Lower Body Strength and Power

Finally, the key aspect of riding a wave that should not be neglected is lower body strength and power. It should be noted that strength is concerned with how much force is exerted, whereas power is the ability to exert force over a set time. With this knowledge, we can look at how these aspects are integrated into surfing.

During the bottom turn, the surfer gains momentum (force) as they reach the base of the wave. This increase of force is transmitted through the lower body to the surfboard, resulting in displacement water in a specified direction to ultimately redirect the board back up toward the top of the wave (Everline, 2007). The primary goal for the surfer now is to position the board in the most critical section of the wave face and once again transmit force through an explosive rotational movement to complete the top turn. To be able to transmit force through the lower body, both strength and power are key. Broadly speaking, the body is moving from a flexed to an extended position. So, to develop lower body strength and power, exercises that facilitate these movements in a similar stance would seem logical (e.g., squat or lunge type movement).

Surfing Australia's High-Performance Centre (HPC) has conducted investigations into how strength and power qualities impact performance (Secomb et al., 2015a, b, c, 2016a, b). Strength was assessed using what is known as an isometric mid-thigh pull (IMPT), where the athlete pulls on a bar that is fixed and placed at their mid-thigh and pushes on a force plate. To measure power, two tests were commonly used: (1) countermovement jump (CMJ), where the athlete places their hands on their waists and bends down and jumps up as high as possible, and (2) squat jump (SJ), where the athlete bends down to a self-selected depth and holds this position for 3 seconds and then jumps as high as possible. The SJ removes the momentum component; therefore, it is a test that quantifies power more closely than the CMJ. Higher IMPT, CMJ, and SJ scores were correlated with the surfers' performance during turning maneuvers (Secomb et al., 2015a, b, c). While strength and power alone won't ensure better turns, they are important training considerations to enhance turning ability. Lower body strength and power training should also be considered given the small amount of time spent standing and riding on a wave and hence the limited ability to adequately train through solely catching waves (see Table 7.6 for training considerations).

Table 7.6 Training considerations for the recreational surfer

Focus	Lower body exercise	Considerations	Progressions
Lower body strength	Squat or lunge	8–12 reps, 4 sets with body weight (BW)	Gradually add weight
Lower body power	Countermovement jump (CMJ)	5 reps, 4 sets, BW	Single leg CMJ/squat jump
Rotational strength	Lunge with upper-body rotation	8–12 reps, 4 sets with BW	Consider adding weight with a medicine ball
Rotational power	Lunge with upper-body rotation	Complete the movement with speed, 5 reps, 4 sets, BW	Consider adding weight with a medicine ball
Force absorption	Drop from 20- to 50-cm box	5 reps, 4 sets. Practice landing soft	Jump up to box and down to ground. Gradually increase height of box

Disclaimer: Please ensure you consult with the appropriate health professional before undertaking any exercise program

Fig. 7.10 A surfer about to begin a floater maneuver on the wave crest. (Photo: James Furness)

A drop and stick test has also been commonly used in surfing research. This test involves the athlete landing after dropping down from a 50-cm high box. An interesting finding was that the relative peak power output (using IMPT) for competitive surfers showed those with greater flexibility in their ankles had a significantly lower relative peak landing force (Secomb et al., 2016a, b). This implies that both mobility and strength are key attributes to improved efficiency when absorbing force during landing task. This finding is very important considering numerous turning maneuvers involve the ability to absorb force. For example, a "floater" is a maneuver that involves the surfer placing the board on the broken part of the wave (Fig. 7.10) and then dropping from a height back down to the bottom of the wave (Fig. 7.11). This is somewhat like a drop and stick test. In this situation, the surfer needs to adequately absorb force as they land to complete the maneuver and limit the chance of an injury.

Fig. 7.11 A surfer in the wave bore having just completed a floater. (Photo: James Furness)

7.7 Injuries in Surfing

7.7.1 Terminology Used Within Injury Epidemiology

When discussing sporting injuries, there is some key terminology that needs to be understood. The first term is injury onset, which refers to how quickly or slowly the injury came on. Injury onset can be classified as either traumatic or a gradual onset event. For example, a shoulder injury caused by direct contact with a surfboard is a traumatic injury, whereas damage that slowly developed from paddling without a specific event is a gradual onset injury. The other consideration is injury duration: how long an injury takes to resolve or heal. This is how injuries are often classified. For example, an acute injury is one that is healed in <3 months, whereas a chronic injury results in healing time >3 months (Jordan et al., 2010). In addition, details about the location, type, and mechanisms of the injury are commonly reported. This includes location (part of the anatomy) and mechanism (how the injury occurred). These terms will be used throughout this section of the chapter to outline injury epidemiology within the sport of surfing.

Table 7.7 provides an initial overview of injury epidemiology research within surfing. There has been a significant growth in research over the past 10 years, with nearly 60% of the publications occurring during this period. Furthermore, it is worthwhile noting the fact that only two of these publications used what is considered a prospective research design. This is an important point, as prospective research designs mean that the data is recorded as the injury occurs, improving the validity of the collected data. Other designs such as cross-sectional surveys rely on the surfer to recall the injury that has previously occurred (often over the previous

Table 7.7 Study characteristics specific to injury epidemiology research in the sport of surfing

Characteristics	Number of studies	References
Year of Publication		
Before 2000	3	Allen et al. (1977) and Lowdon et al. (1983, 1987)
2000–2004	2	Nathanson et al. (2002) and Taylor et al. (2004)
2005–2009	3	Base et al. (2007), Hay et al. (2009), and Nathanson et al. (2007)
2010–2014	4	de Moraes et al. (2013), Furness et al. (2014), Meir et al. (2012), Moran and Webber (2013), Pikora et al. (2012)
2015–2019	10	Bazanella et al. (2017), Burgess et al. (2019), Dimmick et al. (2019), Furness et al. (2015), Inada et al. (2018), Jubbal et al. (2017), Klick et al. (2016), Minghelli et al. (2018), Ulkestad and Drogset (2016), Woodacre et al. (2015)
2020–2024	7	Farì et al. (2021), Furness et al. (2021), Hager et al. (2023), Hohn et al. (2020), Quinn et al. (2022), Remnant et al. (2020), and Szymski et al. (2021)
Total	**29**	
Study design		
Prospective cohort design	2	Inada et al. (2018) and Nathanson et al. (2007)
Cross-sectional or retrospective cohort	27	Allen et al. (1977), Base et al. (2007), Bazanella et al. (2017), Burgess et al. (2019), de Moraes et al. (2013), Dimmick et al. (2019), Farì et al. (2021), Furness et al. (2014, 2015, 2021), Hager et al. (2023), Hohn et al. (2020), Jubbal et al. (2017), Klick et al. (2016), Lowdon et al. (1983, 1987), Meir et al. (2012), Minghelli et al. (2018), Moran and Webber (2013), Nathanson et al. (2002, 2007), Pikora et al. (2012), Quinn et al. (2022), Remnant et al. (2020), Szymski et al. (2021), Taylor et al. (2004), Ulkestad and Drogset (2016), Woodacre et al. (2015)
Location of data collection		
Hospital/medical facility	8	Allen et al. (1977), Dimmick et al. (2019), Hager et al. (2023), Hohn et al. (2020), Inada et al. (2018), Jubbal et al. (2017), Klick et al. (2016), and Quinn et al. (2022)
Online survey/beach/surfing competitions	21	Base et al. (2007), Bazanella et al. (2017), Burgess et al. (2019), de Moraes et al. (2013), Farì et al. (2021), Furness et al. (2014, 2015, 2021), Hay et al. (2009), Lowdon et al. (1983, 1987), Meir et al. (2012), Moran and Webber (2013), Nathanson et al. (2002, 2007), Pikora et al. (2012), Remnant et al. (2020), Szymski et al. (2021), Taylor et al. (2004), Ulkestad and Drogset (2016), Woodacre et al. (2015)

(continued)

Table 7.7 (continued)

Characteristics	Number of studies	References
Nature and duration of injury		
Traumatic/acute injury	27	Allen et al. (1977), Base et al. (2007), Bazanella et al. (2017), Burgess et al. (2019), de Moraes et al. (2013), Dimmick et al. (2019), Farì et al. (2021), Furness et al. (2015, 2021), Hager et al. (2023), Hay et al. (2009), Hohn et al. (2020), Inada et al. (2018), Jubbal et al. (2017), Klick et al. (2016), Lowdon et al. (1983, 1987), Meir et al. (2012), Minghelli et al. (2018), Moran and Webber (2013), Nathanson et al. (2002), Pikora et al. (2012), Quinn et al. (2022), Szymski et al. (2021), Taylor et al. (2004), Ulkestad and Drogset (2016) and Woodacre et al. (2015)
Gradual onset/chronic[a]	8	Bazanella et al. (2017), Furness et al. (2014), Inada et al. (2018), Lowdon et al. (1983, 1987), Nathanson et al. (2002), Remnant et al. (2020), and Taylor et al. (2004)

[a]Some studies looked at both traumatic and gradual onset injuries

12-month period). Thus, the limitation of memory decay becomes apparent, whereby the surfer may remember they had an injury but may not recall the precise details such as injury onset (traumatic or gradual onset) or duration (acute or chronic) or the type (muscle, skin, etc.).

Finally retrospective cohort designs are often employed by hospital-based studies, where researchers obtain patient records and then analyze this data. It also can't calculate injury rates as there is no group of uninjured surfers and therefore doesn't provide data on this metric (this metric is outlined below). You also see a trend of surfers who are injured seeking first aid treatment, and hence injuries are commonly related to the skin such as a laceration, with less frequency of soft tissue sprains and strains as these may be self-managed or managed by an allied health practitioner such as a physiotherapist. Furthermore, the data was not initially collected with the initial intent for a research project and often is limited in what can be reported; for example, these studies lack details of the surfers themselves (experience and skill level, board type, hours surfed per week, etc.).

7.7.2 Injury Prevention Model

In order to discuss injury within the sport of surfing, this section will follow the model of van Mechelen et al. (1992) (Fig. 7.12). The model looks at four key areas which will now be explored.

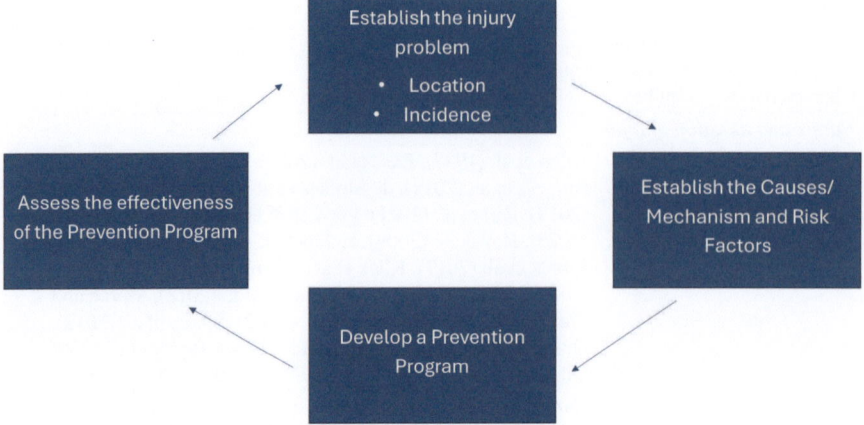

Fig. 7.12 The injury prevention model developed by van Mechelen et al. (1992)

7.7.2.1 Injury Incidence

How often do surfers get injured? This is one of the first questions to ask, and the short answer is not very often if you compare it to mainstream sports such as football and soccer. To fully explore the topic, research investigates injury incidence and prevalence. Incidence looks at the number of injuries over a set period, while prevalence investigates injuries at one time point. For example, incidence is commonly expressed as a rate and uses the denominator of hours (often per 1000 hours). The other term is known as incidence proportion which looks at the number of surfers that has been injured divided by the total number of surfers in the entire group. This is a useful metric as it assists with answering the question, "how likely is a surfer to sustain an injury in specific time?"

Generally, the incidence rate for recreational surfers is one to two injuries per 1000 hours (Furness et al., 2015, 2021; McArthur et al., 2020) and approximately six injuries per 1000 hours for competitive surfers (Inada et al., 2018; Nathanson et al., 2007). This statistic is good for comparative reasons as it is commonly used in research in other sports. For example, when comparing surfing to say a contact sport like rugby union where the incidence rate has been shown to be 69 injuries per 1000 playing hours, surfing begins to be viewed as a relatively safe sport. However, there is a caveat to this statement. One of the only prospective studies conducted by Nathanson et al. (2007) collected data at 32 professional and amateur surfing contests and found the overall injury rate for competitive surfers to be 6.6 injuries per 1000 hours. The same researchers conducted a sub-analysis just looking at the four events held in Hawaii at Pipeline (generally a large wave breaking over shallow reef) and found this incidence rate to increase to 32 injuries per 1000 hours, this being a 385% increase in the incidence rate. This finding really highlights the danger of larger surf over reef and more specifically surfing in Hawaii. It highlights the site-specific nature of injuries in surfing reinforcing how local conditions determine the character and composition of the break which is ridden (Chap. 4).

The more user-friendly metric is incidence proportion, namely, the proportion of surfers who will sustain an injury over a given time. While this is not commonly reported in all the surfing-related injury research, it is approximately one out of three recreational surfers will sustain an injury and one out of two competitive or aerialist surfers will sustain an injury in a 12-month period (Furness et al., 2015, 2021; McArthur et al., 2020).

7.7.2.2 Location

Where on the body do surfers usually get injured? This often depends on where the data was collected. When the data is collected in hospitals and medical facilities, there tends to be a high number of injuries to the head and face, commonly lacerations. A large systematic literature review of injury epidemiology between 1977 and 2018 revealed that if data was collected in healthcare facilities, the frequency of head and neck injuries was 43% as opposed to survey data being 28% (McArthur et al., 2020) (Fig. 7.13) and this trend was also reported in other hospital based studies in the USA (Muhonen et al., 2022) and Australia (Quinn et al., 2022). When the mechanisms of injury are explored, they are often a result of direct trauma with a board. When data is collected via survey methods or directly at surfing events, differences become apparent. The number of lower limb injuries and sprains and strains to muscles and joints increases. These trends are expected, given a minor or moderate ligament injury would often not result in hospitalization.

Traumatic and acute injuries tend to occur in the shoulder, head, and ankle or foot regions in recreational surfers (Fig. 7.14), whereas the knee and the and ankle

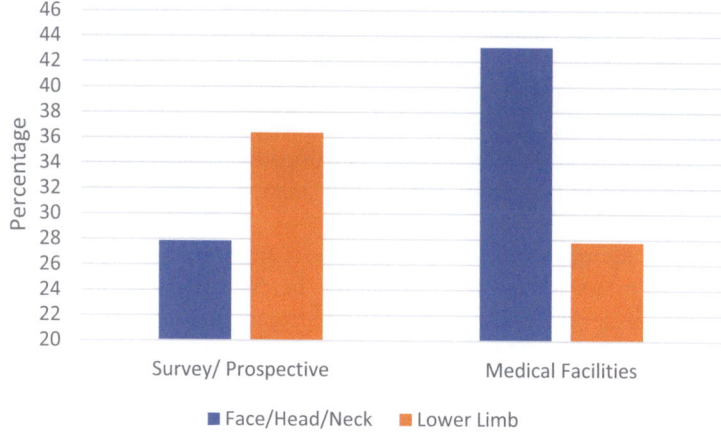

Fig. 7.13 Comparison of injury locations and study locations. (Adapted from McArthur et al. (2020). Reproduced under a CC-BY licence)

Fig. 7.14 Injury location, considering injury onset, duration, and competitive status. (Photo Sebastian Staines on Unsplash. Reproduced under a CC-0 licence)

appear to have high frequencies of injuries in competitive surfers. When looking at gradual onset and chronic injuries, the shoulder and lower back have high frequencies of injuries compared to other regions (Table 7.7).

7.7.3 Traumatic Head Injuries

A concussion also known as a mild traumatic brain injury (MTBI) occurs when an acceleration and deceleration force cause the soft brain tissue to impact the harder skull (Fig. 7.15). This is known as a translational movement. The damage is done by direct impact, bruising the brain. It can also be caused whereby the head is struck on an angle and the brain rotates (rotational movement), and this is believed to be the primary mechanism for concussions. This causes damage to brain cells (neurons) as they rub against each other due to shearing forces. Both of these types of concussion can result in temporary neurological impairment (such as loss of consciousness, balance impairment, etc.); however, MTBI can have longer consequences involving chronic traumatic encephalopathy (degeneration of nerve cells in the brain), cognitive impairment, and mental health issues (Kozminski et al., 2020).

Concussion injury specific to surfing does not appear to be frequently reported. However, this is not to say that it shouldn't be discussed given the severity of repeated and/or severe concussions and the potential long-term consequences (Ledreux et al., 2020). There are three large scale studies that provide some insight into concussion within surfing (Hager et al., 2023; Klick et al., 2016; Kozminski et al., 2020) drawing on emergency department data from

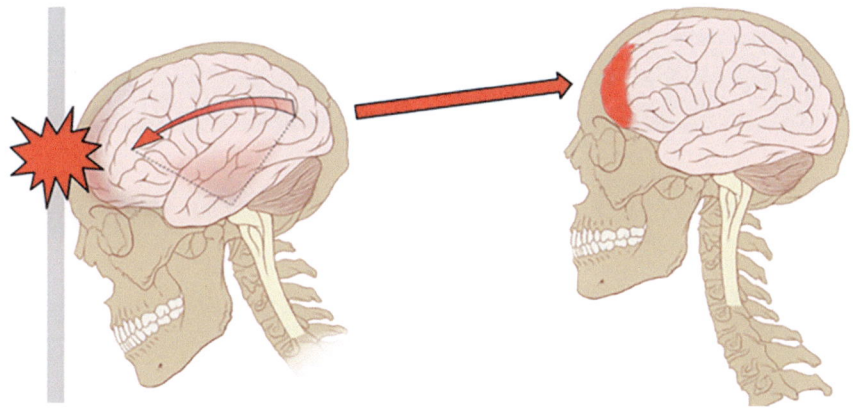

Fig. 7.15 Translation of the brain within the skull resulting in a MTBI. (Image by Max Andrews. Reproduced under a CC BY-SA 4.0 licence)

approximately 100 hospitals within the USA. Combined they found reported concussions ranging from 2.7% to 16.1% with children have approximately double the frequency of such events than adults. It was also found there was a decrease in the rate of lacerations which may be due to the introduction of soft tops surfboards, primarily made from polystyrene foam and fins now being made from pliable plastic.

Given these alarming statistics, the question remains around what could be implemented to minimize concussion within the sport of surfing. Despite a plethora of research papers concluding with a statement that surfers should wear a helmet (soft or hard) to prevent head injuries, it has been estimated that only 1–2% of surfers wear a helmet on a regular basis (Taylor et al., 2005; Ulkestad & Drogset, 2016). While the surfers admit there is a significant risk of head injury, the major reasons for not wearing a helmet are the fact that they restrict performance and influence balance and secondly surfers perceive other sports to have a higher risk of head injury (Taylor et al., 2005).

Researchers at the Queensland Brain Institute, Australia, believe that hard helmets provide protection against more severe brain injury where there is translation of the brain from a direct force. Here the helmet can dissipate force over a larger area. An example of this would be a direct contact with the reef following falling from a wave. A helmet may be less effective from a rotational force due to the inability of the helmet to redistribute or absorb these forces. While helmet use may continue to have low levels of uptake, raising awareness of concussion is needed. This is paramount in the sport of surfing as the surfer may continue to surf following a head injury, therefore increasing the risk of a sequential head injury with potential long-term consequences. There are both signs and symptoms of concussion that surfers should be aware of as depicted in Fig. 7.16.

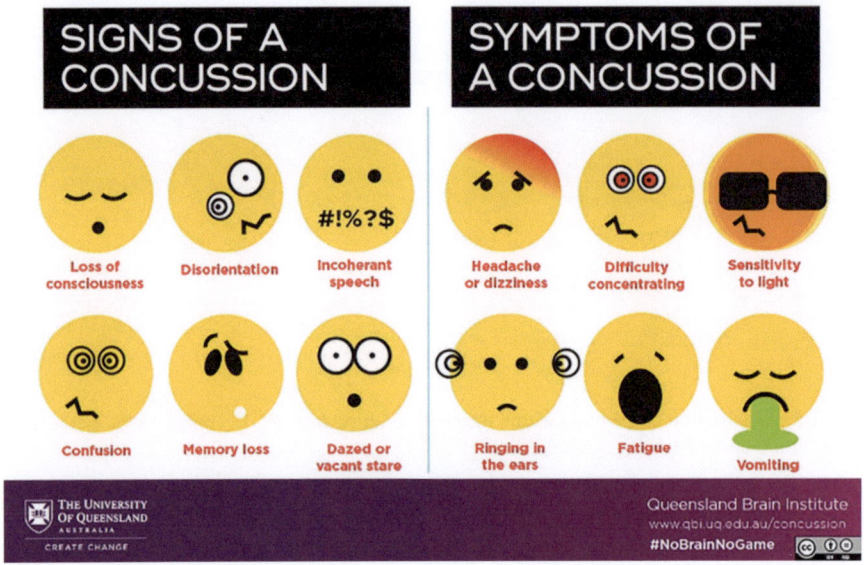

Fig. 7.16 The signs and symptoms of concussion. (Source https://qbi.uq.edu.au/ reproduced under a CC- BY-ND licence)

7.7.3.1 What to Do if a Concussion Is Suspected?

The Australian Sports Commission has provided guidelines and a position statement on concussion, and these should be applied to the sport of surfing. The underpinning principle is that if there is any doubt an athlete (a surfer) is concussed, they should be immediately removed and not allowed to return to sport (surfing) until cleared by a healthcare practitioner (HCP). The slogan "If in doubt, sit them out" is now used (Australian Sports Commission, 2023). Any athlete (surfer) that has a suspected concussion should be referred to a AHPRA registered HCP (such as a medical practitioner or physiotherapist) with the appropriate training and experience in concussion assessment and management (Australian Sports Commission, 2023).

7.7.4 Traumatic Injuries to the Knee

The knee is a common injury site for competitive surfers being primarily due to aggressive turning maneuvers and aerials (Nathanson et al., 2007). When studies have incorporated both recreational and competitive surfers there are often double the rate of knee injuries in the latter group (Furness et al., 2015, 2021). Overall, approximately 50% of the knee injuries relate to the medial collateral ligament (MCL) (Fig. 7.17). Understanding the anatomy of the knee is important as the mechanism of injury is often a valgus force placed on the outside of the knee. This

Fig. 7.17 The ligaments of the knee (in blue). (Source: Jaskirat Singh Benipal)

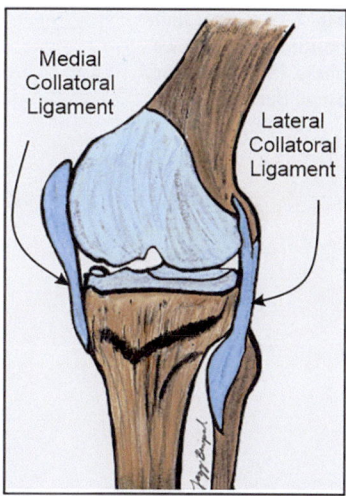

Fig. 7.18 Illustration of the sideways stance promoting the knee to drop inward, on the leg at the rear of the board, as this lowers the center of gravity for the surfer. (Photo: Surfing Croyde Bay on Unsplash. Reproduced under a CC-0 licence)

is often a position the rear leg is placed in during turning maneuvers, aerials, floaters, and barrel riding (Figs. 7.18 and 7.19). The sideways stance promotes the knee to be dropped inward as this lowers the center of gravity for the surfer. A study conducted on professional surfers who underwent medical treatment between 1991 and 2016 found that lower extremity injuries primarily affect the surfers' rear leg (Hohn et al., 2020).

Fig. 7.19 Lower limb position during stance phase. (Source: Jaskirat Singh Benipal)

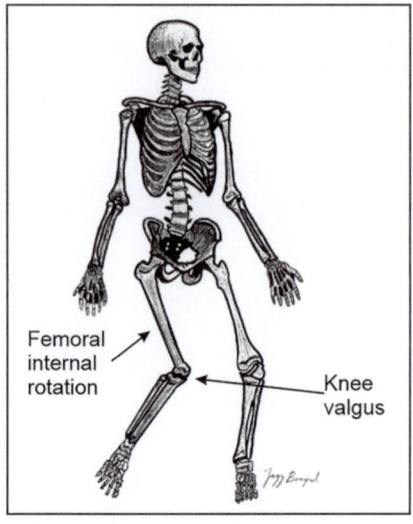

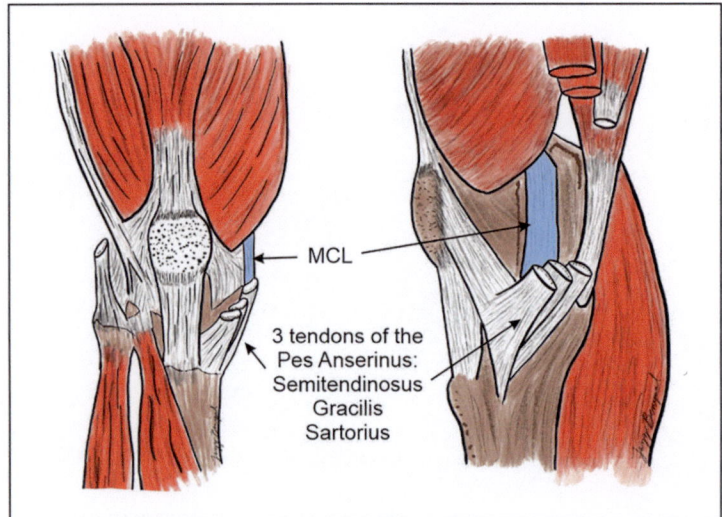

Fig. 7.20 Adductor muscles and their proximity to MCL knee ligament. (Image by Jaskirat Singh Benipal)

7.7.4.1 Injury Reduction Strategies for the Knee

Exposing the lower limbs to movements that are performed in the water with a controlled land environment may also assist in building tolerance in the ligamentous structures (MCL) and the associated musculature. For example, the MCL is covered by the adductor muscles (sartorius tendon, gracilis tendon, and semitendinosus tendon) (Fig. 7.20). Competitive surfers are now training the adductor muscles to

Fig. 7.21 Adductor muscle training. (Photo https://www.fitnessequipmentireland.ie/wp-content/uploads/2019/10/Slide-)

provide additional stability at the medial knee (MCL). A simple sliding exercise is performed to both replicate similar movements in surfing and strengthen the adductor muscles may potentially provide protection to the MCL (Fig. 7.21).

7.7.5 Injuries at the Shoulder

The bone structure of the shoulder joint allows large amounts of movement through the shallow socket and large ball component. The shoulder gains its stability through the musculature that attaches from the scapular (shoulder blade) and the spine and the humerus. It is this balance between mobility and stability that makes the interplay between the muscles and bones at the shoulder joint very complex and prone to injury. As a surfer spends approximately half their time paddling, it is no surprise that the shoulder is a common site for gradual onset injuries. Gradual onset injuries are often a result of repeated microtrauma when the musculature is not given adequate rest to recover and repair. This accumulation of load can result in delayed and poor healing of musculature and consequently pain. Figure 7.22 illustrates that tissue tolerance improves provided there is a period of rest between the loads. When rest is absent and there is cumulative loading, there is a decrease in tissue tolerance, ultimately ending in injury.

This concept of reduction in tissue tolerance is often validated by research on gradual onset and chronic injury in surfing. This type of injury is often related to muscles and tendons within the shoulder, commonly referred to as the rotator cuff

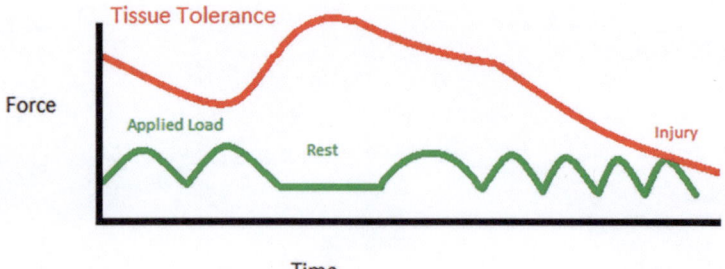

Fig. 7.22 The relationship between load, rest, and injury. (Modified from McGill (2007))

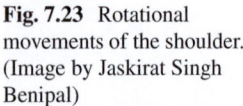

Fig. 7.23 Rotational movements of the shoulder. (Image by Jaskirat Singh Benipal)

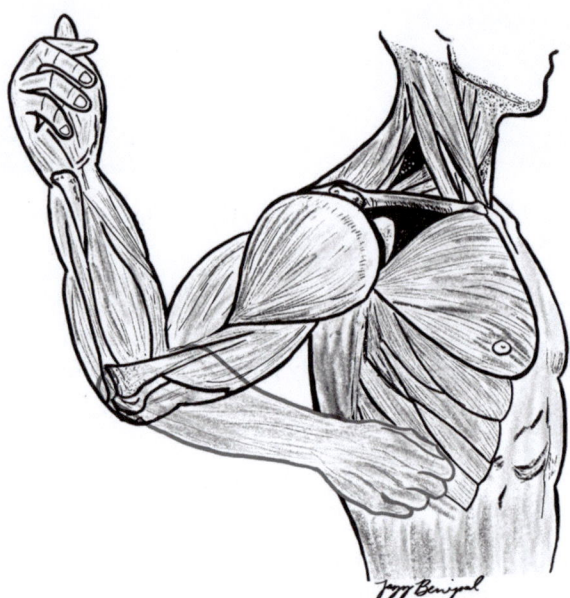

(Hohn et al., 2020; Inada et al., 2018; Remnant et al., 2020). When the mechanism of injury is further explored, prolonged paddling is often a major cause. Gradual onset shoulder injuries are often chronic (64% of injuries will take 3 months or longer to recover) (Remnant et al., 2020) and even when the cause is traumatic with 44% of injuries then become chronic (Furness et al., 2021). As depicted in Fig. 7.22, applied load followed by a period of rest results in improved tissue tolerance.

There are other modifiable risk factors that may contribute to shoulder injury and long recovery periods. For example, the shoulder joint allows for two rotational motions in the sagittal plan—this includes internal and external rotation (Fig. 7.23). A group of muscles known as the rotator cuff muscles assist these movements, including the supraspinatus, infraspinatus, subscapularis, and teres minor (Fig. 7.24). Imbalance in both the range and strength of these motions has been shown to be risk factors for shoulder injury in other sports such as handball (Andersson et al., 2017).

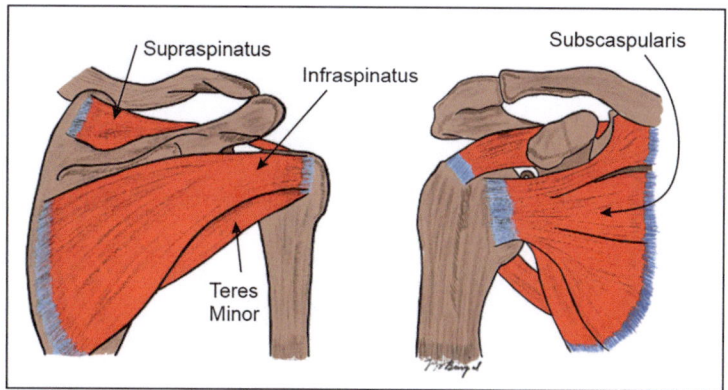

Fig. 7.24 Rotator cuff muscles of the shoulder. (Image by Jaskirat Singh Benipal)

In addition to imbalance in range of motion or strength, poor muscular endurance may also predispose the surfer to shoulder injury. To date, there is a paucity of research investigating these modifiable risk factors in surfing. One published study investigating muscle balance of internal and external rotation strength in 13 competitive surfers identified the muscle balance to be approximately 80%. So, for every 1 unit of internal rotation strength, the surfer had 0.8 units of external rotation strength (Furness et al., 2018a, b).

7.7.5.1 Injury Reduction Strategies for the Shoulder

Adequate recovery is critical to assist in reducing shoulder injury. While this is individualized to each surfer based on underlying experience and surfing frequency, the surfer should be aware of muscular fatigue and gradually increase surfing sessions and frequency over time. In addition to this active recovery methods such as shoulder stretching, specifically the posterior musculature/capsule can be stretched to aid in recovery of the rotator cuff. Figures 7.25, 7.26, and 7.27 and Table 7.8 provide injury reduction considerations for surfers.

7.7.6 Lower Back: Gradual Onset Injury

In studies that have explored gradual onset injuries or injuries of chronic duration, the lower back is frequently represented with frequencies ranging from 16 to 31% (Furness et al., 2014; Inada et al., 2018; Nathanson et al., 2002; Remnant et al., 2020). The spine is made up of three key regions, the cervical, thoracic, and lumbar spine. The primary role of the lumbar spine is to absorb load, and hence the bone structure is large compared to the thoracic and cervical regions. Each segment is

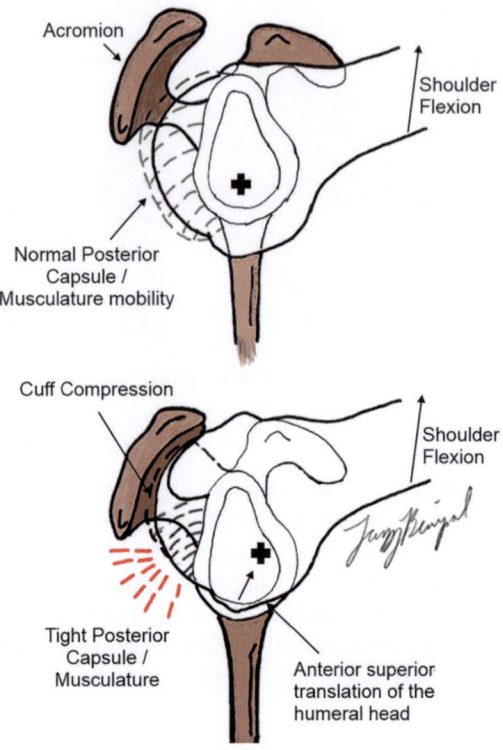

Fig. 7.25 Illustration of the posterior capsule influence on compression of the rotator cuff. Note the movement of the black cross during shoulder flexion. (Image by Jaskirat Singh Benipal)

separated with a spinal disc which is a gel like structure that adapts to various loads (Fig. 7.28). The lumbar spine also has facet joints which are open with flexion and close with extension. The orientation of these joints is also important as they are positioned in the sagittal plane allowing for flexion and extension with minimal rotation. The thoracic spine facets are orientated in the frontal plane and consequently allow for rotation of the spine (Fig. 7.28).

Turning maneuvers and prolonged paddling are common reasons for lower back injuries. When a surfer is in a prone position (chest facing down) and the spine in an extended position, this requires the facet joints to be in proximity with each other (Fig. 7.28). As the surfer rapidly moves from paddling to standing, the spine changes to a flexed position. In addition, explosive turning, cutting, and twisting movements are needed to generate torque and maneuver the board often combining both spine rotation and flexion.

By understanding the anatomy and the activity requirement of surfing, there are several potential causes for gradual lower back onset injuries. Firstly, the prolonged extended position requires the lumbar facets to be held in a sustained closed position along with a sustained muscle contraction; it is very plausible that the pain experienced from a surfer may be due to repeated microtrauma to these structures. Secondly, the explosive spinal movements required during turning maneuvers may compromise the spine with flexion and rotation placing excessive strain on the

Fig. 7.26 Arm across body for posterior capsule. (Photo James Furness)

Fig. 7.27 Sleeper stretch for posterior capsule. (Photo James Furness)

Table 7.8 Injury reduction considerations for the surfer's shoulder

Focus	Exercise	Considerations	Progressions
Shoulder mobility	Posterior capsule stretch	30 second hold. Arm across body.	Sleeper stretch.
Muscular endurance of rotator cuff and shoulder muscles	Consider pool-based or land-based training during low surfing frequency periods	Start with a low to moderate intensity effort for 15–20 minutes	Increase intensity to moderate effort and duration to 30 minutes. During weeks with no surfing, aim to complete these two to three times per week

Disclaimer: Please ensure you consult with the appropriate health professional before undertaking any exercise program

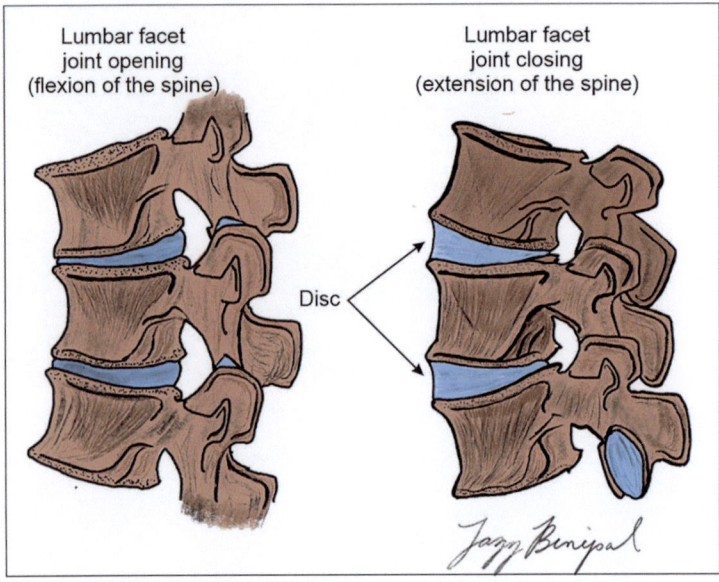

Fig. 7.28 Lumbar spine illustration of facets opening and closing. (Image by Jaskirat Singh Benipal)

intervertebral discs. In addition to this, poor muscular strength and power in the trunk and lower body may also predispose the surfer to injury within the lumbar spine.

7.7.6.1 Practical Applications for the Lumbar Spine

As previously discussed, the surfer should participate in lower limb and trunk exercises (Table 7.5) to ensure adequate strength and power. In conjunction with this, a range of mobility exercise may be appropriate for the surfer with a movement

Table 7.9 Injury reduction considerations for the surfer's lumbar spine

Focus	Exercise	Considerations
Address areas of restriction above and below the lumbar spine	Thoracic rotation mobility exercises	Trial 15–30 second holds Trial dynamic movements with minimal sustained holds
	Hip internal rotation exercises	
Adequate recovery of the lumbar spine	Lumbar mobility exercises	

Disclaimer: Please ensure you consult with the appropriate health professional before undertaking any exercise program

restriction (Table 7.9). For example, if a surfer lacks mobility in the thoracic extension when lying prone (chest down), the lumbar spine may compensate to maintain adequate spinal extension during paddling. The same can occur during turning maneuvers where the thoracic spine and hips provide the rotational component. A reduction in thoracic and hip rotation may increase load at the lumbar spine.

7.7.7 Surfer's Myelopathy

While discussing the lumbar spine and gradual onset injuries, a more serious pathology is what has been coined surfers' myelopathy. This was first described in 2004 after a series of young healthy novice surfers developed nontraumatic spinal cord injuries. A systematic review (Alva-Díaz et al., 2022) found a total of 104 patients who had sustained this injury with 58% occurring after surfing. Furthermore, only 52% reported a complete or moderate recovery. With this condition, the surfer often suddenly (often an hour or less after the activity) develops low back pain, lower limb paraesthesia (the feeling of tingling, numbness, or "pins and needles"), and most commonly bladder and bowel dysfunction. The cause of this serious condition is thought to be a lack of blood supply to the spinal cord.

The spinal cord runs in a canal which sits in front of the vertebrae—at each spinal level, a nerve exits and supplies various parts of our body (Fig. 7.29). Nerves provide movement (motor) and sensation (sensory) within our body. All structures in the body require oxygenated blood, and without this supply, tissue damage occurs (ischemia). Surfers' myelopathy is thought to occur from ischemia (lack of oxygenated blood) from to the artery of Adamkiewicz being kinked or narrowed (vasospasm) related to the novice surfer lying in the prone position for a prolonged period. The novice surfer may exhibit a position of mild hyperextension when waiting for a wave (lying prone with the elbows resting on the board) and may also stay in this position when the wave hits as they lack the ability to duck dive. The prolonged period in this position, as compared to more experienced surfers, promotes a moderate or severe hyperextension of the spine.

Fig. 7.29 The spinal cord and the associated blood supply vessels. (Image by Jaskirat Singh Benipal)

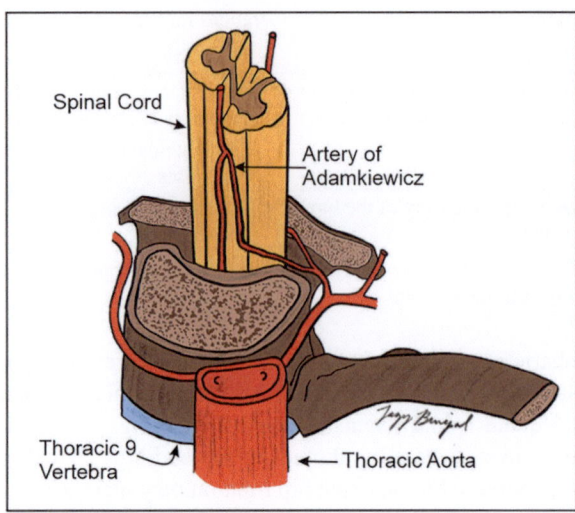

7.7.7.1 Injury Reduction Strategies for Surfer's Myelopathy

Most simply, awareness and education are needed among surf coaches, especially learn to surf schools. Simply being educated on this condition will help ensure novice surfers are not sustaining a hyper-extended position. This could involve teaching the surfer how to sit on the board while waiting for a wave and developing the appropriate technique when a wave hits (either a duck dive or hop off the board) and adding simple warm up strategies involving spinal extension.

7.7.8 Non-musculoskeletal Injuries

7.7.8.1 External Auditory Exostosis, Surfer's Ear

External auditory exostosis (EAE) is a condition which involves bone growth from the temporal bone (skull) into the ear canal, obstructing the function of the ear. As depicted in Fig. 7.30, the red lines (in the ear canal) indicate the abnormal bone growth from the temporal bone (this is one of the major bones on the side of the skull) within the ear canal. The other key anatomy is the ear drum and the inner ear (cochlea).

The external part of the ear (that you can see) acts as a funnel sending sound waves into the ear canal. The sound waves then cause the ear drum to vibrate and make fluid within the cochlea move. This movement of fluid causes tiny hairs within the cochlea to bend and move. The cochlea converts this movement of the hairs to electrical signals that are sent to the brain through the auditory nerve. Our brain is then able to process this information to make sense of what we are hearing.

The current prevalence EAE within the surfing population is estimated to be between 53% and 90% compared with the 3–6% in the general population (Vallée,

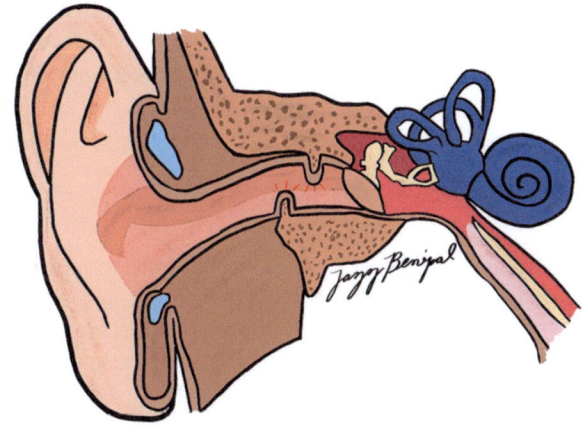

Fig. 7.30 Anatomy of the inner ear, illustrating external auditory exostosis or surfer's ear. (Image by Jaskirat Singh Benipal)

2024). The exact cause of surfer's ear is not known; however, the initial primary hypothesis was cold water immersion. A large latitudinal study conducted in 1984 reviewed frequencies of EAE across the earth. The primary finding was that in latitudes where there was diving for cold water resources (food), there were moderate to high frequencies of EAE; this occurred in the middle latitudes (30–45°) (Kennedy, 1986). Later research has identified the cutoff temperature of water below 19 °C as being a risk factor for EAE with prevalence rates between 61% and 80% of surfers (Attlmayr & Smith, 2015; Chaplin & Stewart, 1998; Hurst et al., 2004; Kroon et al., 2002; Nakanishi et al., 2011; Umeda et al., 1989; Wong et al., 1999). Cold water being the primary risk factor has further been challenged with research recently conducted in warmer climates such as the Gold Coast, Australia, where the water temperature ranges from 20 to 28 °C. The study identified that the current prevalence of EAE was 72% (Simas et al., 2021). Exposure to cold air and wind conditions are other possible causes. Another risk factor for EAE is surfing experience, with the presence and severity of EAE significantly increasing after 5 years of surfing (Alexander et al., 2015; Deleyiannis et al., 1996).

When looking at preventative factors for EAE, the most obvious would be the use of a protective structure such as an earplug or hood which covers the ears. To date, there is some low-level evidence that earplugs reduce the severity of EAE (Alexander et al., 2015; Lambert et al., 2021); however, the designs are cross sectional with no interventional studies to conclusively determine their effect. Given the possible causes of EAE, it would seem reasonable that earplugs and/or a hood would reduce the possibility of developing EAE or reduce the symptoms of EAE such as ear infections or water getting trapped in the ear canal. Surgical intervention is often reserved for severe and symptomatic cases.

Practical Applications for Surfer's Ear

- Promote the understanding of exostosis/surfer's ear.
- Identify surfers with key risk factors:
    - Surfer with greater than 5 years' experience
    - Surfing in cold water climates (<19 °C)

- Early otoscope assessment of a surfer's ear by a general practitioner (GP).
- Consider earplug/hood for symptomatic periods or when surfing in cold climates or to prevent progression of condition if diagnosed.

7.7.9 Skin Cancer

The skin is our largest organ—it is designed to provide protection, control our temperature, and prevent fluid loss. As surfing takes place under direct sunlight and with the reflective properties of water, surfers are at an increased risk of skin cancer compared to other outdoor land-based sports. This condition is not frequently reported in epidemiological research; however, it will be detailed given its serious consequences and its prevalence in surfers compared to the general population.

The first consideration is to understand how skin cancer can occur. Approximately 95% of all skin cancers are a result of exposure to ultraviolet radiation (UVR) (Cancer Council Australia, 2024). The UVR that is emitted from the sun is made up of ultraviolet A (UVA) and ultraviolet B (UVB) rays comprising of 90–95% and 5–10%, respectively, of the total UVR. Most of the UVB is absorbed by the oxygen in the earth's atmosphere which makes up the ozone layer. Areas where the ozone layer is thinner allow for greater UVB to reach our skin which can have a carcinogenic effect on the skin cells. Furthermore, higher levels of UVR are experienced during the periods where the sun is highest in the sky, resulting in the shorter path of the UVR to travel to the earth's surface. Often these peak UVR periods are considered between 9 am and 4 pm depending on the geographic location.

To broadly summarize how UVR damages our skin, we need to understand that skin cells are tightly packed and positioned. There is a process of cell death and development, with skin cells constantly being replaced in an orderly fashion. What a cell does in our body is determined by the genetic information found in the organic chemical known as deoxyribonucleic acid (DNA) in the individual cells. Skin cells that have DNA damaged by UVR may become abnormal and keep growing altering the original well-organized cell structure (Fig. 7.31).

Our skin is made up of three primary layers, the epidermis, dermis, and hypodermis (Fig. 7.32). Each layer contains various structures as depicted in the image. The

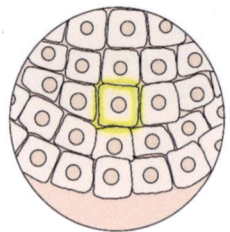
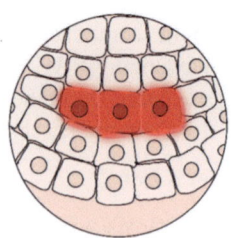
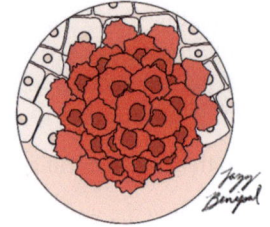

Fig. 7.31 Changes in skin cells due to UVR. (Image by Jaskirat Singh Benipal)

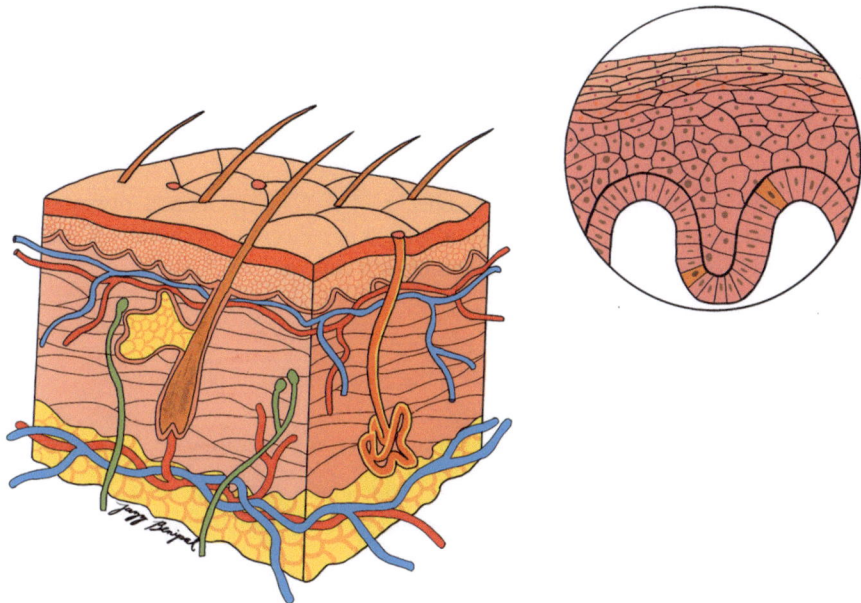

Fig. 7.32 Layers of skin and associated cells. (Image by Jaskirat Singh Benipal)

key cells in each layer are squamous, basal, and melanocytes. All of these correspond to the associated cancers which are squamous cell carcinoma (SCC), basal cell carcinoma (BCC), and melanoma.

Research into surfing and skin cancer has increased over the past decade, and the results reveal that surfers are at particular risk. Research conducted out of Australia initially determined the point prevalence of skin cancers in a group of 116 surfers was 11.2% for BCC, 1.7% for SCC, and 5.2% for melanoma. When this point prevalence was compared with the rate of skin cancers in Australia, surfers were 97 times more likely to have a melanoma, 7 times more likely to have a BCC, and 1.6 times more likely to have a SCC (Climstein et al., 2022).

When looking at location, the skin cancers are commonly located on the face and back (Climstein et al., 2022; Climstein et al., 2016). A very recent study conducted in 2023 also confirmed that surfers were 120 times more likely to be diagnosed with a melanoma when compared to the Australian population. The study found that while lifetime sun exposure was less than other outdoor sports (walkers and runners), surfers were more commonly exposed to peak UVR periods (Miller et al., 2023).

When looking at the sun exposure behaviors within the above studies (Climstein et al., 2016; Miller et al., 2023), surfers spent around 5–7 hours per week exposed to sunlight, and approximately 30% of this time is spent during peak UVR periods (between 10 am and 3 pm). When looking at sun protection strategies, around 90% use sunscreen and 80% use a rash vest, t-shirt, or wetsuit and 20–30% use a surf hat and 30–40% have undergone a skin check in the previous 12 months. Furthermore, only 36% of the surfers completed regular self or partner exam of their own skin.

7.7.9.1 Skin Cancer Protection Strategies for Surfers

Ensure regular self-examination skin checks. The Cancer Council organization within Australia has raised awareness around skin cancers with initiatives such as the ABCDE acronym for self-assessment of skin cancer.

- Asymmetry—of the mole or spot?
- Border—irregularity, jagged, or spreading?
- Colors—are there several colors, or is the color uneven or blotchy?
- Diameter—are the spots getting larger?
- Evolution—is the spot or mole changing or growing over time?

Surfers should be using sun protection strategies (sunscreen, rash vests, etc.). Given the long exposure periods, surfers should be reapplying sunscreen during extended surf sessions.

Surfers need to be aware of the UV index—this could easily be done when checking the surf report from reputable websites (government websites or cancer organizations).

Surfers need to consider minimizing sun exposure during peak UVR periods, commonly between 9 am and 4 pm. Finally, given the high prevalence of skin cancer and excessive UVR exposure during peak UVR periods and the reflective properties of the surf, surfers should have yearly skin checks by a qualified medical practitioner.

Acknowledgments Thank you to Associate Professor Mike Climstein for your valuable feedback and careful editing of this chapter. Your insights have greatly improved the work. Jaskirat Singh Benipal is thanked for the many images he prepared for this chapter.

References

Alexander, V., Lau, A., Beaumont, E., & Hope, A. (2015). The effects of surfing behaviour on the development of external auditory canal exostosis. *European Archives of Oto-Rhino-Laryngology, 272*(7), 1643–1649. https://doi.org/10.1007/s00405-014-2950-5

Allen, R. H., Eiseman, B., Straehley, C. J., & Orloff, B. G. (1977). Surfing injuries At waikiki. *JAMA, 237*(7), 668–670.

Alva-Díaz, C., Rodriguez-López, E., López-Saavedra, A., Metcalf, T., Morán-Mariños, C., Navarro-Flores, A., Velásquez-Rimachi, V., Aguirre-Quispe, W., Shaikh, E. S., Mori, N., Romero-Sanchez, R., & Pacheco-Barrios, K. (2022). Is Surfer's myelopathy an acute hyperextension-induced myelopathy? A systematic synthesis of case studies and proposed diagnostic criteria. *Journal of Neurology, 269*(4), 1776–1785. https://doi.org/10.1007/s00415-021-10775-4

American College of Sports Medicine. (2022). *ACSM's guidelines for exercise testing and prescription* (11th ed.). Wolters Kluwer.

Andersson, S. H., Bahr, R., Clarsen, B., & Myklebust, G. (2017). Preventing overuse shoulder injuries among throwing athletes: A cluster-randomised controlled trial in 660 elite handball players. *British Journal of Sports Medicine, 51*(14), 1073–1080. https://doi.org/10.1136/bjsports-2016-096226

Attlmayr, B., & Smith, I. M. (2015). Prevalence of 'surfer's ear' in Cornish surfers. *The Journal of Laryngology and Otology, 129*(5), 440–444. https://doi.org/10.1017/s0022215115000316

Australian Sports Commission. (2023). *Concussion in Australian sport.* www.ausport.gov.au

Barlow, M. J., Gresty, K., Findlay, M., Cooke, C. B., & Davidson, M. A. (2014). The effect of wave conditions and surfer ability on performance and the physiological response of recreational surfers. *Journal of Strength and Conditioning Research, 28*(10), 2946–2953. https://doi.org/10.1519/jsc.0000000000000491

Base, L. H., Alves, M. A. F., Martins, E. O., & Costa, R. F. d. (2007). Injuries among professional surfers. *Revista Brasileira de Medicina do Esporte, 13*, 251–253.

Bazanella, N. V., Garrett, J. G. Z. D. A., Gomes, A. R. S., Novack, L. F., Osiecki, R., & Korelo, R. I. G. (2017). Influence of practice time on surfing injuries. *Fisioterapia em Movimento, 30*, 23–32.

Borgonovo-Santos, M., Zacca, R., Fernandes, R. J., & Vilas-Boas, J. P. (2021). The impact of a single surfing paddling cycle on fatigue and energy cost. *Scientific Reports, 11*(1), 4566-undefined. https://doi.org/10.1038/s41598-021-83900-y

Burgess, A., Swain, M. S., & Lystad, R. P. (2019). An Australian survey on health and injuries in adult competitive surfing. *The Journal of Sports Medicine and Physical Fitness, 59*(3), 462–468. https://doi.org/10.23736/s0022-4707.18.08381-0

Cancer Council Australia. (2024). *Understanding skin cancer: A guide for people with cancer, their families and friends.* https://www.cancer.org.au/assets/pdf/understanding-skin-cancer-booklet

Chaplin, J. M., & Stewart, I. A. (1998). The prevalence of exostoses in the external auditory meatus of surfers. *Clinical Otolaryngology and Allied Sciences, 23*(4), 326–330. https://doi.org/10.1046/j.1365-2273.1998.00151.x

Clark, J., Glasziou, P., Del Mar, C., Bannach-Brown, A., Stehlik, P., & Scott, A. M. (2020). A full systematic review was completed in 2 weeks using automation tools: A case study. *Journal of Clinical Epidemiology, 121*, 81–90. https://doi.org/10.1016/j.jclinepi.2020.01.008

Climstein, M., Furness, J., Hing, W., & Walsh, J. (2016). Lifetime prevalence of non-melanoma and melanoma skin cancer in Australian recreational and competitive surfers. *Photodermatology, Photoimmunology & Photomedicine, 32*(4), 207–213. https://doi.org/10.1111/phpp.12247

Climstein, M., Doyle, B., Stapelberg, M., Rosic, N., Hertess, I., Furness, J., Simas, V., & Walsh, J. (2022). Point prevalence of non-melanoma and melanoma skin cancers in Australian surfers and swimmers in Southeast Queensland and Northern New South Wales. *PeerJ, 10*, e13243-undefined. https://doi.org/10.7717/peerj.13243

Coyne, J. O., Tran, T. T., Secomb, J. L., Lundgren, L. E., Farley, O. R., Newton, R. U., & Sheppard, J. M. (2017). Maximal strength training improves surfboard sprint and endurance paddling performance in competitive and recreational surfers. *Journal of Strength and Conditioning Research, 31*(1), 244–253. https://doi.org/10.1519/jsc.0000000000001483

de Moraes, G. C., Guimarães, A. T., & Gomes, A. R. (2013). Analysis of injuries' prevalence in surfers from Paraná seacoast. *Acta Ortopédica Brasileira, 21*(4), 213–218. https://doi.org/10.1590/s1413-78522013000400006

Deleyiannis, F. W., Cockcroft, B. D., & Pinczower, E. F. (1996). Exostoses of the external auditory canal in Oregon surfers. *American Journal of Otolaryngology, 17*(5), 303–307. https://doi.org/10.1016/s0196-0709(96)90015-0

Dimmick, S., Gillett, M., Buchan, C., Sheehan, P., Franks, M., Ratchford, A., Porges, K., Day, R., Milne, T., & Anderson, S. (2019). Prospective analysis of surfing and bodyboard injuries. *Trauma, 21*(2), 113–120.

Eurich, A. D., Brown, L. E., Coburn, J. W., Noffal, G. J., Nguyen, D., Khamoui, A. V., & Uribe, B. P. (2010). Performance differences between sexes in the pop-up phase of surfing.

Journal of Strength and Conditioning Research, 24(10), 2821–2825. https://doi.org/10.1519/JSC.0b013e3181f0a77f

Everline, C. (2007). Shortboard performance surfing: A qualitative assessment of maneuvers and a sample periodized strength and conditioning program in and out of the water. *Strength & Conditioning Journal, 29*(3), 32–40. https://search.ebscohost.com/login.aspx?direct=true&db=s3h&AN=106135906&site=ehost-live&scope=site

Farì, G., Notarnicola, A., Paolo, S. D. I., Covelli, I., & Moretti, B. (2021). Epidemiology of injuries in water board sports: Trauma versus overuse injury. *The Journal of Sports Medicine and Physical Fitness, 61*(5), 707–711. https://doi.org/10.23736/s0022-4707.20.11379-3

Farley, O., Harris, N. K., & Kilding, A. E. (2012). Anaerobic and aerobic fitness profiling of competitive surfers. *Journal of Strength and Conditioning Research, 26*(8), 2243–2248. https://doi.org/10.1519/JSC.0b013e31823a3c81

Frank, M., Zhou, S., Bezerra, P., & Crowley, Z. (2009). Effects of long-term recreational surfing on control of force and posture in older surfers: A preliminary investigation. *Journal of Exercise Science & Fitness, 7*(1), 31–38. https://doi.org/10.1016/s1728-869x(09)60005-8

Furness, J. (2016). *Musculoskeletal and physiological profile of elite and recreational surfers: Injuries and sports specific screening*. Bond University.

Furness, J., Hing, W., Abbott, A., Walsh, J., Sheppard, J. M., & Climstein, M. (2014). Retrospective analysis of chronic injuries in recreational and competitive surfers: Injury location, type, and mechanism. *International Journal of Aquatic Research & Education, 8*(3), 277–287. https://search.ebscohost.com/login.aspx?direct=true&db=s3h&AN=97671206&site=ehost-live&scope=site

Furness, J., Hing, W., Walsh, J., Abbott, A., Sheppard, J. M., & Climstein, M. (2015). Acute injuries in recreational and competitive surfers: Incidence, severity, location, type, and mechanism. *The American Journal of Sports Medicine, 43*(5), 1246–1254. https://doi.org/10.1177/0363546514567062

Furness, J. W., Hing, W. A., Sheppard, J. M., Newcomer, S. C., Schram, B. L., & Climstein, M. (2018a). Physiological profile of male competitive and recreational surfers. *Journal of Strength and Conditioning Research, 32*(2), 372–378. https://doi.org/10.1519/jsc.0000000000001623

Furness, J., Schram, B., Cottman-Fields, T., Solia, B., & Secomb, J. (2018b). Profiling shoulder strength in competitive surfers. *Sports (2075–4663), 6*(2), 52-undefined. https://search.ebscohost.com/login.aspx?direct=true&db=s3h&AN=131001581&site=ehost-live&scope=site. https://mdpi-res.com/d_attachment/sports/sports-06-00052/article_deploy/sports-06-00052.pdf?version=1527663155

Furness, J., McArthur, K., Remnant, D., Jorgensen, D., Bacon, C. J., Moran, R. W., Hing, W., & Climstein, M. (2021). Traumatic surfing injuries in New Zealand: A descriptive epidemiology study. *PeerJ, 9*, e12334. https://doi.org/10.7717/peerj.12334

Hager, M., Leavitt, J., Carballo, C., Gratton, A., & Yon, J. (2023). Surfing injuries: A US epidemiological study from 2009–2020. *Injury, 54*(6), 1541–1545. https://search.ebscohost.com/login.aspx?direct=true&db=s3h&AN=163848331&site=ehost-live&scope=site

Hay, C. S., Barton, S., & Sulkin, T. (2009). Recreational surfing injuries in Cornwall, United Kingdom. *Wilderness & Environmental Medicine, 20*(4), 335–338. https://doi.org/10.1580/1080-6032-020.004.0335

Hohn, E., Robinson, S., Merriman, J., Parrish, R., & Kramer, W. (2020). Orthopedic injuries in professional surfers: A retrospective study at a single orthopedic center. *Clinical Journal of Sport Medicine, 30*(4), 378–382. https://doi.org/10.1097/jsm.0000000000000596

Hurst, W., Bailey, M., & Hurst, B. (2004). Prevalence of external auditory canal exostoses in Australian surfboard riders. *The Journal of Laryngology and Otology, 118*(5), 348–351. https://doi.org/10.1258/002221504323086525

Inada, K., Matsumoto, Y., Kihara, T., Tsuji, N., Netsu, M., Kanari, S., Yakame, K., & Arima, S. (2018). Acute injuries and chronic disorders in competitive surfing: From the survey of professional surfers in Japan. *Sports Orthopaedics and Traumatology, 34*(3), 256–260.

Jordan, J. L., Holden, M. A., Mason, E. E., & Foster, N. E. (2010). Interventions to improve adherence to exercise for chronic musculoskeletal pain in adults. *Cochrane Database of Systematic Reviews, 2010*(1), Cd005956. https://doi.org/10.1002/14651858.CD005956.pub2

Jubbal, K. T., Chen, C., Costantini, T., Herrera, F., Dobke, M., & Suliman, A. (2017). Analysis of surfing injuries presenting in the acute trauma setting. *Annals of Plastic Surgery, 78*(5 Suppl 4), S233–s237. https://doi.org/10.1097/sap.0000000000001026

Kennedy, G. E. (1986). The relationship between auditory exostoses and cold water: A latitudinal analysis. *American Journal of Physical Anthropology, 71*(4), 401–415. https://doi.org/10.1002/ajpa.1330710403

Klick, C., Jones, C. M., & Adler, D. (2016). Surfing USA: An epidemiological study of surfing injuries presenting to US EDs 2002 to 2013. *The American Journal of Emergency Medicine, 34*(8), 1491–1496. https://doi.org/10.1016/j.ajem.2016.05.008

Kozminski, B. U., Ahmed, N., Cautela, F. S., Shah, N. V., Shangguan, X., Doran, J. P., Newman, J. M., Horowitz, E. H., Gonzales, A. S., 3rd, Lee, C. J., Persaud, C. S., Urban, W. P., & Stickevers, S. M. (2020). Surfing-related head injuries presenting to United States emergency departments. *Journal of Orthopaedics, 19*, 184–188. https://doi.org/10.1016/j.jor.2019.11.042

Kroon, D. F., Lawson, M. L., Derkay, C. S., Hoffmann, K., & McCook, J. (2002). Surfer's ear: External auditory exostoses are more prevalent in cold water surfers. *Otolaryngology and Head and Neck Surgery, 126*(5), 499–504. https://doi.org/10.1067/mhn.2002.124474

LaLanne, C. L., Cannady, M. S., Moon, J. F., Taylor, D. L., Nessler, J. A., Crocker, G. H., & Newcomer, S. C. (2017). Characterization of activity and cardiovascular responses during surfing in recreational male surfers between the ages of 18 and 75 years old. *Journal of Aging and Physical Activity, 25*(2), 182–188. https://doi.org/10.1123/japa.2016-0041

Lambert, C., Marin, S., Esvan, M., & Godey, B. (2021). Impact of ear protection on occurrence of exostosis in surfers: An observational prospective study of 242 ears. *European Archives of Oto-Rhino-Laryngology, 278*(12), 4775–4781. https://doi.org/10.1007/s00405-021-06609-8

Ledreux, A., Pryhoda, M. K., Gorgens, K., Shelburne, K., Gilmore, A., Linseman, D. A., Fleming, H., Koza, L. A., Campbell, J., Wolff, A., Kelly, J. P., Margittai, M., Davidson, B. S., & Granholm, A. C. (2020). Assessment of long-term effects of sports-related concussions: Biological mechanisms and exosomal biomarkers. *Frontiers in Neuroscience, 14*, 761. https://doi.org/10.3389/fnins.2020.00761

Locquet, M., Beaudart, C., Durieux, N., Reginster, J.-Y., & Bruyère, O. (2019). Relationship between the changes over time of bone mass and muscle health in children and adults: A systematic review and meta-analysis. *BMC Musculoskeletal Disorders, 20*(1), 429. https://doi.org/10.1186/s12891-019-2752-4

Loveless, D., & Minahan, C. (2010). Peak aerobic power and paddling efficiency in recreational and competitive junior male surfers. *European Journal of Sport Science, 10*(6), 407–415. https://search.ebscohost.com/login.aspx?direct=true&db=s3h&AN=55053273&site=ehost-live&scope=site

Lowdon, B. J., Pateman, N. A., & Pitman, A. J. (1983). Surfboard-riding injuries. *The Medical Journal of Australia, 2*(12), 613–616.

Lowdon, B. J., Pateman, N. A., Pitman, A. J., & Ross, K. (1987). Injuries to international competitive surfboard riders. *Traumatismes subis par des athletes pratiquant le surfing au niveau international, 27*(1), 57–63. https://search.ebscohost.com/login.aspx?direct=true&db=s3h&AN=SPH203970&site=ehost-live&scope=site

McGill, S. (2007). Low back disorders: evidence-based prevention and rehabilitation (Third edition. ed.). Human Kinetics.

McArthur, K., Jorgensen, D., Climstein, M., & Furness, J. (2020). Epidemiology of acute injuries in surfing: Type, location, mechanism, severity, and incidence: A systematic review. *Sports (2075–4663), 8*(2), 25-undefined. https://search.ebscohost.com/login.aspx?direct=true&db=s3h&AN=142091310&site=ehost-live&scope=site

McLeod, M., Breen, L., Hamilton, D. L., & Philp, A. (2016). Live strong and prosper: The importance of skeletal muscle strength for healthy ageing. *Biogerontology, 17*(3), 497–510. https://doi.org/10.1007/s10522-015-9631-7

Meir, R. A., Zhou, S. H. I., Rolfe, M. I., Gilleard, W. L., & Coutts, R. A. (2012). An investigation of surf injury prevalence in Australian surfers: A self-reported retrospective analysis. *New Zealand Journal of Sports Medicine, 39*(2), 52–58. https://search.ebscohost.com/login.aspx?direct=true&db=s3h&AN=84509846&site=ehost-live&scope=site

Meir, R., Duncan, B., Crowley-McHattan, Z., Gorrie, C., & Sheppard, J. (2015). Water, water, everywhere, nor any drop to drink: Fluid loss in Australian recreational surfers. *Journal of Australian Strength & Conditioning, 23*(6), 16–20. https://search.ebscohost.com/login.aspx?direct=true&db=s3h&AN=112820925&site=ehost-live&scope=site

Mendez-Villanueva, A., & Bishop, D. (2005). Physiological aspects of surfboard riding performance. *Sports Medicine, 35*(1), 55–70. https://doi.org/10.2165/00007256-200535010-00005

Miller, I. J., Stapelberg, M., Rosic, N., Hudson, J., Coxon, P., Furness, J., Walsh, J., & Climstein, M. (2023). Implementation of artificial intelligence for the detection of cutaneous melanoma within a primary care setting: Prevalence and types of skin cancer in outdoor enthusiasts. *PeerJ, 11*, e15737-undefined. https://doi.org/10.7717/peerj.15737

Minghelli, B., Nunes, C., & Oliveira, R. (2018). Injuries in recreational and competitive surfers: A nationwide study in Portugal. *The Journal of Sports Medicine and Physical Fitness, 58*(12), 1831–1838. https://doi.org/10.23736/s0022-4707.17.07773-8

Moran, K., & Webber, J. (2013). Surfing injuries requiring first aid in New Zealand, 2007–2012. *International Journal of Aquatic Research & Education, 7*(3), 192–203. https://search.ebscohost.com/login.aspx?direct=true&db=s3h&AN=89639149&site=ehost-live&scope=site

Muhonen, E. G., Kafle, S., Torabi, S. J., Abello, E. H., Bitner, B. F., & Pham, N. (2022). Surfing-related craniofacial injuries: A NEISS database study. *The Journal of Craniofacial Surgery, 33*(8), 2383–2387. https://doi.org/10.1097/scs.0000000000008769

Nakanishi, H., Tono, T., & Kawano, H. (2011). Incidence of external auditory canal exostoses in competitive surfers in Japan. *Otolaryngology and Head and Neck Surgery, 145*(1), 80–85. https://doi.org/10.1177/0194599811402041

Nathanson, A., Haynes, P., & Galanis, D. (2002). Surfing injuries. *The American Journal of Emergency Medicine, 20*(3), 155–160. https://doi.org/10.1053/ajem.2002.32650

Nathanson, A., Bird, S., Dao, L., & Tam-Sing, K. (2007). Competitive surfing injuries: A prospective study of surfing-related injuries among contest surfers. *The American Journal of Sports Medicine, 35*(1), 113–117. https://doi.org/10.1177/0363546506293702

Nessler, J. A., Ponce-Gonzalez, J. G., Robles-Rodriguez, C., Furr, H., Warner, M., & Newcomer, S. C. (2019). Electromyographic analysis of the surf paddling stroke across multiple intensities. *Journal of Strength and Conditioning Research, 33*(4), 1102–1110. https://doi.org/10.1519/jsc.0000000000003070

O'Neill, B., Leon, E., Furness, J., Schram, B. E. N., & Kemp-Smith, K. (2021). The effects of a 2-hour surfing session on the hydration status of male recreational surfers. *International Journal of Exercise Science, 14*(6), 1388–1399. https://search.ebscohost.com/login.aspx?direct=true&db=s3h&AN=173328987&site=ehost-live&scope=site

Pikora, T. J., Braham, R., & Mills, C. (2012). The epidemiology of injury among surfers, kite surfers and personal watercraft riders: Wind and waves. *Medicine and Sport Science, 58*, 80–97. https://doi.org/10.1159/000338583

Queensland Brain Institute. *Do helmets protect against concussion?* https://qbi.uq.edu.au/concussion/do-helmets-protect-against-concussion

Quinn, J., Salmon, L., Ngo, D., Taylor, F., & Platt, S. (2022). The epidemiology of surfing injuries in a major Australian Centre – A ten year clinical audit. *Injury, 53*(6), 1887–1892. https://doi.org/10.1016/j.injury.2022.03.045

Redd, M. J., & Fukuda, D. H. (2016). Utilization of time motion analysis in the development of training programs for surfing athletes. *Strength & Conditioning Journal (Lippincott Williams & Wilkins), 38*(4), 1–8. https://search.ebscohost.com/login.aspx?direct=true&db=s3h&AN=117201043&site=ehost-live&scope=site

Remnant, D., Moran, R. W., Furness, J., Climstein, M., Hing, W. A., & Bacon, C. J. (2020). Gradual-onset surfing-related injuries in New Zealand: A cross-sectional study. *Journal of Science and Medicine in Sport, 23*(11), 1049–1054. https://doi.org/10.1016/j.jsams.2020.05.010

Sallis, R. (2015). Exercise is medicine: A call to action for physicians to assess and prescribe exercise. *The Physician and Sportsmedicine, 43*(1), 22–26. https://doi.org/10.1080/00913847.2015.1001938

Secomb, J. L., Farley, O. R. L., Lundgren, L., Tran, T. T., King, A., Nimphius, S., & Sheppard, J. M. (2015a). Associations between the performance of scoring manoeuvres and lower-body strength and power in elite surfers. *International Journal of Sports Science & Coaching, 10*(5), 911–918. https://search.ebscohost.com/login.aspx?direct=true&db=s3h&AN=111954335&site=ehost-live&scope=site

Secomb, J. L., Lundgren, L. E., Farley, O. R., Tran, T. T., Nimphius, S., & Sheppard, J. M. (2015b). Relationships between lower-body muscle structure and lower-body strength, power, and muscle-tendon complex stiffness. *Journal of Strength and Conditioning Research, 29*(8), 2221–2228. https://doi.org/10.1519/jsc.0000000000000858

Secomb, J. L., Nimphius, S., Farley, O. R. L., Lundgren, L. E., Tai, T. T., & Sheppard, J. M. (2015c). Relationships between lower-body muscle structure and, lower-body strength, explosiveness and eccentric leg stiffness in adolescent athletes. *Journal of Sports Science & Medicine, 14*(4), 691–697. https://search.ebscohost.com/login.aspx?direct=true&db=s3h&AN=111246110&site=ehost-live&scope=site

Secomb, J. L., Sheppard, J. M., & Dascombe, B. J. (2015d). Time-motion analysis of a 2-hour surfing training session. *International Journal of Sports Physiology and Performance, 10*(1), 17–22. https://doi.org/10.1123/ijspp.2014-0002

Secomb, J. L., Nimphius, S., Farley, O. R., Lundgren, L., Tran, T. T., & Sheppard, J. M. (2016a). Lower-body muscle structure and jump performance of stronger and weaker surfing athletes. *International Journal of Sports Physiology and Performance, 11*(5), 652–657. https://doi.org/10.1123/ijspp.2015-0481

Secomb, J. L., Parsonage, J., Dowse, R., Ferrier, B., Sheppard, J., & Nimphius, S. (2016b). The importance of mobility in combination with strength for force absorption. *Journal of Australian Strength & Conditioning, 24*(6), 55–56. https://search.ebscohost.com/login.aspx?direct=true&db=s3h&AN=120775201&site=ehost-live&scope=site

Sheppard, J. M., McNamara, P., Osborne, M., Andrews, M., Oliveira Borges, T., Walshe, P., & Chapman, D. W. (2012). Association between anthropometry and upper-body strength qualities with sprint paddling performance in competitive wave surfers. *Journal of Strength and Conditioning Research, 26*(12), 3345–3348. https://doi.org/10.1519/JSC.0b013e31824b4d78

Simas, V., Hing, W., Rathbone, E., Pope, R., & Climstein, M. (2021). Auditory exostosis in Australian warm water surfers: A cross-sectional study. *BMC Sports Science, Medicine & Rehabilitation, 13*(1), 1–7. https://search.ebscohost.com/login.aspx?direct=true&db=s3h&AN=150304229&site=ehost-live&scope=site

Szymski, D., Achenbach, L., Siebentritt, M., Simoni, K., Kuner, N., Pfeifer, C., Krutsch, W., Alt, V., Meffert, R., & Fehske, K. (2021). Injury epidemiology of 626 athletes in surfing, wind surfing and kite surfing. *Open access Journal of Sports Medicine, 12*, 99–107. https://search.ebscohost.com/login.aspx?direct=true&db=s3h&AN=152843051&site=ehost-live&scope=site

Taylor, D. M., Bennett, D., Carter, M., Garewal, D., & Finch, C. F. (2004). Acute injury and chronic disability resulting from surfboard riding. *Journal of Science and Medicine in Sport, 7*(4), 429–437. https://doi.org/10.1016/s1440-2440(04)80260-3

Taylor, D. M., Bennett, D., Carter, M., Garewal, D., & Finch, C. (2005). Perceptions of surfboard riders regarding the need for protective headgear. *Wilderness & Environmental Medicine, 16*(2), 75–80. https://doi.org/10.1580/1080-6032(2005)16[75:Posrrt]2.0.Co;2

Ulkestad, G.-E., & Drogset, J. O. (2016). Surfing injuries in Norwegian Arctic waters. *The Open Sports Sciences Journal, 9*(1), 153–161.

Umeda, Y., Nakajima, M., & Yoshioka, H. (1989). Surfer's ear in Japan. *Laryngoscope, 99*(6 Pt 1), 639–641. https://doi.org/10.1288/00005537-198906000-00012

Vallée, A. (2024). External auditory exostosis among surfers: A comprehensive and systematic review. *European Archives of Oto-Rhino-Laryngology, 281*(2), 573–578. https://doi.org/10.1007/s00405-023-08258-5

van Mechelen, W., Hlobil, H., & Kemper, H. C. G. (1992). Incidence, severity, aetiology and prevention of sports injuries. *Sports Medicine, 14*(2), 82–99. https://doi.org/10.2165/00007256-199214020-00002

Wong, B. J., Cervantes, W., Doyle, K. J., Karamzadeh, A. M., Boys, P., Brauel, G., & Mushtaq, E. (1999). Prevalence of external auditory canal exostoses in surfers. *Archives of Otolaryngology – Head & Neck Surgery, 125*(9), 969–972. https://doi.org/10.1001/archotol.125.9.969

Woodacre, T., Waydia, S. E., & Wienand-Barnett, S. (2015). Aetiology of injuries and the need for protective equipment for surfers in the UK. *Injury, 46*(1), 162–165. https://doi.org/10.1016/j.injury.2014.07.019

Chapter 8
SheShaka: Surfing Is a Feminist Practice That Promotes Spatial Justice

Gemma Tarpey-Brown, Fran Edmonds, Natalie Galea, Georgina Sutherland, Cathy Vaughan, and Karen Block

> *"Just as none of us is outside or beyond geography, none of us is completely free from the struggle over geography."*
> –Said, 1994, p. 7
>
> *"Pipeline is for the fucking girls".*
>
> –Simmers, 2024

8.1 Introduction

For centuries, surfing was an art mastered by Polynesian Royalty including queens who were celebrated for their grace and agility riding the raging waves of the Pacific (Kempton, 2021) (Chap. 2). This history may come as a surprise to many in Australia where, today, the local lineup is often a space defined by a hierarchy of gender power relations (Olive et al., 2015), governed by hegemonic masculinity. In surf spaces, contemporary hegemonic masculinity is performed in similar ways to how it is performed onshore. Surf spaces refer to places where surfing and surf culture are practised such as the ocean, the sea, and other waterways where people surf, as well as places like beachside carparks and surf shops (Anderson, 2014) (Chap. 9). Hegemonic masculinity is a dominant form of masculinity, culturally celebrated and marked by its subordination of women and its marginalization of men who do not embody or enact its key traits (Connell & Messerschmidt, 2005). A growing body of evidence has demonstrated the nuanced ways hegemonic masculinity and other hierarchies of power permeate surf environments (Burtscher & Britton, 2022;

G. Tarpey-Brown · F. Edmonds · G. Sutherland · C. Vaughan · K. Block (✉)
The University of Melbourne, Parkville, VIC, Australia
e-mail: g.tarpeybrown@unimelb.edu.au; edmondsf@unimelb.edu.au; georgina.sutherland@unimelb.edu.au; c.vaughan@unimelb.edu.au; keblock@unimelb.edu.au

N. Galea
The University of Sydney Business School, Darlington, NSW, Australia
e-mail: natalie.galea@unsw.edu.au

© The Author(s), under exclusive license to Springer Nature Switzerland AG 2025
D. M. Kennedy (ed.), *The Science and Culture of Surfing*,
https://doi.org/10.1007/978-3-031-80979-8_8

Comer, 2010; Comley, 2016; Crellin, 2022; Evers, 2009; Franklin & Carpenter, 2018; lisahunter, 2018; Olive, 2019; Olive et al., 2018; Thorpe et al., 2017; Waitt, 2007).

In this chapter, we use an intersectional approach to explore how multiple processes of gendered, racial, and social exclusion are enacted in surf spaces to create and maintain an environment of spatial injustice. To do so, we will draw on a blend of autoethnography, findings from the SheShaka project and existing literature. SheShaka was a participatory action research project that used photovoice to identify gendered barriers to surfing experienced by women who lived in a coastal town in regional Australia. The photographs included in this chapter are from the SheShaka project and provide a visual representation of the nuanced ways power relations operate not only on land but also in surf spaces. The stories shared by all the women who participated, alongside the photographs, explore how surfing can be a subversive practice for many women, whereby rigid gender norms are transgressed. We argue that by harnessing this transgressive power, surfing has the potential to transform the ocean into a site of spatial justice.

8.1.1 A Note on Positionality and Language

We, the authors, are a group of white-settler feminist women with a shared interest in promoting spatial justice for women, girls, and other groups that may be marginalized or excluded from participating in sport. Though not all of us are surfers, we each have a special connection with human movement, sport, and embodied practices.

8.2 Intersectional Spatial Justice and Storying: A Conceptual Framework

Our analysis of women's experiences of surfing is informed by intersectionality and Soja's (2009) theory of spatial justice. Intersectionality (Crenshaw, 1991) refers to the ways in which social experiences and inequalities are shaped by attributes such as gender, race, Indigeneity, class, ability/disability, and age. These experiences intersect within specific contexts, structures, and systems of power, which create overarching and interdependent forms of privilege and oppression through gender inequality, racism, homophobia, and ableism. Intersectionality is a critical concept that highlights how, for example, a white woman may feel more welcome in surf spaces than an Aboriginal man due to the way racism and colonialism interact to potentially supersede gendered exclusion.

We use spatial justice theory to further explore how different power relations alter who has rights and access to surf spaces. For Soja (2009), spatial justice allows us to explore the relationship between space and power to produce greater social justice at all geographical scales, from the local to the global. Soja (2009) posits:

> Spatial (in)justice refers to an intentional and focused emphasis on the spatial or geographical aspects of justice and injustice. As a starting point, this involves the fair and equitable distribution in space of socially valued resources and the opportunities to use them. (p. 2)

At its core, surfing is a spatial practice that relies on multiple geomorphic and hydrological spatial movements. It's (mostly) a sport practised in the ocean—a space historically occupied by men. The male occupation and subsequent domination of the beach continue to this day and have created an environment of gender inequality and spatial injustice. Pavlidis (2018) argues it is crucial to examine how gender inequity relates to spatial inequity if we are to effectively challenge gender inequity in sport and, by extension, society. She goes on to demonstrate how, through sport, everyday public spaces become key sites of feminist resistance: it is in public spaces such as the beach that dominant relations of power are challenged, negotiated, and ultimately transformed.

Exploring how women counter spatial injustice in the surf has led us to consider how we may address unequal power relations in the production of scientific knowledge. Phillips and Bunda (2018) use storying as a tool to effectively deconstruct colonialist conventions of research. Storying as a research method brings to life the agency and choices of the storyteller in constructing meaning.

In the following sections, we provide an outline of how modern surf spaces have been sites of gendered and racialized exclusion. We then use storying to present the lived experiences of women who surf.

8.3 Background

8.3.1 Rigid Gender Norms in the Surf

Modern surfing culture has a misogynistic recent history, which has positioned women as passive observers routinely denied agency and the opportunity to participate (Booth, 2002). In Booth's historical analysis of how the beach shaped neocolonial Australian masculinity and contributed to white nation-building, he argues modern surfing culture was informed by a fraternal social structure that sexualized women in order to reenforce male dominance in the surf. The representation of the passive bikini-clad girl on the beach waiting for her surfer boyfriend is demonstrative of how modern surfing is a "patriocolonial" practice framed by patriarchal and colonial hierarchies of gender, sexuality, and race (lisahunter, 2016). In patriocolonial surf culture, if a woman's body does not meet stereotypical beauty standards associated with the surfer girl ideal, her ability to participate is considerably limited. Comer (2010) examined the rapid neoliberalization of surfing in the 1990s where the surf industry saw women and girls as a largely untapped consumer market. Comer's analysis makes clear that these stereotypical beauty standards are tightly linked with Western colonial framings of beauty. At the beginning of the twenty-first century, the white, blue-eyed, blonde, able-bodied heteronormative image of the surfer dude remained the same; however, these features had simply been transposed onto the surfer girl. Women were now allowed to surf, but only if they wore a tiny bikini when doing so (lisahunter, 2018).

8.3.2 Postfeminism and Hypersexualization in Surfing

From a commercial perspective, surfing is a multibillion dollar global industry that profits on the hypersexualization of women. To successfully operate within this industry, women are positioned to celebrate this hypersexual landscape as a form of individual liberation and empowerment, the product of agentic third-wave feminism and postfeminist discourses. These discourses are underpinned by neoliberal logics whereby women are considered active agents in building and maintaining their own embodied hypersexual representation (Gill, 2007). Through a combination of theoretical and empirical analysis, Thorpe et al. (2017) unpack how surfers internalize, embody, and practise current discourses of third-wave feminism and postfeminism. In professional surfing, women are high performing athletes operating in a world where postfeminist sensibilities suggest gender equality has more or less been achieved. Arguably, women are empowered to choose the way their bodies are represented in the media and advertising landscape. Unconstrained by patriarchal regulations and expectations, they are expected to celebrate events such as being selected as one of Playboy's "top 21 hottest surfer babes," as it demonstrates women are finally being promoted in well-established media (Stab Magazine, 2017). The celebration of women's gains in surfing is intricately linked to maintaining a particular understanding of how sex, gender, and sexuality interact and are embodied (lisahunter, 2018). Surfing's narrow conception of how these identities may exist and interact severely restricts, rather than expands, the possibilities of transforming the ocean into a site where harmful hierarchies of domination are challenged and spatial justice achieved.

8.3.3 How Gendered Harms Are Enacted in Surf Spaces

Whether increased participation and visibility of women in the surf challenges patriarchal hierarchies of power is an open question. Empirical evidence suggests that rather than challenging existing gender power relations, women who surf may actively participate in the production of harmful misunderstandings of gender as a binary construct (Olive, 2019). In an ethnographic study conducted in southern

Spain, Sanz-Marcos (2021) critiques male dominance in modern surfing, noting that popular surf breaks are typically androcentric places where women are encouraged to mimic men's behavior, surfing style, and technique. Androcentrism positions men as the standard, whereby their needs, bodies, values, priorities, and ways of being are seen as the social norm. In research conducted in southern California, Comley (2016) identified a number of strategies women use in order to feel comfortable in the surf despite the male-dominated environment. One strategy she found was that women who were highly skilled surfers mirrored men's behavior by displaying assertiveness and aggression when competing for waves. This supported Sanz-Marcos' findings where there was a hypercompetitive and aggressive culture among the few women in the surf as they strove to emulate the patriocolonial behavior of male surfers. This leads to the normalization of aggressive behavior, as both men and women take part in it.

8.3.4 How Racialized Harm Is Enacted in Surf Spaces

In Australia, the beach operates as a public space defined by colonial possession which seeks to exclude and include depending on a subject's (racial and gender) identity (Moreton-Robinson, 2011). By extension, the beach is a physical manifestation of a sovereign international border established to protect the white-settler-colonial state. During colonization, Aboriginal and Torres Strait Islander peoples were erased from the beach, an erasure that persists to this day. Moreton-Robinson (Goenpul/Quandamooka) outlines how the beach has become a space almost exclusively managed by white men who hold a possessive claim that extends out to the surf, where they carve, conquer, and charge over the waves as a way to continually lay sovereign claim to these spaces. In surf spaces, this ownership is performed through the practice of localism. Olive (2019) defines localism as the process surfers use to claim authority over certain beaches and protect surf breaks from outsiders. Previously, critical surf studies have viewed localism as a practice typically performed by white men and focused on how this ultimately excludes cisgendered white women. By using an intersectional lens to analyze localism as a form of settler-colonialism in Australia, Olive explored how women who surf are complicit in maintaining surfing hierarchies of social exclusion that ultimately uphold spatial injustice.

Women's acceptance in the surf has become contingent on meeting and maintaining the status quo—a status quo that is fundamentally underpinned by spatial injustice whereby white men dominate surf spaces. If women enter the surf and challenge established hierarchies of gender and race, the ocean thus becomes a socially contested space.

8.4 How Women Have Carved Out a Place for Themselves in the Surf

8.4.1 The 1980s Onward: "Things Can Only Get Better"[1]

In the early 1980s, I (Fran) was kind of in love with the idea of surfing, so I thought I'd try my hand at making a surfboard. Although I did make a surfboard, loving the idea of surfing didn't really translate into being a good surfer. Perhaps starting surfing later in life affected my surfing ability? Many years later, I now understand why surfing in those days was more of an idea rather than a reality.

I decided to make a surfboard, as part of my year 12 Art portfolio. I was fascinated with the intersection between art and sport; surfing seemed to fit perfectly. Also, I'd always been a water-baby. During my final school year, I spent three weeks riding my bike from our family beach house to the relatively new Rip Curl factory on the Geelong Road, now the Surf Coast Highway, in Torquay. In those days, once summer had finished, Torquay was like a ghost town. Rip Curl was still a small company, the factory was nothing like the huge corporate structure it is today. Under the careful guidance of the (now) renowned board-shaper Russell Graham,[2] I was guided in the art of surfboard making. Russell was a generous teacher and mentor for such a young novice, who had very little surf knowledge.

8.4.2 Space Invaders[3]

Although Torquay was quiet, it didn't mean the surf was empty. There were still plenty of surfers, albeit nothing like the crowds of today. At that time, there were few girls surfing, and the equipment left a lot to be desired. The board I made was short (not great if you're learning to surf). It was the era of the short board, and you rarely saw anyone, let alone a woman on a longboard.[4] The water in Victoria is notoriously cold and the '80s wetsuits weren't exactly user-friendly. During the school holidays, I'd scour the relatively new surfboard factories trying to buy a second-hand wetsuit. These were never the right fit (made for boys) and were so restrictive that you could barely raise your arm higher than your elbow. Board shorts revealed the same issues, and rash vests were non-existent. There was very little choice in women's surf wear. So, while I'm reluctant to admit, in those days I felt intimidated to enter the line-up. I was barely a beginner surfer, I didn't wear a bikini, have blonde hair or blue eyes, and felt safest when most of my body was covered up. Since then, the surf industry has made some inroads with more women surfers involved in designing functional surf wear and equipment for diverse body types. Though the industry continues to hypersexualise women, perpetuating the surfer girl aesthetic.

---

[1] D:Ream, 1993, "Things Can Only Get Better"

[2] Russell died in March 2022 (Matthews, 2022).

[3] Space Invaders also refers to a seminal text by Puwar (2004) examining how certain bodies are entitled to certain spaces, while other bodies are not.

[4] A myth still persists that short boards are for those with above average surfing proficiency and that long boards are for "girls" (see Sanz-Marcos, 2021).

8.4.3 Country Is Kin: Colonization and Waterways (Fig. 8.1)

Not long after I graduated from university, I was living nowhere near the sea. I spent a few years in Central Australia working and studying on remote Aboriginal communities. Being 'out bush' taught me many things, which were later consolidated when I worked in native title, compiling historical and anthropological evidence for Aboriginal land and sea claims. One central lesson from these experiences was that Country, which is inclusive of waterways, is not 'owned' by people. For Aboriginal people Country is kin and despite over 200 years of colonisation this connection remains.

Aboriginal people have created sturdy and beautiful water vessels for thousands of years, revealing ongoing connections to waterways and a deep understanding of water and its resources.[5] This knowledge is both men's and women's business.[6] Later, as a PhD student I worked with the artist Maree Clarke (Mutti Mutti/Wamba Wamba/Yorta Yorta/Boonwurrung), who was my field supervisor. While Maree doesn't surf, her work is about her connections to Country across southeast Australia from the Murray River (Dhungala) to

Fig. 8.1 Maree Clarke, 2022. Remember Me. Printed calico adhered to trees as scar tree installation, Lorne Sculpture Biennale 2022. (Photograph Maree Clarke. Image courtesy of Vivien Anderson Gallery)

---

[5] See Gapps and Smith (2015) for a detailed history of Indigenous watercraft.
[6] Cater, 2014

Port Phillip Bay and beyond.[7] These connections are evident in her knowledge of canoe-making, revivifying her Ancestral connections to culture through artmaking.[8]

Maree's glass canoe, made from pulled strips of coloured glass that mimic microscopic images of river reeds, evoke images of the ocean. She made a series of ephemeral canoe shape designs from calico and adhered them to trees along the coastal track at Lorne. The calico shapes paid homage to the forty-four language groups of Victoria and symbolised the markings left on trees when bark is removed to make canoes.

At Lorne Point, where Maree's scar tree installations loomed large, I hear the sounds of young girls' voices as they surf out the back, and am reminded that not so long ago, their laughter and happy chatter were scarce in this space. As Maree's artwork reminds us, Aboriginal women have always had connections to the sea. Today, there are more women and girls participating in surfing, which is changing the line-up and the culture. However, there is still a way to go before women from all cultural backgrounds are able to negotiate this space on equal terms.

8.4.4 White and Masculine Privilege

During COVID-19, I was privileged to live on the Surf Coast and was able to surf most days. The water became increasingly crowded with locals working from home. One day, a relative novice riding a short board took off on what looked like an almost un-surfable wave. He clumsily changed direction and instead of avoiding me, the pointy nose of his board sliced across my forehead to the skull-bone. I managed to body board back into shore; beach goers rallied around, an ambulance was called, and I spent the night in hospital awaiting plastic surgery.

Soon after, I was looking on Facebook and noticed that my accident had attracted a post. There was a call out for people to be vigilant and surf safely, as crowds were increasing due to lockdown easing. Most comments were supportive. However, one stood out as the writer attempted to dissect what had happened and apportion blame, wondering why I was in the 'impact zone'. The comment revealed the privilege this 'experienced' male surfer felt in being able to analyse a situation that he had not witnessed. He did not know me or my surfing ability, but nevertheless was confident to provide his opinion.

8.4.5 A Final Observation

My initial encounters with the surf industry, provided a safe, happy and respectful insight into the world of surfing, what it could offer and what it could become. Today, while aware of my surfing limitations, I continue to remain alert to the male presence in the ocean. I'm still dropped in on by men, with no apologies forthcoming, and anticipate being given unsolicited surfing advice. While there is more diversity in the waves, it remains predominantly

---

[7] These areas of southeast Australia were some of the most devastated regions during the British colonial incursion (see Thorner et al., 2018).
[8] A monograph of Maree's major work has been curated by Russell-Cook (2021).

male and white. As more women, Indigenous surfers and people of colour participate in the sport surf spaces are transforming for the better.[9]

8.5 Creating Spaces to Subvert Gender Stereotypes and Challenge Spatial Injustice

A couple of years after having my son, I booked a surf camp in South Australia. I went by myself as no friends could come with me. Most of them had young families, couldn't leave the kids for that long, felt too guilty or were breastfeeding.

The camp was incredible, I surfed amazing waves in these epic remote locations. At night we surf skated around a big shed to disco music. We went exploring caves, did offroad driving and I met some really fun people.

When I got back, three different men, on three separate occasions made the comment that I was having a "mid-life crisis". All of these men were in their 50's and all of these men surf themselves and go on regular surf trips. Were these comments "jokes" or were they passive aggressive digs. Digs that I'm not behaving the way a woman in her late 30's should be behaving.

Would they have made any comments if I had spent my holiday fund on a shopping spree, spa weekend, wine tour in the Yarra Valley or a trip to a clinic to get my face injected with filler? All this would be far more socially acceptable, far more normal. I turn 40 this year and will unapologetically continue to surf, skate and be wild.

Chrissie (SheShaka participant)

Chrissie found surfing later in life. Living in a coastal town in regional Australia, Chrissie told us stories about how at times she felt she lost her "toughness" in the surf. She would feel too intimated to paddle out due to the seriousness of the lineup, permeated with anger, aggression, and a fiercely competitive energy. She would regularly be the only woman in the lineup which contributed to a sense of intimidation. To alleviate these feelings, Chrissie would remind herself that she was a woman from the northeast coast of Scotland, where winters were long, cold, and dark. In order to make it through the dark winter, you have to cultivate toughness, and it was this toughness that would push her to paddle out in spite of the intimidating environment. Chrissie explained that on land, it wasn't so much intimidation she experienced when expressing her desire to surf. Rather, it was condescending remarks from her partner's male friends such as those featured in her "midlife crisis" story above.

In suggesting Chrissie's surf trip was an indication she was going through a midlife crisis, despite those men doing the exact same thing multiple times a year,

---

[9] Regarding the urgency to change surf culture, I saw a social media post recently by black surfers speaking out against ongoing racism in surf spaces by white men. A local white surfer proudly said he was known as the "enforcer" because he didn't say anything when he was on the wave—he just gave people "haircuts." Apparently that means "shaving someone's head with the board fin" (blackgirlssurf & solidarity.in.surf, 2024).

the double standard of aging is made clear. The double standard of aging is well established and therefore well ingrained, contributing to harmful gender stereotypes that position women as aging with a "special severity" that progressively "destroys" them (Sontag, 1997). For men, hegemonic masculinity frames middle age as a process that merely enhances. Through the double standard of aging, we come to understand the importance of the body not only as a sexed object but as a gendered subject expected to continuously perform what it means to be woman (Butler, 1986). The activities and practices a woman participates in are influenced by gender norms and associated stereotypes. From a male perspective, surfing has been described as an activity where the body becomes enmeshed in complex class and spatial relations, often related to localism (Evers, 2004). For Chrissie, these relations are also gendered and typically restricted by harmful gender norms that shape stereotypes around what a woman's body should and should not be doing. In Fig. 8.2, Chrissie is not only identifying herself as a surfer, but she is also proudly mocking the label of being middle-aged. Simultaneously, as she skates boldly down the middle of the road, she demonstrates how feminist resistance to forms of injustice contains a spatial element (Pavlidis, 2018).

Fig. 8.2 Chrissie occupying public space on a surf-skate board as she holds up the middle finger. Image taken as part of the SheShaka project

Stories of how women have been using surfing to explore feminist resistance to spatial injustice and challenge gender stereotypes are not solely based in the Global North. In the past decade, there has been an expansion in the number of women and girls taking to the surf across the Global South in countries such as Sri Lanka (Burtscher & Britton, 2022), Papua New Guinea (Britton, 2018), Iran, Senegal, and Ghana. Building on this momentum is Sea Sisters, a surfing and development organization based in eastern Sri Lanka, established with the goal of creating safer spaces for women and girls on the beach and in the ocean. Sea Sisters implements programs to build women's swimming and surfing skills to enhance women's empowerment. In Sri Lanka, the ocean has traditionally been viewed as a dangerous place. Male-dominated ocean-based industries such as fishing, maritime trade, and the navy significantly contribute to the Sri Lankan economy and shape cultural perceptions of the sea as a space that is unsafe for women and girls.

> "I grew up by the ocean. But **I have never seen it as a place for me.** They say, Sri Lankan girls don't surf. It's hard to believe what you don't see."
> —Sanu, Sea Sisters (Burtscher & Britton, 2022)

Sea Sisters works with Sri Lankan women like Sanu who never perceived the ocean as a safe place. Sea Sisters aims to challenge these perceptions, which are deeply embedded in patriarchal gender norms. Sea Sisters and other surfing and development organizations strive to create an ocean environment underpinned by principles of gendered spatial justice, where women in the Global South feel empowered to enter a space they once felt excluded from. Being supported to access and be in the ocean encourages women who have been excluded from surf spaces to establish their presence among the waves, challenging and ultimately resisting community concerns and harmful gender stereotypes.

8.6 Queering the Waves

> A significant barrier for me when I first started to get interested in surfing again was how male dominated it felt. I remember going to spots along the surf coast for a surf. I would see a group of men together getting ready to go out. I'd suddenly feel too anxious and embarrassed. I would just drive back out of the car park and head home.
>
> I did that a fair few times, until I found a group of women to surf with, and those feelings of anxiety, fear and embarrassment fell away immediately. I just love the friendships that can be struck up and the camaraderie between women that changed my experience of surfing dramatically. They are all my close mates still today…
>
> I also don't feel I fit into the "norm" as someone that identifies as queer and a part of the LGBTQIA+ community. I feel passionate about challenging heteronormative ideas and beliefs that are placed on women in our society. But I realised with SheShaka, I'm much more comfortable doing that in spaces outside of body and beauty because I still in some ways feel uncomfortable in my own body. I finally found a photo that I really like of myself, because I can see and feel the joy and solace that being out in the ocean and sunshine brings me.
>
> Ellie (SheShaka participant)

Ellie, a queer woman who participated in the SheShaka project, spoke about how surfing was a practice that cultivated joy. Surfing provided Ellie the opportunity to challenge heteronormative constructions of gender that position the feminine body as an object that must align with normative beauty standards. By surfing with a group of women, Ellie was able to overcome feelings of anxiety, fear, and embarrassment, feelings that had previously prevented her from being able to surf. For Ellie, surf culture's promotion of hegemonic masculinity and associated gender dynamics severely impacted her ability to paddle out. Waitt (2007) highlights the legitimacy of Ellie's feelings of anxiety upon seeing a group of men standing in the car park in front of the surf. In participant observations conducted in New South Wales, Waitt noted an interaction between a group of male surfers and a young woman about to paddle out. The woman, slim, confident, and meeting normative beauty standards, was unable to pass the men without comment. As she entered the water, one man mentioned the size of her breasts while the rest of the group broke into laughter and began to wolf whistle. As a queer woman passionate about challenging heteronormativity and the associated expectations placed on women's bodies to consistently remain desirable to the male gaze, the ocean for Ellie become a site of spatial injustice where dimensions of gender and sexuality interacted to create a distinctive barrier (Fig. 8.3).

Despite global LGBTQIA+ advocacy movements promoting queer rights and inclusion, surfing has remained a harshly heteronormative arena where misunderstandings of gender as a binary construct are perpetually reinforced. Community surf spaces such as car parks and other places adjacent to the surf are often teeming with semi-naked men's bodies half dressed in wet rubber (wetsuits), a scene flourishing with homoeroticism (Evers, 2009). Yet Evers explains there are a series of

Fig. 8.3 Ellie finding joy and solace in the surf as the sun sets. Image taken as part of the SheShaka project

unwritten heteronormative expectations which function to prevent surf spaces becoming sites of homoerotic relations. The significance of the car park as a site used by male surfers to maintain their heterosexuality is again made clear, further contributing to the tensions which routinely caused Ellie to turn her car around.

Research based in the UK examined the relationship between gender, sexuality, and the body (Roy, 2016). For the lesbian surfers Roy interacted with, surfing in a group with other women who identified as queer offered an effective avenue through which they could challenge harsh heteronormative surf culture. The way Ellie overcame fear, anxiety, and embarrassment is reflected in Roy's findings. Ellie found confidence to surf and be in surfing spaces once she connected with a group of women who surfed. When surfing with other women, Ellie felt the ocean transform into a safe space for her. The ease with which this transformation occurred, whereby surfing went from an anxiety inducing practice into one she used to find "solace" and "joy" in the ocean, indicates surfing spaces have the potential to become spaces used to challenge heteronormative social norms.

For queer Hawaiian surfer Keala Kenneally, the entrenched homophobia in surfing meant living in a perpetual state of fear. In the late 1990s, Keala spent most of a decade in the top 10 in the women's World Surf League. However, when she made the decision to be more open about her sexuality, her career notably suffered. Keala recently stated that "to come out as gay in the surfing industry is to step out of the bounds of what is considered 'marketable'" (Coldwell, 2014). As knowledge of her queerness grew, her sponsorship deals diminished. Prior to Keala's marginalization, Australian surfers Pauline Menczer and Jodie Cooper experienced a similar fate. Pauline did not meet the hypersexualized surfer chick aesthetic. She had rheumatoid arthritis, grew up in poverty at Sydney's Bondi Beach, and was a queer woman who felt forced to hide her sexuality from the surfing world for most of her career for fear of repercussions. Pauline believes it was because of these factors that she struggled to find sponsorship deals throughout her career.

Today, Tyler Wright remains the only openly queer surfer on the world tour and has publicly spoken about her own struggle with internalized homophobia. These queer histories outline the overt homophobia and gender discrimination within the surf industry, culture, and space. Yet, as lisahunter (2016) asks, if surfing is a tool that enhances fluidity between human and nature, may it also be used as a practice to enhance fluidity between gendered constructions of man/woman/person?

8.7 Being in Sacred Blue Spaces as a White-Settler Woman

Tyler's story begins in childhood, in a small town on the southern New South Wales coast. Before she became a two-time world champion, Tyler Wright was a girl leading the charge and resisting gender stereotypes in ways that older men in the surf found "confronting." This resulted in men flouting surf etiquette to drop in on her, breaking of the surfing code that is both dangerous and deeply disrespectful as it suggests the surfer who should be entitled to the wave may waste it. Alongside these sexist microaggressions, Tyler experienced gendered structural discrimination. This discrimination was most acute in the significant gender pay gap, where women's prize money was less than half the amount awarded to professional men. In her interview for the SheShaka project, Tyler explored how these compounding experiences of gender-based discrimination led her to internalize narratives associated with women not deserving equal pay or recognition as they "weren't as good" as men. Tyler remembered being "angry" and "mad" and developing "an attitude." As a younger surfer, she didn't know why these feelings of fury were surfacing in her. It wasn't until she was older that she began to understand how unequal gender power relations were affecting her experiences. Learning more about intersectionality allowed Tyler to reshape the ways she saw gender in surfing. In doing so, Tyler recognized the "influence of patriarchy" in surf spaces and also noted how racism functions in the same spaces (Fig. 8.4).

Using intersectionality to contextualize her experiences of discrimination in surfing spaces, Tyler discussed how surfing can be a practice that reveals violent colonial histories. For decades, surfing has been a practice dominated and managed by white male settlers. Yet for Tyler, surfing has promoted a richer understanding and recognition of First Nations sovereignty.

> Surfing is for sure a powerful outlet that can support mental wellbeing and connection to community. I think for me, it's not so much surfing itself because sometimes it's kind of a work orientated space for me.

Fig. 8.4 Tyler Wright at the 2024 Olympic Games where she represented Australia in the surfing at Teahupo'o, Tahiti, in French Polynesia. Image taken as part of the SheShaka project

> What I do find is my space is just being in the ocean and being around people that love and respect the ocean. That's where I feel really connected. One of the other things is that [as surfers] we are incredibly privileged to be able to travel around the world and be connected to Indigenous cultures and First Nations people wherever we go. To be connecting to community in that way, to be connecting to the land and the stories in the ocean… I think for me that part of surfing has been incredibly healing and incredibly restoring for me over the years.
>
> Tyler (SheShaka participant)

Tyler outlines how she has been able to cultivate healing and restorative power through recognizing the role of blue spaces in First Nations knowledges and cultures. What we today refer to as blue spaces have long been sacred spaces for Indigenous peoples. Sacred blue spaces are not a site of extraction nor domination. Exploring the notion of how spatial justice may be used to enhance Indigenous peoples' protection of sacred places, June Lorenzo (2017), a Laguna Pueblo/Navajo legal scholar, emphasizes that within Indigenous knowledges, sacred spaces are living active participants in the lives of Indigenous people. In Australia, surfing has become a cultural practice used to uphold settler-colonial power relations that exclude racialized bodies from accessing sacred blue spaces (Moreton-Robinson, 2011; Wheaton, 2017). Therefore, the possibility of surfing being used to resist the very culture of racialized colonial exclusion it was used to maintain becomes questionable.

Yet Tyler is beginning to ask this question with increasing intensity. In 2020, during a World Surf League event, Tyler knelt for 439 s with "Black Lives Matter" emblazoned on her surfboard. Each of those 439 s represented an Aboriginal person who had died in police custody since 1991 in Australia. Tyler's solidarity may be seen as a demonstration that surfers have the capacity to challenge the dominant culture of the sport. How this attempt to transform existing systems of racial exclusion and settler-colonialism generates support at grassroots and community levels of surfing remains unknown. Emerging evidence from New Zealand suggests that surfing and bodyboarding provides a space for Māori men and women to create a distinct form of cultural capital (Nemani & Thorpe, 2016). Surfing communities in Australia and the rest of the Global North must take note to build more effective forms of resistance against racialized (and gendered) forms of spatial injustice, which currently contribute to sacred blue spaces being dominated and defined by white-settler power relations.

8.8 Gendered Violence in the Water Is a Health Equity Issue

> In the surf, I remove myself from any situation that I don't feel comfortable in. I'll get out of the water…There's been a time when there's been an altercation, where there's been words. There's been a couple of remarks from men in the surf like "did you bring the balloons? Didn't know it was a party." That sort of thing… Me and my friends, we just get out of the water when it happens. We do it to put people back in their place. We're at a beach break and we're all just having fun. Nobody's sponsored and we're not having a competition. We're not going to win a trophy. So, there's been a couple of little altercations like that, but I've never let myself get into a position of full-on confrontation in the surf. I'm not good

with confrontation. Why poke the bear? If you call a dickhead a dickhead, they're not going to stop being a dickhead. They're going to keep being a dickhead so you may as well go. I live at the beach; I can go anywhere. So, I tend to remove myself. But **you're consciously changing your behaviour and I think most women here would have done that in the surf, at least once.** Changing your behaviour like that can make you sometimes feel a bit shit, sometimes a bit frustrated. But when you step back… it's a power grab. So let them have power.

At certain surf breaks, the energy in the water is just not great. My friend was surfing there [a world famous surf break in Australia] a few months ago with her husband. A male surfer pushed her off her surfboard! He came over and pushed her off. It's just horrible! He was all, "you've dropped in on me. You've ruined my wave." There was nowhere she could go. There was nothing she could do. He pushed her off her board. Of course, she got out and had a cry and felt a bit yuck and then came to us. We were like, yeah, he's a dick. He'll always be a dick and you'll always be a legend.

Kate (SheShaka participant)

Kate's story highlights how women manage the threat of confrontation in everyday surfing spaces and how quickly confrontation can escalate into physical violence. Kate outlines instances where she was forced to remove herself from the surf in order to avoid confrontation. Here, spatial injustice is demonstrated in its most literal form, in that harmful gender power dynamics prevent women from staying in the surf and transmute surfing spaces into sites of violence.

When surfing is understood as a practice that takes place in public blue spaces, and growing evidence outlines how spending time in blue spaces contributes to positive mental and physical health outcomes, men's intimidation and resistance to women's inclusion in these spaces becomes an issue of health equity. It is not only men's resistance to women's inclusion that may impact women's health outcomes. Experiencing gendered violence has serious negative short-, medium-, and long-term health consequences (Vives-Cases et al., 2011). Chrissie, a participant in SheShaka, described an incident of verbal abuse and intimidation where her friend "copped" it. Chrissie explained how when out in the surf "this guy just went crazy-psycho at my friend. All the other guys were not really getting involved, cause this guy was a psycho… she was really intimidated."

Other instances of gendered violence in the surf have resulted in serious injury. In 2018, Jodie Cooper, one of the world's first openly queer professional surfers, feared for her life when she was assaulted by a man in the surf. The man was a local surfer who was subsequently prosecuted for assaulting Jodie. In her statement to police and the court, Jodie described the assault:

"He was trying to provoke me off the wave… he just grabbed me with two hands and just forced me under the water." Jodie was able to make her way to the surface three times, however the man repeatedly held her head underwater and pulled her hair. Cooper was able to escape by pretending to be unconscious, stating that "when I went limp, thinking I was dead or drowned or something, that's when he released [me]." (Kahn, 2019)

The stories shared by Kate, Chrissie, and Jodie demonstrate that different forms of violence against women persist across all levels of surfing, from the professional to grassroots. The impacts that gendered violence perpetrated in the surf has on the

Fig. 8.5 A photo composed as part of the SheShaka project to represent the 1 in 3 women worldwide who have experienced physical and/or sexual violence in their lifetime. Image taken as part of the SheShaka project

health and well-being of women and girls remain largely under-researched, though evidence indicates that violence against women and girls is a leading contributor to women's burden of disease and mortality globally (Fig. 8.5).

8.9 Whose Break Is It Anyway? Surfing in Lombok with Local Women (Fig. 8.6)

The island of Lombok had a reputation amongst surfers as being home to some of the best waves in the world. Everyone I (Gemma) spoke to about going surfing in Lombok followed their beautiful description of the island with an equivocation about how much it had changed in the past decade, how it had been "totally ruined" by the crowds who clog the line-up – typically Europeans who can barely surf. When I heard white settlers in Australia bemoan the busy line-ups in Lombok, I would feel a gnawing sensation in my gut. My fellow surfers spoke about the ocean in Lombok, Bali, and other Indonesian islands as a space they were **entitled** to enter, manage, control and ultimately dominate, indicating that practices of localism somehow extended beyond Australian borders. There was no acknowledgement of how Lombok, and the people of Lombok, existed prior to the island's "discovery" as a global surfing destination. The Islamic and pre-Islamic systems of power and knowledge on the island simply did not exist. Lombok was viewed through its utility as

Fig. 8.6 Gemma and Rana on the beach in southern Lombok before their first surf together. Image taken as part of the SheShaka project

a surfing destination, a utility that had rapidly declined due to the number of westerners clogging up the surf. And I was about to be another member of the herd, clogging up the already "ruined" ocean.

On my first day in Lombok we didn't surf. Instead, we watched from the shore as shirtless white men on shortboards tiptoed their way across jagged reef out towards the break. Despite Indonesia being the world's largest Muslim majority country, there was no one out there wearing a hijab, All the surfers were westerners, and most were men. Some were joined by Indonesian surf guides and teachers, though many were alone. I was under the impression Rana had been surfing with her husband, and would be somewhat comfortable on the board. It wasn't until later that night I learnt she had never been surfing before. I was initially surprised, though I understood that even for a Lombok local, access to surf spaces could be unjust.

The next morning, we chose a slow-moving reef break about a kilometre offshore. Rana and I jumped on the fishing boat with two white men and their surf guide. Rana's daughter waved from the shore with her father as the wobbly wooden boat made its way towards the breaking waves. The surf was already crowded with groups of westerners learning to surf, Indonesian instructors by their sides. Rana positioned herself on the right-hand sand of the breaking waves away from the crowd while I paddled over to the peak. More fishing boats arrived bringing more surfers. There was a sense of fun and frivolity amongst the group.

As I paddled back over to find Rana, I saw her zooming through the white water towards me. Her face was beaming with joyous abandon as she experienced her body being propelled through space by oceanic energy. An Indonesian surf instructor was sitting in her path with one of his western tourist students. She came racing towards the pair who safely moved out of her way. As Rana surfed past me, she rolled off the wave and began to make the long paddle back, her head down as she beat against the lulling waves one arm at a time. The instructor who she had surfed past paddled over to me. I initially thought he was celebrating Rana for catching her first wave. It took me a moment to realise he was actually lambasting me for leaving her out in the surf "unsupervised". He screamed that she was "dangerous" and shouldn't be out there. He continued to yell as he told me it was my responsibility to stay by her side at all times lest she cause injury to one of his students. Rana had considerately positioned herself a safe distance from the other surfers. She was not a danger to anyone, nor was it my responsibility to supervise her. I was her guest in Lombok and would not have known about that specific surf spot without her. Was it not she who was supervising me, a visitor to her homeland?

After the incident, Rana shared that she was elated by catching her first ever wave. Yet at the same time, she felt embarrassed. She was worried she had caused harm by deciding to paddle for the broken wave. I assured her that she had done absolutely nothing wrong. She was within her rights to be taking up space in the ocean that bordered her home. If she was not permitted to learn how to surf in Lombok, was she allowed to learn how to surf anywhere? In an attempt to prevent any further confrontation, we decided to head back to shore. As we made the long paddle back, I started to wonder how much her status as a "danger" to the westerners in the surf were inscribed in land-based tropes of Islamophobia, where women in hijab are interpellated as a security threat. Despite this experience, Rana started surf lessons and continues to paddle out every week building her confidence and practising her right to exist in surf spaces.

8.10 Fighting for Spatial Justice in Surf Spaces

In this chapter, we have explored how women who surf experience distinct forms of spatial injustice that make them feel unsafe, unwelcome, and excluded in surf spaces. While the stories have highlighted the gendered nature of spatial injustice, we are cognizant of the importance of including intersectional perspectives that highlight how gendered discrimination in the surf often interacts with other power relations based on race, sexuality, and ability/disability. We have shown how spatial injustice affects "other" women (Carter-Francique, 2013; Ratna & Samie, 2019) by including perspectives from queer women and women from culturally and racially marginalized backgrounds. We have not included perspectives from Aboriginal and Torres Strait Islander women, and we note this is a key limitation.

Women's inclusion in surf spaces remains contingent on maintaining dominant power relations that privilege white, able-bodied heteronormative men as the spatial managers of the Australian beach. There is a dire need to move beyond current understandings of gender equality in surfing, and sport more broadly, which has focused on the inclusion of white heterosexual cis-women. Through the experiences of women who participated in SheShaka, surfing can be understood as a practice that allows women to fight for the creation of fairer surf spaces where place and resources to the ocean are equally distributed across gender, race, ability, sexuality, and class. While women experienced intersecting barriers that at times were complex, and even violent, women continued to surf as a way to envision the ocean as a site of spatial justice.

Acknowledgments The SheShaka project was supported by a grant from the Melbourne Social Equity Institute at the University of Melbourne and approved by the University of Melbourne Human Research Ethics Committee (project ID 25946). We thank the women who participated in SheShaka for sharing their experiences in the surf and for creating such beautiful photographs. This research was conducted on the stolen lands and waterways of the Gunditjmara and Wurundjeri Woi Wurrung Peoples of the Eastern Maar and Kulin Nations.

References

Anderson, J. (2014). Exploring the space between words and meaning: Understanding the relational sensibility of surf spaces. *Emotion, Space and Society, 10*, 27–34. https://doi.org/10.1016/j.emospa.2012.11.002

Black Girls Surf & Solidarity in Surf. (2024, July 7). Solidarity in surf: Last year, @arte.tv came to Santa Cruz, CA to capture the real essence of racism in surfing. Instagram. https://www.instagram.com/reel/C9IsAvsS7XN/

Booth, D. (2002). *Australian beach cultures: The history of sun, sand and surf*. Routledge.

Britton, E. (2018). 'Be like water': Reflections on strategies developing cross-cultural Programmes for women, surfing and social good. In L. Mansfield, J. Caudwell, B. Wheaton, & B. Watson (Eds.), *The Palgrave handbook of feminism and sport, leisure and physical education* (pp. 793–807). Palgrave Macmillan. https://doi.org/10.1057/978-1-137-53318-0_50

Burtscher, M., & Britton, E. (2022). "There was some kind of energy coming into my heart": Creating safe spaces for Sri Lankan women and girls to enjoy the wellbeing benefits of the ocean. *International Journal of Environmental Research and Public Health, 19*(6), 3342. https://doi.org/10.3390/ijerph19063342

Butler, J. (1986). Sex and gender in Simone de Beauvoir's second sex. *Yale French Studies, 72*, 35. https://doi.org/10.2307/2930225

Carter-Francique, A. (2013). Intersections of race, ethnicity, and gender in sport. In *Gender relations in sport* (pp. 73–93). Brill. https://scholar.google.com/citations?view_op=view_citation&hl=en&user=4RpZ5GYAAAAJ&citation_for_view=4RpZ5GYAAAAJ:IjCSPb-OGe4C

Cater, G. (2014). History: Aboriginal rafts and canoes from 1770. https://www.surfresearch.com.au/index.html

Coldwell, W. (2014, October 10). Caught on camera: The homophobic world of surfing. *The Guardian*. https://www.theguardian.com/lifeandstyle/2014/oct/10/homophobic-surfing-documentary-gay-surfers

Comer, K. (2010). *Surfer girls in the new world order*. Duke University Press.

Comley, C. (2016). "We have to establish our territory": How women surfers 'carve out' gendered spaces within surfing. *Sport in Society, 19*(8–9), 1289–1298. https://doi.org/10.1080/17430437.2015.1133603

Connell, R. W., & Messerschmidt, J. W. (2005). Hegemonic masculinity: Rethinking the concept. *Gender & Society, 19*(6), 829–859. https://doi.org/10.1177/0891243205278639

Crellin, D. (2022). *Troubled waters: The ocean as contested space in California surf culture*. University of California.

Crenshaw, K. (1991). Mapping the margins: Intersectionality, identity politics, and violence against women of color. *Stanford Law Review, 43*(6), 1241–1299. https://doi.org/10.2307/1229039

Evers, C. (2004). Men who surf. *Cultural studies review, 10*(1), Article 1. https://doi.org/10.5130/csr.v10i1.3519

Evers, C. (2009). 'The point': Surfing, geography and a sensual life of men and masculinity on the Gold Coast, Australia. *Social & Cultural Geography, 10*(8), 893–908. https://doi.org/10.1080/14649360903305783

Franklin, R., & Carpenter, L. (2018). Surfing, sponsorship and sexploitation: The reality of being a female professional surfer. In *Surfing, sex, genders and sexualities*. Routledge.

Gapps, S., & Smith, M. (2015). Nawi—Exploring Australia's indigenous watercraft: Cultural resurgence through museums and indigenous communities. *AlterNative: An International Journal of Indigenous Peoples, 11*(2), 87–102. https://doi.org/10.1177/117718011501100201

Gill, R. (2007). Postfeminist media culture: Elements of a sensibility. *European Journal of Cultural Studies, 10*(2), 147–166. https://doi.org/10.1177/1367549407075898

Kahn, T. (2019). "I was fearful for my life:" Pro Surfer Jodie Cooper Wins Criminal Case. *Sea Maven Magazine*. https://www.seamavenmagazine.com/blog/pro-surfer-jodie-cooper-wins-criminal-case

Kempton, J. (2021). *Women on waves*. Simon & Schuster.

lisahunter. (2016). Becoming visible: Visual narratives of 'female' as a political position in surfing: The history, perpetuation, and disruption of Patriocolonial pedagogies? In H. Thorpe & R. Olive (Eds.), *Women in action sport cultures* (pp. 319–347). Palgrave Macmillan UK. https://doi.org/10.1057/978-1-137-45797-4_16

lisahunter. (2018). The long and short of (performance) surfing: Tightening patriarchal threads in boardshorts and bikinis? *Sport in Society, 21*(9), 1382–1399. https://doi.org/10.1080/17430437.2017.1388789

Lorenzo, J. L. (2017). Spatial justice and indigenous peoples' protection of sacred places: Adding indigenous dimensions to the conversation. *Justice Spatiale | Spatial Justice, 11*, 1–17.

Matthews, G. (2022, June 11). Torquay and surf industry icon a master surfboard-maker. The Sydney Morning Herald. https://www.smh.com.au/national/torquay-and-surf-industry-icon-a-master-surfboard-maker-20220611-p5at0p.html

Moreton-Robinson, A. (2011). Bodies that matter: Performing white possession on the beach. *American Indian Culture and Research Journal, 35*(4), 57–72. https://doi.org/10.17953/aicr.35.4.41936g11158r78n4

Nemani, M., & Thorpe, H. (2016). The experiences of 'Brown' female Bodyboarders: Negotiating multiple axes of marginality. In H. Thorpe & R. Olive (Eds.), *Women in action sport cultures: Identity, politics and experience* (pp. 213–233). Palgrave Macmillan UK. https://doi.org/10.1057/978-1-137-45797-4_11

Olive, R. (2019). The trouble with newcomers: Women, localism and the politics of surfing. *Journal of Australian Studies, 43*(1), 39–54. https://doi.org/10.1080/14443058.2019.1574861

Olive, R., McCuaig, L., & Phillips, M. G. (2015). Women's recreational surfing: A patronising experience. *Sport, Education and Society, 20*(2), 258–276. https://doi.org/10.1080/13573322.2012.754752

Olive, R., Roy, G., & Wheaton, B. (2018). Stories of surfing: Surfing, space and subjectivity/intersectionality. In *Surfing, sex, genders and sexualities*. Routledge.

Pavlidis, A. (2018). Making "space" for women and girls in sport: An agenda for Australian geography: Making "space" for women in sport. *Geographical Research, 56*(4), 343–352. https://doi.org/10.1111/1745-5871.12302

Phillips, L. G., & Bunda, T. (2018). *Research through, with and as storying*. Routledge.

Puwar, N. (2004). *Space invaders: Race, gender and bodies out of place*. https://research.ebsco.com/linkprocessor/plink?id=aca94bb6-edf1-3bdb-b62a-095911b96b7d

Ratna, A., & Samie, S. F. (Eds.). (2019). *Race, gender and sport: The politics of ethnic "other" girls and women*. Routledge. https://doi.org/10.4324/9781315637051

Roy, G. (2016). Coming together and paddling out: Lesbian identities and British surfing spaces. In H. Thorpe & R. Olive (Eds.), *Women in action sport cultures: Identity, politics and experience* (pp. 193–211). Palgrave Macmillan UK. https://doi.org/10.1057/978-1-137-45797-4_10

Russell-Cook, M. (Ed.). (2021). *Maree Clarke: Ancestral memories/edited by Myles Russell-Cook with contributors*. Council of Trustees, National Gallery of Victoria.

Said, E. W. (1994). *Culture and imperialism*. (1st vintage books ed. Vintage Books.

Sanz-Marcos, P. (2021). Can women escape the male influence when surfing? An ethnographic study of female surfers in southern Spain. *Sport in Society, 26*(4), 605–617. https://doi.org/10.1080/17430437.2021.1996347

Soja, E. W. (2009). The City and spatial justice. *Justice Spatiale | Spatial Justice, 1*, 1–5.

Sontag, S. (1997). The double standard of aging. In *The other within us*. Routledge.

Stab Magazine. (2017, December 14). *Playboy ranked the 21 "Hottest Surfer Babes On Instagram."* StabMag.https://stabmag.com/girls/playboy-ranked-the-21-hottest-surfer-babes-on-instagram/

Thorner, S., Edmonds, F., Clarke, M., & Balla, P. (2018). Maree's backyard: Intercultural collaborations for indigenous sovereignty in Melbourne. *Oceania, 88*(3), 269–291. https://doi.org/10.1002/ocea.5206

Thorpe, H., Toffoletti, K., & Bruce, T. (2017). Sportswomen and social media: Bringing third-wave feminism, postfeminism, and neoliberal feminism into conversation. *Journal of Sport and Social Issues, 41*(5), 359–383. https://doi.org/10.1177/0193723517730808

Vives-Cases, C., Ruiz-Cantero, M. T., Escriba-Aguir, V., & Miralles, J. J. (2011). The effect of intimate partner violence and other forms of violence against women on health. *Journal of Public Health, 33*(1), 15–21. https://doi.org/10.1093/pubmed/fdq101

Waitt, G. (2007). (Hetero)sexy waves: Surfing, space, gender and sexuality. In *Rethinking gender and youth sport*. Routledge.

Wheaton, B. (2017). 8. Space invaders in Surfing's white tribe: Exploring surfing, race, and identity. In *8. Space invaders in surfing's white tribe: Exploring surfing, race, and identity* (pp. 177–195). Duke University Press. https://doi.org/10.1515/9780822372820-010

Chapter 9
Surfing Economics: Understanding, Managing and Protecting the Value of Surfing Ecosystems

Ana Manero

9.1 Introduction

"Surfing" and "economics" may, at first, seem two contrasting concepts. Surfing, whether regarded as a sport, a pastime, or a lifestyle, is often portrayed as a counter-culture, anti-capitalist endeavor. Surfing originated as an integral part of Polynesian culture, but it slowly weaved itself into Western society, primarily as a leisure activity, starting in the early 1900s (Chap. 2). By the mid-twentieth century, particularly in places like California, Hawaii, and Australia, surfing had become a symbol of a new form of life, a new way of understanding and experiencing the world, which starkly contrasted with the prevailing postwar discourses of industrial productivity and consumerism. Economics has been through even longer and more complex transformations over the course of its thousands of year history (Poitras, 2003), to be stereotyped today in popular culture as the dogma of free markets and profit maximization. So what do the two—surfing and economics—have in common? Why would surfers be interested in economics, and why would an economist care about waves?

To start with, economics can be an ally to surfing to understand waves as valuable resources. Let's look at how waves are formed in the natural environment. At a minimum, we need a "surf break," which allows waves to break or peel under a delicate set of conditions dependent on winds, tides, currents (Chap. 3), shoreline, and the seabed, among others (Chaps. 4 and 5) (Reineman, 2016). Any changes to these factors, namely, those affecting the coastline, can dramatically impact the processes needed for wave formation. Like many coastal areas around the world, surf-rich

A. Manero (✉)
The University of Western Australia, Crawley, WA, Australia

The Australian National University, Canberra, ACT, Australia
e-mail: ana.maneroruiz@uwa.edu.au

© The Author(s), under exclusive license to Springer Nature Switzerland AG 2025
D. M. Kennedy (ed.), *The Science and Culture of Surfing*,
https://doi.org/10.1007/978-3-031-80979-8_9

locations are increasingly subject to growing pressures, including coastal erosion, sand loss, declining water quality, competing land and ocean uses, and expanding urban developments (Scott & Rogers, 2018). At the same time, surfing's appeal can be such that the influx of new residents, visitors, economic growth, and media attention can play a decisive role in changing local traditions and ways of life (Fig. 9.1).

The dual but often conflicting goals of socioeconomic growth and ocean sustainability are brought together under so-called blue economy, that is, the use of ocean resources for economic growth and social prosperity, while preserving the health of ocean ecosystems (Lee et al., 2020). To gather ocean stakeholders behind a common set of goals, the United Nations declared the 10 years from 2021 to 2030 as the "Decade of Ocean Science for Sustainable Development" (McKinley et al., 2023). The "Oceans Decade" is articulated around ten key "challenges," and, to address these, it supports a range of global initiatives, including a growing number of projects targeted at the intersection of surfing and science (e.g., A Liquid Future, 2024).

Amidst the complexity of competing goals and growing pressures, decision-makers, from town planners to engineers, ought to understand the value provided by natural features to inform processes that guarantee the sustainability of our cherished surf coasts. When faced with multiple issues and limited resources, decision-makers need to make difficult trade-offs that will, ideally, deliver the greatest collective benefit, at the lowest cost. But how would coastal planners know what the

Fig. 9.1 "A past of fishermen, A future of surfers"—street mural in Baleal, Portugal. (Photo: Ana Manero)

best option is? How can they determine the worth of priceless assets, such as waves? And how can they know how big a loss it would be, if a surf break was indeed to disappear? Economics, and, in particular, environmental economics, offers a suitable approach to answer these and many other questions (Hanley et al., 2019).

Although money is indeed central to economic science (Ekstedt, 2012), economics is much more than the study of moneymaking. Economics is a broad discipline concerned about how humans—both at the individual and the collective level—make choices, especially when allocating scarce resources to meet their wants and needs (Backhouse & Medema, 2009). For anyone who has ever attempted to catch a wave, it is evident that surfing waves are, a scarce resource: they are valuable to those who seek them, and demand often exceeds supply. Indeed, there are usually more people trying to catch waves than waves available (Mixon, 2018). Beyond competition for wave-riding, contestation for coastal spaces expands at a broader scale. For millennia, coastal areas have been valued as strategic zones, where a variety of activities take place, within a limited space, thus placing pressures both on the environment and among users (Tuda et al., 2014). Today, with over one-third of the world's population living within 100 km from the coast (Reimann et al., 2023), governments worldwide are increasingly managing the access and use of coastal resources through purposely defined strategies, such as integrated coastal zone management or marine protected areas (McKenna et al., 2008; Touron-Gardic & Failler, 2022).

While there is a growing understanding about natural processes and human-induced changes to the coast, these are rarely examined in light of their impacts on surf breaks and their ability for wave formation (Manero, 2023). The lack of consideration contrasts with surfing's documented influence on a variety of socioeconomic outcomes, such as faster economic growth (McGregor & Wills, 2017), higher real estate prices (Scorse et al., 2015), public infrastructure investments (Earhart, 2015), job creation (Mills & Cummins, 2015), and skill development (Mach, 2019). Bearing these effects in mind, the field of economics offers a suite of approaches to conceptualize and communicate the multiple social, economic, and environmental values associated with the practice of surfing. When appropriately employed, tools such as expenditure analysis and nonmarket valuation can generate the necessary evidence to inform decision-making processes potentially affecting surfing amenity (Monteferri et al., 2020). These may include a range of interventions, from large projects such as the construction of a new marina to local management practices, like sand nourishment programs. Importantly, economics doesn't stand alone in the quest to understand people's values and behaviors. Fields as diverse as Indigenous studies, anthropology, ethnography, history, geography, psychology, and engineering offer unique lenses and, collectively, contribute to the holistic study of the importance of surfing to society (Aramoana Waiti & Awatere, 2019; Lanagan, 2002; Olive et al., 2023; Olive & Wheaton, 2020; Usher & Kerstetter, 2015).

This chapter covers essential economics concepts through the lens of surfing as a recreational activity but also from the perspective of surf breaks and their surrounding areas as valuable components of the natural environment. The chapter will first address the financial impacts derived from surf participation—one of the most commonly used approaches to gauge the "worth" or "value" of surfing. Further, an overview of ecosystem services and nonmarket valuation principles will guide an examination of surfing's so-called intangible benefits. Finally, a discussion will be given to widespread rival and exclusion attitudes among surfers and how these transform waves from a (theoretically) public resource into (almost) a private good.

9.2 Surfing's Impact on Local, Regional, and Global Economies

9.2.1 Accounting for Market-Based Economic Impacts

The practice of surfing, the mere act of riding a wave with the use of a craft, or even your own body, is free. Waves have no price and there is no charge to ride a wave. However, substantial efforts are typically involved in getting surfers to that point. One needs, at a minimum, some form of equipment, starting with bathing suits, probably a wetsuit if surfing in waters under 24 °C, a board, and all the accessories that go with it: fins, tail pads, leash, bags, and roof racks, to name a few. In sunny locations, one may also choose some good-quality sunscreen and a hat, whereas cold spots may call for booties and a hood. While some surfers may walk to their local breaks, most will need to get into a car, before they get onto their boards. Those living seaside are probably already paying a premium to rent or buy close to the coast. Of course, not all surfers have the perfect wave at their doorstep, so they need to factor in elements needed to travel further away: flights, accommodation, insurance, etc. the list goes on. On aggregate, all of these surf-motivated financial transactions can have a substantial impact on local and regional economies.

One of the most common approaches to measuring the impact of surfing on "the economy" is by estimating direct expenditures. It is understood that money spent in gear, travel, and other surf-related activities would not have occurred without the pursuit of catching waves. A small part, however, may be considered a transfer from other exchanges that would have occurred elsewhere in the market economy, for example, people taking up alternative hobbies or traveling to other locations (Nelsen et al., 2007). Direct expenditures offer the advantage of being relatively easy to estimate, as they are based on observable market transactions and consistent with standard accounting systems, namely, the System of National Accounts or SNA (European Commission et al., 2008). The SNA is an internationally recognized framework, developed by the United Nations Statistics Division, that provides standardized recommendations for compiling measures of economic activity, based on globally agreed-upon accounting rules (United Nations, 2024a).

In a practical example of a common hobby, under the SNA, someone learning to dance could be seen as contributing to gross domestic product (GDP), through the purchase of new shoes and lessons. In turn, the dance teacher may further input into "the economy" and job creation, by deciding to expand the studio and hire new staff. Something that sets surfing aside from dancing is its entire dependence on the natural environment. Without people's access to the natural environment, surfing would not take place, and neither would the financial transactions derived from it. These are referred to as monetary flows from the environment into "the economy" (Pelletier et al., 2021), and also occur through other forms of outdoor recreation, such as rock-climbing or mountain-biking. In recognition of these complex relationships, the System of Environmental-Economic Accounting (SEEA) expands the concepts and methods of the SNA to specifically account for the interrelationships between the economy and the environment and how environmental assets generate benefits to people (United Nations, 2024b).

The SEEA framework can be used to quantify the monetary contributions emerging from a particular component of the ecosystem (e.g., surf break) or, more broadly, an activity that is dependent on the ecosystem (e.g., recreational surfing at the national level). These types of tools and data can be used, for example, by local governments to understand how surfing affects market dynamics and their implications for questions such as tax revenues, labor market, incomes, and economic inequality, as a result of surf-related activities (Murphy & Bernal, 2008). Importantly, the SEEA framework, in particular the SEEA Ecosystem Accounting (or SEEA EA) component, can help account for what would happen if a surf break would disappear and, with it, all the economic activity it generates. For instance, a surf school may lose its business, and the flow-on economic activity, if the local break becomes severely compromised or even disappears, as a result of coastal threats such as climate change or development-induced sand erosion. These changes in the extent and condition of natural features (also referred to as assets; see Sect. 9.3.1) can be documented within the SEEA EA framework, thus allowing for the systematic accounting of gains and losses and determining long-term trends (United Nations, 2021). Moreover, SEEA EA's monetary accounts can serve to record changes in economic activity and attribute these to particular changes in the environment.

Such information can be fed into decision-support tools such as economic impact analysis (EIA), which aims to understand how a particular activity (e.g., recreation) or intervention (e.g., infrastructure development or environmental protection) will impact market outputs and dynamics (Branigan & Ramezani, 2018). In addition to direct effects, EIA accounts for indirect and induced impacts, which consider secondary and "ripple" effects, through subsequent rounds of spending within the market economy. A cautionary critique of EIA is that it can be misused to justify decisions on the basis of overinflated benefits, through the improper application of economic multipliers (Joseph et al., 2020). Another limitation of EIA, and more broadly of market-based accounting systems, is the exclusion of benefits that are not the result of an economic production process, as defined by the SNA (Sylla et al., 2021). These include benefits that cannot be measured through observable market transactions, such as the pleasure of riding a wave, or the cultural significance of surfing to local communities.

SEEA EA aims to recognize non-SNA benefits by recording stocks or flows of environmental assets, but not their equivalency in monetary terms (United Nations, 2021). For example, if a surf break was to disappear due to the construction of a new marina, as it occurred in 2021, in Western Australia (Manero, 2023), the SEEA EA could be used to record the loss of the ecosystem asset, as well as the losses associated with reduced surf-derived economic activity. By contrast, non-SNA benefits in the form of community connectedness and physical and mental well-being would not be accounted for, despite these translating into millions of dollars through better quality of life (Buckley & Cooper, 2023). A suite of techniques known as nonmarket valuation, or NMV (Champ et al., 2017), aim to systemically estimate the value of goods and services that are not bought and sold in markets—a crucial step toward understanding the total economic value of natural resources and ecosystem services (see Sect. 9.3.2).

9.2.2 Surfing's Economic Impact in Numbers

Across the world, a number of studies have quantified the economic impact of surfing through direct expenditures (Table 9.1). In a global study of surf tourism, Mach and Ponting (2021) estimated that international surf travel generated between US$32 billion and US$64 billion through direct expenditure. Further, survey participants indicated they would be prepared to pay an additional 10–20% for future international trips (equivalent to US$2 billion to US$4 billion, across the whole sample), if the trip was demonstrably more sustainable. Over 90% of sampled surfers rated themselves as having a skill level of intermediate or above, thus suggesting they also practice the sport regularly in their home countries, in addition to international trips. Because the study did not account for domestic economic impacts, it is likely that figures reported by Mach and Ponting (2021) are a lower-bound estimate of surf-driven input into the global market economy.

Few studies exist of surf-driven expenditure at the national levels, with Manero et al. (2024) and Mills and Cummins (2015) providing estimates for Australia and the UK, respectively. Aggregating across the national surfing population, and adjusting for inflation and currency conversions, the annual monetary impact of recreational surfing is estimated at U$1.85 billion in Australia and U$3.11 billion in the UK. A number of studies have estimated monetary impacts from surf tourism in popular destinations, such as Uluwatu in Bali, Indonesia, and Mundaka in the Basque Country, Spain (see Table 9.1). The figures vary largely, as a result of differences in sampling approaches, items included, and the nature and size of the surfing population. Generally speaking, reported expenditures exceed US$200 per person per day, adjusting for inflation into 2024 prices.

In addition to users' expenditure, professional sponsorships and competitions drive important fluxes of money across the globe and into local economies. While no official figures are publicly available, some estimates put the cost of running a large championship tour event at US$3 million (Bousquette, 2024). Accounting for 10 of them per year, and another 100+ World Surf League (WSL) events (such as qualifying series, big wave, longboard, and junior tours), the overall economic impact of professional surfing contests is in the order of dozens, if not hundreds, of

Table 9.1 Summary of studies estimating direct expenditure derived from recreational surfing

Area and source	Direct expenditure per person	Population group	Aggregated expenditure across the surfing population
Australia (Manero et al., 2024)	A$3719/year	Australian surfers	A$2.71 (direct)–A$4.8 (with multiplayers) billion/year
Global (Mach & Ponting, 2021)	US$2150/year	International travelers	US$31.5–64.9 billion/year
Guarda do Embaú, Santa Catarina, Brazil (Bosquetti & de Souza, 2019)	US$61/day	Surfer tourists	US$4.2 million/year
Wrightsville, Carolina, and Kure, North Carolina, USA (Hritz & Franzidis, 2018)	US$76–154/day	Local surfers and visitors	N/A
UK (Mills & Cummins, 2015)	UK£2014/year	UK surfers	UK£1.8 (direct)–UK£4.8 (with multiplayers) billion/year
Huanchaco, Trujillo, Peru (Hodges, 2015)	US$45/day	Surfer tourists	US$0.30 million/year
Todos Santos, Baja California, Mexico (Hodges, 2014)	US$111/day	Surfer tourists	US$/0.75–0.97 million/year
Uluwatu, Bali, Indonesia (Margules et al., 2014)	US$150/day	Surfer tourists	US$35.3 million year
Mundaka, Basque Country, Spain (Murphy & Bernal, 2008)	US$120/day	Local surfers and visitors	US$1–4.5 million/year
Pichilemu, Cardenal Caro, Chile (Wright et al., 2014)	US$168/day	Surfer tourists	US$2–8 million/year
Gold Coast, Queensland, Australia (Lazarow, 2009)	A$1942 /year	Local surfers and visitors	A$126–233 million/year
Trestles, California, USA (Nelsen et al., 2007)	US$25–55/visit	Local surfers and visitors	US$8–13 million/year
South Stradbroke Island, Queensland, Australia (Lazarow, 2007)	A$1775/year	Local surfers and visitors	A$20 million/year
Bastion Point, Victoria, Australia (Lazarow, 2007)	A$3078/year	Local surfers and visitors	A$230,850 /year

millions of dollars every year. Local figures for the Margaret River Pro event, held in April in Australia's south west, estimate the competition, in 2023 alone, attracted 3500 visitors and A$8 million into the region. While the event organization is largely supported by large sponsorship partnerships, substantial economic flows also originate from side activities and spillover effects into local hospitality businesses (Fig. 9.2).

Professional surfing is likely to generate even greater impacts through its declaration as a permanent Olympic Sport in 2022, following a first (temporarily) inclusion in the Tokyo 2020 games. Being part of French Polynesia, the island of Tahiti

Fig. 9.2 Young fans meet WSL Men's Championship Tour professional surfers during the 2024 Margaret River Pro event. Left: Eleven times World Champion Kelly Slater, at event site in Surfers Point. Right (from left to right): Cole Houshmand Kade Matson, CT staff member, Griffin Colapinto, and Crosby Colapinto, at popular winery and restaurant, Xanadu Wines. (Photos: Ana Manero)

hosted the 2024 Olympic surfing event, in what is known as the world's "heaviest, most perfect" wave (Walsh, 2024). Teahupo'o has long been a mecca for expert surfers and keen observers alike, generating an economy of its own. Nevertheless, the extraordinary global attention and infrastructure developments deployed to host the Olympics have raised concerns for irreversibly altering the small village's delicate balance between the surfing economy and traditional ways of life (Hand Studio, 2024).

The surfing industry may become a continuous and long-lasting contributor to local economies, not only through tourism but also through manufacturing and retail. One of the most iconic examples is the town of Torquay, located on the Surf Coast of Victoria, Australia, some 100 km southwest of Melbourne. A 2014 report indicated that surfing contributed to one-quarter of the region's jobs and economic added value, making it the single most important activity sector (AECGroup, 2014). Established in the 1960s by local Torquay entrepreneurs, Rip Curl and Quicksilver have grown to become some of the world's largest and most recognized surfing brands. In 2019, exactly 50 years after its founding, Rip Curl was acquired by major outdoor retailer Kathmandu in a US$350 million deal. Beyond direct financial metrics, iconic surf brands are often associated with a détendu, outdoor lifestyle that can fuel consumer demand and influence broader industry trends, including among a wide non-surfer customer base.

9.3 Understanding Surf Breaks and Waves as Natural Resources

9.3.1 Surf Breaks, Surfing Resources, and Surfing Ecosystems

When thinking of surfing from an environmental and geographical perspective, one can frame the activity at different scales. An avid surfer may focus on a particular surf break and the specific set of local conditions that allow it to generate the best possible wave. Conversely, an officer working for the regional coastal planning authority might be more concerned with the growing number of tourists attracted to the area, while also having to address worsening coastal erosion threatening built infrastructure. Alternatively, an ecologist might wonder how surf charter boats are affecting marine life and how both recreational and biodiversity benefits can be effectively preserved (Touron-Gardic & Failler, 2022). Different scopes call for different understandings and a purposeful definition of surf-related terms. Based on the comprehensive framework developed by Manero and Mach (2023), we provide a brief definition of surf breaks, surfing resources, and surfing ecosystems as follows:

Surf breaks are the discrete locations where waves break and become ridable (Scarfe et al., 2003), typically characterized by specific bathymetric and topographic features, such as sand bars, reefs, or headlands. Surf breaks are consistent with the SEEA EA definition of "ecosystem assets," as the sources of services that support human health and well-being (Hein et al., 2016).

Surfing resources are understood as surf breaks and the related physical factors that allow the practice of the activity (Atkin & Greer, 2019). These include spaces and processes within the adjacent coastline and the "swell corridor" (ocean areas through which swells travel). Importantly, the notion of "surfing resources" sits within the broader framing of "natural resources," i.e., assets present in the environment from which humans benefit with no or few modifications (Keith et al., 2017).

Surfing ecosystems can be defined as surf breaks and their surrounding environments, including physical features, living organisms, humans, and the interactions occurring between all of them (Manero & Mach, 2023). From an environmental economics perspective, "surfing ecosystems" fit within the "ecosystem services" framework (Millennium Ecosystem Assessment, 2005).

These three dimensions are schematically represented in Fig. 9.3, where surf breaks, surfing resources, and surfing ecosystems are represented in a nested model, in alignment with key environmental economics frameworks and concepts.

Through the ecosystem services framework, surfing can be understood as an ecosystem service, that is, an activity that necessarily depends on the natural environment and from which humans derive benefits (Millennium Ecosystem Assessment, 2005). Typically, ecosystem services are classified into four main categories: provisioning (i.e., products obtained from ecosystems, such as food, water,

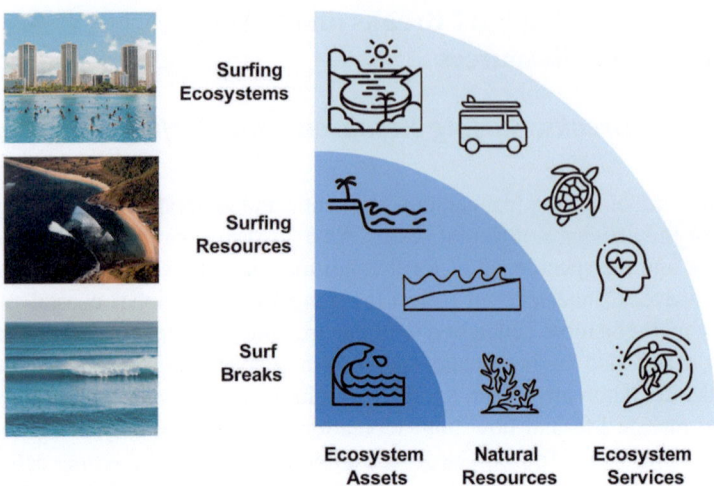

Fig. 9.3 Surf breaks, surfing resources, and surfing ecosystems. (Photos: Jess Loiterton and Nick Wehrli from Pexels)

and raw materials), regulating (i.e., processes that regulate natural phenomena, such as water filtration or carbon sequestration), supporting (i.e., services necessary for the formation of other beneficial processes, such as nutrient cycling or habitat provision), and cultural services (i.e., nonmaterial/intangible benefits derived from interactions with ecosystems, such as recreation and spirituality) (Millennium Ecosystem Assessment, 2005). Surfing ecosystems may provide ecosystem services consistent with all four categories (Manero & Mach, 2023). For instance, reefs yielding surfable waves are often also important habitats for marine species, while large reef formations may help mitigate extreme weather events by dissipating wave energy during severe storms. However, most surf scholars focus on surfing as cultural ecosystem services, for its many "nonmaterial" benefits, including tourism, recreation, health and well-being, social connectedness, and cultural identity, among several others (Román et al., 2022).

9.3.2 Nonmarket Values

In Sect. 9.2.2, we looked at market-based values associated with surfing, which are often relatively easy to calculate as they are based on financial transactions. A different set of values emerges when understanding surfing as a cultural ecosystem service, given that this includes benefits that cannot be bought or sold, such as culture, community connectedness, and personal well-being. These can be understood as our *held values*, i.e., the qualities or end-states that people consider desirable and render the environmental good or service important to us (Brown, 1984). For example, think of the pleasure you obtain while surfing an uncrowded wave in a pristine surf destination. In order to attain those *held values* (personal well-being and environmental preservation), resource users—in our case surfers—will typically be prepared to make some sacrifices in terms of money and time. At every occasion a surfer goes out to catch a wave, whether it is at their local break or one far away, they are making trade-offs, giving up money that could be spent elsewhere and time that could be dedicated to other activities, including working more to earn more money. Surfers consider those sacrifices to be worthwhile, in exchange for the benefits they will obtain once they get in the water. Thus, by quantifying how big or small those sacrifices are, one can infer how much the surf is "worth." This is referred to as *assigned values,* i.e., the worth or relative importance users place on a good or service, within a given context (Brown, 1984). When the goods and services in question cannot be bought or sold, their worth is referred as "nonmarket values."

Nonmarket values vary on a number of factors, including why and how individuals derive benefits from that particular environmental good or service. A common approach to classifying nonmarket values is the "total economic value" framework or TEV (Segerson, 2017). Under the TEV (Fig. 9.4), values are divided into "use" and "nonuse," depending on whether goods or services are being utilized by the person doing the valuation, or not. "Direct use values" are obtained through immediate interaction and can be "consumptive" or "nonconsumptive," depending on whether there is an extraction from the environment (e.g., fishing) or not (e.g., riding or contemplating waves). "Indirect use values" are those whereby benefits occur through an additional process, such as surf breaks mitigating the impact of storms

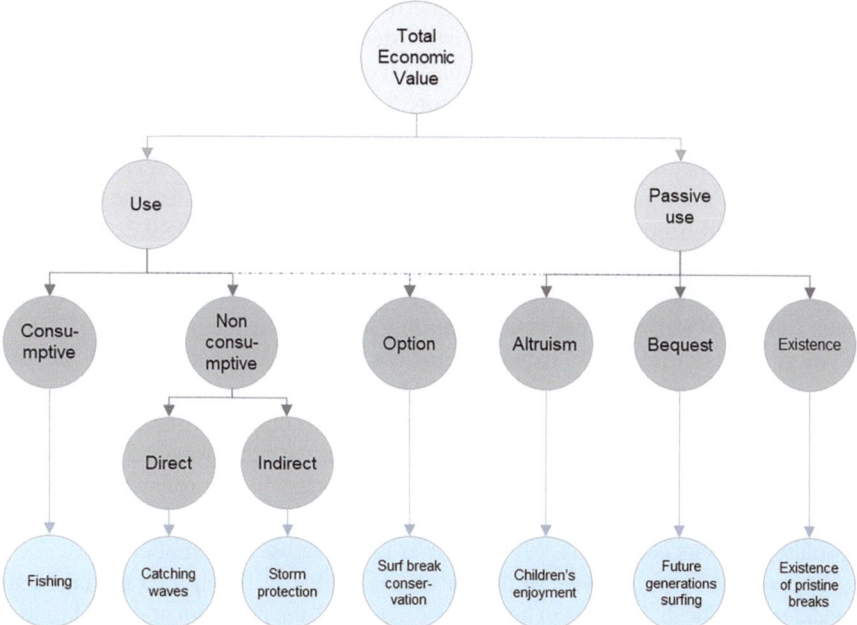

Fig. 9.4 Total economic value framework: surfing nonmarket values with examples

on coastal erosion. "Passive-use" (or "nonuse") values provide individuals with satisfaction from the enjoyment of the resource by others ("altruistic" or "bequest" values) or simply by knowing that natural feature exists ("existence" value). Further, "option" values reflect the worth of avoiding irreversible damages, to allow for possible future benefits for oneself or others (Manero et al., 2022).

To infer the assigned value of goods and services that have no market, economists apply a suite of techniques known as "nonmarket valuation" or NMV (Champ et al., 2017). The fundamental goal is to understand how changes to environmental features (e.g., a wave quality being hindered by sand sediment movement) would result in changes to users' "utility," i.e., the satisfaction or pleasures people derive from using or consuming a particular good and service.

One of the most common methods in nonmarket valuation of recreational surfing is the travel cost model, which estimates the benefit (i.e., utility) surfers derive from a recreation site, based on the travel time and any monetary costs needed to access it, such as petrol or parking fees. Surfers choosing to visit the site would experience a benefit that is greater than what they are actually paying for it (including time and money)—otherwise, the visit would not take place. If the conditions at a local break are not good enough, one may decide to travel farther away to reach a location that has better waves or is less crowded. It will get to a point where the travel effort will be greater than the benefit the surfer expects to get—at which point, the surfer will decide not to take the trip—it is *not worth* the effort. This maximum level of "effort" is referred to as "willingness-to-pay" and can be understood as the equivalent amount of money a user would be prepared to trade, in exchange for the "utility" they derive from the good or service in question (Flores, 2017).

Of course, many surfers choose to live close to desirable surf locations. A surfer living 5 min from the beach may only make an effort equivalent to a few dollars, but the benefit they derive is worth much more to them. In welfare economics, this is referred to as "consumer surplus," i.e., the difference between a user's (theoretical) willingness-to-pay and what they are actually paying for the good or service (Pascoe, 2019). Using the travel cost method, a few studies have estimated the consumer surplus derived from visits to surf breaks, as summarized in Table 9.2. Taking the example of Burnett et al. (2020), the results can be interpreted such that the average surfer visiting Mavericks, in California, experiences a net benefit, in consumer surplus, equivalent to US$57 per visit. According to economic theory (Parsons, 2017), if you gave this surfer US$57 to stay home (as a compensation for foregoing the surf session), he or she would be equally happy. To the extent of our knowledge, no studies have tested this in practice. Asking surfers to stay at home in exchange for payment could be a controversial experiment, yet a very insightful one. Surf scholars would likely be interested in knowing how willingness-to-pay (or willingness-to-accept) changes with the quality of the surf. Would there be a point, perhaps under absolutely perfect conditions, where a surfer would refuse any amount?

In practical terms, a caveat of the travel cost method is that it does not take into consideration the differences in house prices, namely, the premium of coastal living. For seaside residents, time and monetary costs may be low given they only need to walk, or drive a few minutes, and perhaps not even pay for parking, thanks to their

Table 9.2 Summary of studies estimating nonmarket values derived from recreational surfing

Area and source	Method	Individual nonmarket values	Aggregate nonmarket values
Mavericks, California, USA (Burnett et al., 2020)	Travel costs	US$57/person/visit	US$23.8 million/year
New South Wales, Australia (Pascoe, 2019)	Travel costs	A$9–30/person/visit	N/A
Costa da Caparica, Setúbal, Portugal (Silva & Ferreira, 2014)	Travel costs	€47/person/visit	€1.0 m/year
Trestles, California, USA (Nelsen, 2012)	Travel costs	US$114–US$232/person/visit	US 14.6 million/year
Pleasure Point, California, USA (Tilley, 2001)	Travel costs	N/A	US$6.2 m/year
Santa Cruz, California, USA (Scorse et al., 2015)	Hedonic pricing	US$106,000/house/mile	N/A
Portugal (Ramos et al., 2019)	Contingent valuation	€1.6/person/visit (user fee) €15–22/person (one-off payment)	N/A

local permit. However, higher real estate prices in proximity of desirable natural features are a well-documented phenomenon (Brander & Koetse, 2011), although only one published study has evaluated the effect of surf breaks. Scorse et al. (2015) compared differences in house prices in Santa Cruz, California, depending on their distance to the surf break, as well as numerous other characteristics. This approach is known as "hedonic pricing" and aims to estimate values of environmental features that are capitalized in market prices, in this case, house prices (Taylor, 2017). The study by Scorse et al. (2015) found that, controlling for other factors, the benefit that Californian home buyers derive from living nearby the Santa Cruz surf break was equivalent to US$106,000, for each mile (1.6 km) closer the break.

Both travel cost and hedonic pricing methods are based on observable behaviors, such as trips or home purchases. Thus, these methods are jointly known as "revealed preferences," because users' priorities in the face of trade-offs are made explicit through their real-life choices. A different set of nonmarket valuation tools comprises approaches based on reported preferences, typically collected through surveys where participants are asked what their choices would be, when faced with hypothetical scenarios. These techniques are known as "stated preferences" (Champ et al., 2017). Ramos et al. (2019) employed a stated preferences approach, asking Portuguese surfers whether they would agree to pay an access fee to finance the preservation of the area and/or a one-off contribution toward a beach cleanup in the supposed case of an oil spill. When questions are posed as having binary outcomes (e.g., yes or no), the approach is commonly referred to as "contingent valuation" (Boyle, 2017). When multiple options are possible, the method is commonly referred to as "choice models" or "choice experiments," typically offering participants choices among three or more bundles of goods and services, at varying levels of hypothetical changes (Holmes et al., 2017).

In Ramos et al. (2019), respondents who said "yes" were offered payment levels ranging from €1 to €100. This allows for estimating the average "willingness-to-pay" (or WTP), which, in the case of the beach-use fee, was €1.6 per person per visit. Although these figures are based on hypothetical scenarios and non-enforceable payments (participants are not charged at the end of the survey), stated preferences provide decision-makers with important insights into people's priorities and attitudes. For instance, the fee Portuguese surfers stated they would be willing to pay is reasonable and comparable to what one could expect to pay for 1 h of public parking. Knowing users are ready to contribute an additional amount to preserve surfing amenity, local governments could consider rolling out new financing mechanisms. The money collected could be invested in management initiatives to deliver benefits for users and the environment, such as the creation of marine protected areas (Orchard et al., 2023; Touron-Gardic & Failler, 2022; Whitelaw et al., 2014). WTP measures also provide an indication of the potential loss that people would experience if those natural features became affected, thus helping justify the preservation of natural assets so as to safeguard users' utility.

In addition to welfare estimates, a growing body of literature is documenting health and well-being outcomes, with surfing as a form of therapy (Britton et al., 2020; Olive et al., 2023) and among the general population (Manero et al., 2024; Suendermann, 2015), but few do so in economic terms. Buckley and Cooper (2023) calculated the global mental health benefits of surfing at US$0.38–1.30 trillion per year, as an approximation using baseline data sourced from visitations to national parks (Buckley et al., 2019). Priorities for future research could include robust quantification, in monetary terms, of surf-derived health outcomes (Manero et al., 2024), which could serve to evaluate cost-effectiveness of health prevention and promotion initiatives (Le et al., 2021).

Finally, it is important to acknowledge that surfing can also entail negative consequences, both for individuals (e.g., risk of injury) (Chap. 7) and the environment. Numerous accounts exist documenting exacerbated pressures from surf tourism in, otherwise, relatively undisturbed locations (Buckley et al., 2017; Ponting & O'Brien, 2013). Further, the ever-growing demand for equipment and travel inevitably results in growing amounts of nonrecyclable waste and carbon emissions. Although surfers broadly see themselves as nature stewards, significant contradictions arise when examining the sport's environmental footprint, albeit some recent technological innovations toward greater sustainability (Wheaton, 2020).

9.4 Waves as a Common-Pool Resource

In resource economics and environmental management, a key concept lies at the intersection of excludability and rivalry (Adams & McCormick, 1987). Excludability refers to the ability to prevent users from accessing a service or resource, typically enforced through monetary payment or some form of authorization. Thus, a service or resource is considered excludable if it is possible to control and stop people from accessing it. The ocean itself is generally non-excludable, because it is difficult to restrict access to it and typically there are no fees or rules in accessing public coastal spaces. However, some surf breaks are located too far away from the coast to paddle out to, meaning surfers must hire a boat or jet ski to reach them. The need to source and pay for transportation to reach these waves acts as a form of excludability, which could be, potentially, regulated or taxed by local authorities. Buckley et al. (2017) present an illustrative case study of excludability linked to surf tourism in the Maldives. In 2012, two laws were introduced that granted island resorts greater control over access to adjacent surf breaks, including via live-aboard charter vessels, as well as traditional boats used for fishing and transport. As a result, renowned surf breaks such as Pasta Point and Lohis became practically inaccessible to anyone other than the limited number of guests staying at the exclusive resorts laying claim to each of the breaks. While proponents hailed the rules as necessary instruments of surf management plans to control over-tourism, critics viewed these regulations as an avenue toward privatization of public goods, often benefiting foreign investors, rather than local islanders (Mach & Ponting, 2018). Other forms of excludability include localism and territorialism, whereby local surfers purposely behave in such way to discourage others from taking part. This may take the form of "stealing waves" (e.g., breaking surfing etiquette dictating right of way), physical aggression, intimidation (Mach & Ponting, 2018), and even "surf gangs," purposely organized to stop outsiders from surfing at their claimed local spots (Mixon, 2018).

Complementary to "excludability" is the notion of "rivalry." Rivalry explains a relationship between resource users when the use by one individual diminishes the ability of another individual to benefit from the same resource. Surfing is a prime illustration of rivalry, given that when a surfer is riding a wave, that specific wave cannot be enjoyed by another surfer at the same time, without causing interference and even accidents. There are some rare exceptions when some people may agree to (or tolerate) "sharing waves," but even in these cases, most surfers would prefer ride alone. While each wave in itself is a rivalrous resource, a surf break, generating a continuous flow of waves may or may not be rivalrous, depending on how many surfers are competing for the same waves, at any given time.

As Adams and McCormick (1987) elaborate, there is a degree to rivalry, meaning that a resource can be more or less rivalrous depending on whether the demand for the resource (e.g., number of surfers) is greater or smaller than its supply (e.g., number of waves). Competition has been found to be more ferocious in good quality and "big-wave" surf breaks, requiring a greater level of skill and posing greater dangers, compared to regular surf breaks (Kaffine, 2009; Mixon & Caudill, 2017). Within the context of surfing, this is what Ponting and O'Brien (2015) refer to as "social carrying capacity," i.e., the maximum number of surfers at which they are comfortable and the line-up feels uncrowded. Most surfers would relate to a situation where they have been surfing with one or two good friends, perhaps in a secluded location, with "waves for everyone." In the presence of a couple of friends, the surf break would likely be non-rivalrous. However, a few minutes later, the idyllic balance is disrupted by a crowd of local surfers, with little appetite for wave-sharing. The surf break has now become rivalrous. Adams and McCormick (1987) define this kind of resources as "congestible," i.e., those where there is a limit to the number of users who can share a resource in a non-rival manner.

Surf breaks can become congested, even when they are somehow excludable, for instance, in the form of payment for transportation. This is the case of popular destinations, such as the beginner-friendly bay of Gerupuk, in Lombok, Indonesia, where surf-tour boats offload scores of newcomers onto a collection of gently rolling waves, making them perilously overcrowded. At the opposite end of the spectrum in terms of easy waves is Nazaré (Portugal), home to the largest waves on the planet, where, towed by jet skis, professional and skilled amateur surfers fiercely contented for the ride of a lifetime.

The combination of excludability and rivalry results in four main types of resources (or goods): private goods, club goods, common-pool resources, and public goods (see Fig. 9.5 for classification examples):

Private goods are excludable and rivalrous. An obvious example of a private good is a surf board, which one would need to buy or rent and whose use cannot be shared by two people at one time (unless you are tandem surfing). Exclusive, by-invitation only wave pools, such as Kelly Slater's Surf Ranch in California, USA, and Surf Lakes, in Queensland, Australia, could also be understood as private surf goods.

Club goods are excludable and non-rivalrous, such as an artificial wave park (Ponting, 2017). To access these facilities, users pay a fee, which in Melbourne, Australia, range from A$89 upward for an intermediate, 55-min session (URBNSURF, 2024). Interestingly, there is no rivalry at a wave pool because the

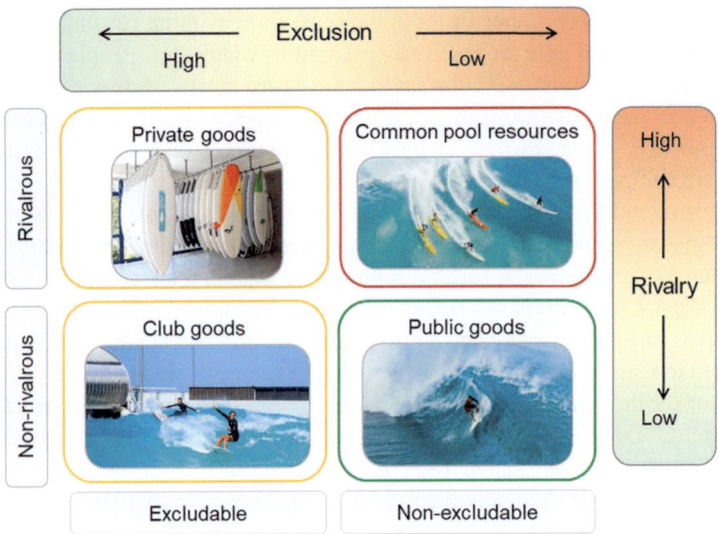

Fig. 9.5 A classification of goods and resources based on exclusion and rivalry. (Image sources Common-pool-resource and public good images by Jess Loiterton from https://www.pexels.com; Private and club goods: David M. Kennedy)

number of users is capped, and all surfers are scheduled to have (practically) the same number of waves within their designated session.

Public goods are non-excludable and have low or no rivalry, such as an uncrowded wave, within a short paddle distance from an open-access beach. Realistically, not many places around the world still offer this setting, with surfers willing to pay large sums to holiday or live near uncrowded lineups. It could be argued that the remoteness and costs associated with such idyllic conditions, by definition, turn those surf breaks into club goods.

Common-pool resources, also referred to as "commons," are non-excludable and rivalrous. The all-too-familiar busy lineup, where every surfer aims to maximize their own benefit, with little respect for etiquette or the safety of others, is a textbook example of the "tragedy of the commons" (Rider, 1998).

The "tragedy of the commons" is a concept dating back to the time of Aristotle but popularized in modern times by Hardin's (1968) essay bearing that title, published in the prestigious journal *Science*. In economics, the "tragedy of the commons" refers to a situation where individuals share access to a non-excludable good or resource (known as "commons") yet decide to act according to their own self-interest, ultimately leading to the degradation of the resource for all users. Within the context of ocean management, a classic case of the "tragedy of the commons" is overfishing (Berkes, 1985). As each fisherman gains direct benefit from every additional catch, the reduction of fish stocks is felt by all, until the entire fish population collapses, to the detriment of all fishermen. Solutions to this conundrum have been proposed, most notably by Ostrom (1990), who was awarded The Sveriges Riksbank Prize in Economic Sciences in Memory of Alfred Nobel 2009 for her analysis of governance of the commons, making her the first woman to win the prestigious prize.

Ostrom (2008) proposed eight key principles for robust, long-term management of the "commons." These include (i) clearly defined boundaries, (ii) proportional equivalence between benefits and cost, (iii) collective-choice arrangements, (iv) monitoring, (v) graduated sanction, (vi) conflict-resolution mechanism, (vii) minimal recognition of rights to organize, and (viii) nested enterprises. While these eight principles have been found to contribute to effective governance of a variety of commons, from grasslands to forests, fisheries, and irrigation systems, their application to surfing resources is inadequate. As a starting point, Ostrom's principles presuppose the existence of a governing institution, such as a fishermen's association acting as a custodian of the shared resources. This is not the case for most waves across the world, which also lack clear boundaries and few robust mechanisms for enforcing rule compliance.

Surf scholars observe that the *surf etiquette* rules, determining who has *right of way,* offer some form of guidance to sharing waves. In a sense, this mirrors Ostrom's proportionality between cots and benefits, arguing that the surfer waiting the longest and positioning him- or herself at the right spot deserves the catch the next wave. However, in congested surf breaks, confusion around "whose turn is it?" and individuals' drive to maximize their own satisfaction often leads to socially suboptimal outcomes (Rider, 1998). These chaotic situations have been found to give rise to informal governance structures, such as surf gangs, aimed at enforcing boundaries and access rights, often based on the principles of localism and camaraderie (Mixon, 2018).

9.5 Conclusion

This chapter explores the intersection of surfing and economics, illuminating a complex and multifaceted linkage and how surfing interacts with local and global economies. Surfing, sometimes perceived merely as a leisure activity, has shown itself to be a significant economic driver, influencing local real estate markets and attracting international tourism. These economic impacts are driven by direct expenditures and flow-on effects catalyzed by surf-related activities. Furthermore, based on an environmental economics approach, this chapter explains a framework through which the nonmarket values of surfing—such as its contributions to community well-being and cultural identity—can be recognized and quantified. This is crucial in today's world where economic rationale often dictates policy decisions. By applying a suite of tools like expenditure analysis and nonmarket valuation, economics can offer evidence-based insights that assist policymakers and community planners in making informed decisions that acknowledge both the financial and "intangible" values provided by surfing ecosystems. The discussion extends to examine how surf breaks and waves are, in fact, a common-pool resource, i.e., excludable and rivalrous, but escaping standard recommendations for effective governance of the commons. Moreover, exacerbated environmental impacts caused by over-visitation and waste from the ever-growing surf industry highlight a contradiction with the, otherwise, environmentally friendly attitudes that characterize the surfing psyche.

The insights provided in this chapter may help surfers better understand the potential that economics offers, as an ally in understanding, managing, and protecting the multiple values of surfing. By knowing how economic valuation tools work, surfers can be better informed to participate in community consultation processes and gain greater agency around how their knowledge is leveraged and for what purpose. Ultimately, being able to effectively translate the feeling of "stoke" into a "tangible" measure may facilitate communications between surfers and coastal planners. Elevating the voices of surfers as key ocean stakeholders could bring due attention to their rights and interests and toward more holistic and sustainable care for our cherished ocean. On the flip side, surfing offers a unique opportunity for economists to study something that is an important driver of macroeconomic activity, such as retail and tourism, but at the same time, a valuable natural asset, resource, and ecosystem, generating billions of dollars' worth in both market and nonmarket values. From a behavioral standpoint, the analysis how surfers make choices on where to go and which waves to paddle for offers a remarkably unique natural experiment of resource allocation and utility maximization, all against the backdrop of breathtaking natural beauty.

Table 9.3 Glossary of key surfing economic terms

Term	Definition	Relevant surf-related references
Common-pool resources (CPR)	Environmental goods or services that are shared among many individual in a non-excludable, rivalrous manner	Rider (1998)
Consumer surplus	The benefit consumers perceive when they pay less for a product or service than what they are actually prepared to pay (i.e., difference between their willingness-to-pay and the price paid)	Scorse and Hodges (2017)
Monetary flows	Financial transactions associated with environmental goods or services, in line with the SEEA EA framework	Manero et al. (2024)
Ecosystem services	Goods and services provided by nature from which humans derive benefits	Aguiar-Quintana et al. (2022)
Environmental footprint	Total impact that surfing activities have on the environment, including resource consumption, waste generation, and changes to coastal ecosystems	Wheaton (2020)
Nonmarket values	Values associated with ecosystem services (e.g., surfing) that are not reflected in market transactions, such as the pleasure of riding a wave or the cultural significance to local communities	Manero and Mach (2023)
Nonmarket valuation	Suite of techniques aimed at quantifying, often in monetary terms, the value assigned to goods and services that cannot be bought or sold	Scorse and Hodges (2017)
SEEA EA	System of Environmental-Economic Accounting—Ecosystem Accounting—a UN-backed, international accounting framework that integrates environmental data with economic accounts, helping to measure the contributions of ecosystems to the economy and how economic activities impact those ecosystems	Manero et al. (2024)
Social carrying capacity	Maximum number of surfers at which they still feel comfortable and uncrowded	Ponting and O'Brien (2015)
Surf breaks	Discrete locations where waves break and become ridable typically characterized by specific bathymetric and topographic features, such as sand bars, reefs, or headlands	Scarfe et al. (2003)
Surfing ecosystems	Surf breaks and their surrounding environments, including physical features, living organisms, humans, and the interactions occurring between all of them	Manero and Mach (2023)
Surfing resources	Surf breaks and the related physical processes that allow the practice of the activity	Atkin and Greer (2019)
Total economic value	A systematic framework for classifying nonmarket values, divided into use and nonuse values	Nelsen (2012)

Glossary (Table 9.3) References

A Liquid Future. (2024). *Surfer scientists*. https://aliquidfuture.org/projects

Adams, R. D., & McCormick, K. (1987). Private goods, club goods, and public goods as a continuum∗. *Review of Social Economy, 45*(2), 192–199.

AECGroup. (2014). Economic value of the surf industry to surf coast shire.
Aguiar-Quintana, T., Román, C., & Gubisch, P. M. M. (2022). The post-COVID-19 tourism recovery led by crisis-resistant tourists: Surf tourism preferences in the Canary Islands. *Tourism Management Perspectives, 44*, 101041.
Aramoana Waiti, J. T., & Awatere, S. (2019). Kaihekengaru: Māori surfers' and a sense of place. *Journal of Coastal Research, 87*(SI), 35–43.
Atkin, E. A., & Greer, D. (2019). A comparison of methods for defining a surf break's swell corridor. *Journal of Coastal Research, 87*(sp1), 70–77.
Backhouse, R. E., & Medema, S. G. (2009). Retrospectives: On the definition of economics. *Journal of Economic Perspectives, 23*(1), 221–233.
Berkes, F. (1985). Fishermen and 'the tragedy of the commons'. *Environmental Conservation, 12*(3), 199–206.
Bosquetti, M. A., & de Souza, M. A. (2019). Surfonomics. Guarda do Embaú, Brazil. The economic impact of surf tourism on the local economy. S. t. W. Coalition. https://www.savethewaves.org/wp-content/uploads/2020/07/GuardaDoEmbau_SurfonomicsStudy.pdf
Bousquette, I. (2024). As surfing makes its olympic debut, Billionaire-owned world surf league hopes to catch a wave of opportunities. Forbes. https://www.forbes.com/sites/isabellebousquette/2021/07/27/as-surfing-makes-its-olympic-debut-billionaire-owned-world-surf-league-hopes-to-catch-a-wave-of-opportunities/
Boyle, K. J. (2017). Contingent valuation in practice. In P. A. Champ, T. C. Brown, & K. Boyle (Eds.), *A primer on nonmarket valuation* (pp. 83–131). Springer.
Brander, L. M., & Koetse, M. J. (2011, October 1). The value of urban open space: Meta-analyses of contingent valuation and hedonic pricing results. *Journal of Environmental Management, 92*(10), 2763–2773.
Branigan, J., & Ramezani, F. (2018). Assessing the value of public infrastructure at a regional level: Cost benefit analysis supplemented by economic impact analysis. *The Australasian Journal of Regional Studies, 24*(2), 147–167.
Britton, E., Kindermann, G., Domegan, C., & Carlin, C. (2020, Feb 1). Blue care: A systematic review of blue space interventions for health and wellbeing. *Health Promotion International, 35*(1), 50–69.
Brown, T. C. (1984). The concept of value in resource allocation. *Land Economics, 60*(3), 231–246.
Buckley, R. C., & Cooper, M.-A. (2023). Mental health contribution to economic value of surfing ecosystem services. *npj Ocean Sustainability, 2*(1), 20.
Buckley, R. C., Guitart, D., & Shakeela, A. (2017). Contested surf tourism resources in the Maldives. *Annals of Tourism Research, 64*, 185–199.
Buckley, R., Brough, P., Hague, L., Chauvenet, A., Fleming, C., Roche, E., Sofija, E., & Harris, N. (2019, November 12). Economic value of protected areas via visitor mental health. *Nature Communications, 10*(1), 5005.
Burnett, P., Vardon, M., Keith, H., King, S., & Lindenmayer, D. (2020). Measuring net-positive outcomes for nature using accounting. *Nature Ecology & Evolution, 4*(3), 284–285.
Champ, P. A., Boyle, K., & Brown, T. C. (Eds.). (2017). *A primer on nonmarket valuation* (2nd ed.). Springer.
Earhart, N. H. W. (2015). *The effects of surf-driven development on the local population of Playa Gigante, Nicaragua (Publication Number 10017883)*. (M.A., University of Denver). ProQuest One Academic.
Ekstedt, H. (2012). *Money in economic theory*. Taylor & Francis Group.
European Commission, International Monetary Fund, Organisation for Economic Co-operation and Development, United Nations, & World Bank. (2008). System of National Accounts.
Flores, N. E. (2017). Conceptual framework for nonmarket valuation. In P. A. Champ, K. Boyle, & T. C. Brown (Eds.), *A primer on nonmarket valuation* (2nd ed., pp. 27–54). Springer.
Hand Studio. (2024). *Children of Teahupo'o* https://www.youtube.com/watch?v=2m_kXi1w3YI
Hanley, N., Shogren, J., & White, B. (2019). *Introduction to environmental economics* (3rd ed.). Oxford University Press.

Hardin, G. (1968). The tragedy of the commons. *Science, 162*(3859), 1243–1248.

Hein, L., Bagstad, K., Edens, B., Obst, C., de Jong, R., & Lesschen, J. P. (2016). Defining ecosystem assets for natural capital accounting. *PLoS One, 11*(11), e0164460.

Hodges, T. (2014). Impacto Económico del Surf en la Bahía de Todos Santos, Baja California, México. S. t. W. Coalition.

Hodges, T. (2015). The economic impact of surfing in Huanchaco world surfing reserve, Peru. S. t. W. Coalition. https://www.savethewaves.org/wp-content/uploads/2020/07/HuanchacoSurfonomicsStudy_SaveTheWaves.pdf

Holmes, T. P., Adamowicz, W., & Carlsson, F. (2017). Choice experiments. In P. A. Champ, K. J. Boyle, & T. C. Brown (Eds.), *A primer on nonmarket valuation*. Springer.

Hritz, N., & Franzidis, A. F. (2018). Exploring the economic significance of the surf tourism market by experience level. *Journal of Destination Marketing & Management, 7*, 164–169.

Joseph, C., Gunton, T., Knowler, D., & Broadbent, S. (2020). The role of cost-benefit analysis and economic impact analysis in environmental assessment: The case for reform. *Impact Assessment and Project Appraisal, 38*(6), 491–501.

Kaffine, D. T. (2009). Quality and the commons: The Surf Gangs of California. *The Journal of Law and Economics, 52*(4), 727–743.

Keith, H., Vardon, M., Stein, J. A., Stein, J. L., & Lindenmayer, D. (2017, November 1). Ecosystem accounts define explicit and spatial trade-offs for managing natural resources. *Nature Ecology & Evolution, 1*(11), 1683–1692.

Lanagan, D. (2002). Surfing in the third millennium: Commodifying the visual argot. *The Australian Journal of Anthropology, 13*(3), 283–291.

Lazarow, N. (2007). The value of coastal recreational resources: A case study approach to examine the value of recreational surfing to specific locales. *Journal of Coastal Research, 50*(50), 12–20.

Lazarow, N. (2009). Using observed market expenditure to estimate the value of recreational surfing to the Gold Coast, Australia. *Journal of Coastal Research, 56*(Special Issue 56), 1130–1134.

Le, L. K.-D., Esturas, A. C., Mihalopoulos, C., Chiotelis, O., Bucholc, J., Chatterton, M. L., & Engel, L. (2021). Cost-effectiveness evidence of mental health prevention and promotion interventions: A systematic review of economic evaluations. *PLoS Medicine, 18*(5), e1003606.

Lee, K.-H., Noh, J., & Khim, J. S. (2020). The blue economy and the United Nations' sustainable development goals: Challenges and opportunities. *Environment International, 137*, 105528.

Mach, L. (2019). Surf-for-development: An exploration of program recipient perspectives in Lobitos, Peru. *Journal of Sport and Social Issues, 43*(6), 438–461.

Mach, L., & Ponting, J. (2018, November 2). Governmentality and surf tourism destination governance. *Journal of Sustainable Tourism, 26*(11), 1845–1862.

Mach, L., & Ponting, J. (2021). Establishing a pre-COVID-19 baseline for surf tourism: Trip expenditure and attitudes, behaviors and willingness to pay for sustainability. *Annals of Tourism Research Empirical Insights, 2*(1), 100011.

Manero, A. (2023, June 13). A case for protecting the value of 'surfing ecosystems'. *npj Ocean Sustainability, 2*(1), 6.

Manero, A., George, P., Yusoff, A., Olive, L., & White, J. (2024). Understanding surfing as a 'blue space' activity for its contributions to health and wellbeing. *npj Ocean Sustainability, 3*(1), 37. https://doi.org/10.1038/s44183-024-00076-4.

Manero, A., & Mach, L. (2023). Valuing surfing ecosystems: An environmental economics and natural resources management perspective. *Tourism Geographies, 25*(6), 1602–1629.

Manero, A., Taylor, K., Nikolakis, W., Adamowicz, W., Marshall, V., Spencer-Cotton, A., Nguyen, M., & Grafton, R. Q. (2022). A systematic literature review of non-market valuation of Indigenous peoples' values: Current knowledge, best-practice and framing questions for future research. *Ecosystem Services, 54*, 101417.

Manero, A., Yusoff, A., Lane, M., & Verreydt, K. (2024). A national assessment of the economic and wellbeing impacts of recreational surfing in Australia. *Marine Policy, 167*(September 2024), 106267.

Margules, T., Ponting, J., Lovett, E., Mustika, P., & Pardee Wright, J. (2014). Assessing direct expenditure associated with ecosystem services in the local economy of Uluwatu, Bali, Indonesia. S. t. W. Coallition. https://www.savethewaves.org/wp-content/uploads/Bali_Surfonomics_Final%20Report_14_11_28_nm.pdf.

McGregor, T., & Wills, S. (2017). *Surfing a wave of economic growth* (CAMA Working Paper No. 31/2017.

McKenna, J., Cooper, A., & O'Hagan, A. M. (2008). Managing by principle: A critical analysis of the European principles of Integrated Coastal Zone Management (ICZM). *Marine Policy, 32*(6), 941–955.

McKinley, E., Burdon, D., & Shellock, R. (2023). The evolution of ocean literacy: A new framework for the United Nations Ocean Decade and beyond. *Marine Pollution Bulletin, 186*, 114467.

Millennium Ecosystem Assessment. (2005). *Ecosystems and human well-being: Current state and trends, volume 1*. Island Press. https://www.millenniumassessment.org/documents/document.766.aspx.pdf

Mills, B., & Cummins, A. (2015). An estimation of the economic impact of surfing in the United Kingdom. *Tourism in Marine Environments, 11*(1), 1–17.

Mixon, F. G. (2018). Camaraderie, common pool congestion, and the optimal size of surf gangs. *Economics of Governance, 19*(4), 381–396.

Mixon, F. G., & Caudill, S. B. (2017). Guarding giants: Resource commons quality and informal property rights in big-wave surfing. *Empirical Economics, 54*(4), 1697–1715.

Monteferri, B., Scheske, C., & Muller, M. R. (2020). *The legal protection of surf breaks: An option for conservation and development* (Vol. 1, 1st ed., pp. 149–162). Routledge.

Murphy, M., & Bernal, M. (2008). The impact of Surfing on the local economy of Mundaka, Spain. S. t. W. Coalition. https://savethewaves.org/wp-content/uploads/2020/07/MundakaSpain_SurfonomicsStudy.pdf

Nelsen, C. E. (2012). Collecting and using economic information to guide the management of coastal recreational resources in California University of California. http://public.surfrider.org/files/nelsen/Nelsen_2012_CA_beachsurfecon_dissertation.pdf

Nelsen, C., Pendleton, L., & Vaughn, R. (2007). A socioeconomic study of surfers at trestles beach. *Shore and Beach, 75*. https://ref.coastalrestorationtrust.org.nz/site/assets/files/8007/nelsen_2007_trestleseconimpact.pdf

Olive, R., & Wheaton, B. (2020). Understanding blue spaces: Sport, bodies, wellbeing, and the sea. *Journal of Sport and Social Issues, 45*(1), 3–19.

Olive, L., Dober, M., Mazza, C., Turner, A., Mohebbi, M., Berk, M., & Telford, R. (2023, March 1). Surf therapy for improving child and adolescent mental health: A pilot randomised control trial. *Psychology of Sport and Exercise, 65*, 102349.

Orchard, S., Reiblich, J., & dos Santos, M. D. (2023). A global review of legal protection mechanisms for the management of surf breaks. *Ocean & Coastal Management, 238*, 106573.

Ostrom, E. (1990). *Governing the commons: The evolution of institutions for collective action*. Cambridge University Press.

Ostrom, E. (2008). Design principles of robust property-rights institutions: What have we learned. In D. H. Cole & M. D. McGinnis (Eds.), *Elinor Ostrom and the Bloomington school of political economy: Resource governance* (Vol. 2, pp. 215–248). Lexington Books/Fortress Academic.

Parsons, G. R. (2017). Travel cost models. In P. A. Champ, T. C. Brown, & K. J. Boyle (Eds.), *A primer on nonmarket valuation* (2nd ed.). Springer.

Pascoe, S. (2019). Recreational beach use values with multiple activities. *Ecological Economics, 160*, 137–144.

Pelletier, M.-C., Heagney, E., & Kovač, M. (2021). Valuing recreational services: A review of methods with application to New South Wales National Parks. *Ecosystem Services, 50*, 101315.

Poitras, G. (2003). *A Brief History of Economics: Artful Approaches to the Dismal Science* (Vol. 65). Blackwell Publishing.

Ponting, J. (2017). Simulating nirvana: Surf parks, surfing spaces, and sustainability. In G. Borne & J. Ponting (Eds.), *Sustainable surfing* (pp. 219–237). Routledge.

Ponting, J., & O'Brien, D. (2013). Liberalizing Nirvana: an analysis of the consequences of common pool resource deregulation for the sustainability of Fiji's surf tourism industry. *Journal of Sustainable Tourism, 22*(3), 384–402.

Ponting, J., & O'Brien, D. (2015). Regulating "Nirvana": Sustainable surf tourism in a climate of increasing regulation. *Sport management review, 18*, 99–110.

Ramos, P., Pinto, L. M. C., Chaves, C., & Formigo, N. (2019). Surf as a driver for sustainable coastal preservation—An application of the contingent valuation method in Portugal. *Human Ecology, 47*(5), 705–715.

Reimann, L., Vafeidis, A. T., & Honsel, L. E. (2023). Population development as a driver of coastal risk: Current trends and future pathways. *Cambridge Prisms: Coastal Futures, 1*(e14), 1–12.

Reineman, D. (2016). The utility of surfers' wave knowledge for coastal management. *Marine Policy, 67*, 139–147.

Rider, R. (1998). Hangin' ten: The common-pool resource problem of surfing. *Public Choice, 97*(1/2), 49–64.

Román, C., Borja, A., Uyarra, M. C., & Pouso, S. (2022). Surfing the waves: Environmental and socio-economic aspects of surf tourism and recreation. *Science of the Total Environment, 826*, 154122.

Scarfe, B., Elwany, M. H. S., Mead, S. T., & Black, K. (2003). The science of surfing waves and surfing breaks – A review. *Integrative Oceanography Division.* https://escholarship.org/uc/item/6h72j1fz

Scorse, J., & Hodges, T. (2017). The non-market value of surfing and its body policy implications. In G. Borne & J. Ponting (Eds.), *Sustainable surfing* (1st ed., pp. 137–143). Taylor & Francis Group.

Scorse, J., Reynolds, F., & Sackett, A. (2015). Impact of surf breaks on home prices in Santa Cruz, CA. *Tourism Economics, 21*(2), 409–418.

Scott, S. Q., & Rogers, S. H. (2018). Surf's up? How does water quality risk impact surfer decisions? *Ocean & Coastal Management, 151*, 53–60.

Segerson, K. (2017). Valuing environmental goods and services: An economic perspective. In P. Champ, K. Boyle, & T. Brown (Eds.), *A primer on nonmarket valuation. The economics of non-market goods and resources* (Vol. 13). Springer.

Silva, S. F., & Ferreira, J. C. (2014). The social and economic value of waves: An analysis of Costa de Caparica, Portugal. *Ocean and Coastal Management, 102*(PA), 58–64.

Suendermann, S. (2015). *Beyond the waves: Exploring the social value of surfing to the Surf Coast community.* http://www.actionsportsfordev.org/assets/Uploads/Item-44-Appendix-1-Social-Value-of-Surfing-1.pdf

Sylla, M., Harmáčková, Z. V., Grammatikopoulou, I., Whitham, C., Pártl, A., & Vačkářová, D. (2021, August 1). Methodological and empirical challenges of SEEA EEA in developing contexts: Towards ecosystem service accounts in the Kyrgyz Republic. *Ecosystem Services, 50*, 101333.

Taylor, L. O. (2017). Hedonics. In P. A. Champ, K. J. Boyle, & T. C. Brown (Eds.), *A primer on nonmarket valuation* (pp. 235–292). Springer.

Tilley, C. F. (2001). A valuation of the pleasure point surf-zone in Santa Cruz, CA using travel cost modeling California State University. https://digitalcommons.csumb.edu/caps_thes_restricted/174

Touron-Gardic, G., & Failler, P. (2022). A bright future for wave reserves? *Trends in Ecology & Evolution, 37*(5), 385–388.

Tuda, A. O., Stevens, T. F., & Rodwell, L. D. (2014, January 15). Resolving coastal conflicts using marine spatial planning. *Journal of Environmental Management, 133*, 59–68.

United Nations. (2021). System of environmental-economic accounting—Ecosystem accounting (SEEA EA). White cover publication, pre-edited text subject to official editing. https://seea.un.org/ecosystem-accounting

United Nations. (2024a). *National accounts publications.* Retrieved 06 June 2024 from https://unstats.un.org/unsd/nationalaccount/pubs.asp

United Nations. (2024b). *System of Environmental-Economic Accounting (SEEA)*. Retrieved 06 June 2024 from https://seea.un.org/

URBNSURF. (2024). *Intermediate surf sessions*. Retrieved June 04 from https://support.urbnsurf.com/hc/en-us/categories/360002244411-Intermediate-Surf-Sessions

Usher, L. E., & Kerstetter, D. (2015, January 1). Re-defining localism: an ethnography of human territoriality in the surf. *International Journal of Tourism Anthropology, 4*(3), 286–302.

Walsh, D. (2024, May 24). 'Eight-storey building onto a four-foot reef': Why Teahupo'o is so perfect ... and terrifying. Sydney Morning Herald. https://www.smh.com.au/sport/eight-storey-building-onto-a-four-foot-reef-why-teahupo-o-is-so-perfect-and-terrifying-20240520-p5jf59.html

Wheaton, B. (2020). Surfing and environmental sustainability. In B. Wilson & B. Millington (Eds.), *Sport and the environment* (Vol. 13, pp. 157–178). Emerald Publishing Limited.

Whitelaw, P. A., King, B. E. M., & Tolkach, D. (2014, May 19). Protected areas, conservation and tourism—Financing the sustainable dream. *Journal of Sustainable Tourism, 22*(4), 584–603.

Wright, J., Hodges, T., & Sadrpour, N. (2014). Economic impact of surfing on the economy of Pichilemu, Chile. S. t. W. Coalition. https://www.savethewaves.org/wp-content/uploads/2020/07/Pichilemu_SurfonomicsStudy_SaveTheWaves.pdf

Chapter 10
Surf Tourism

Danny O'Brien

10.1 Introduction

Surfing has always been deeply intertwined with the phenomenon of travel (Chaps. 9 and 13). Warshaw (2017) noted that the relationship between surfing and travel is as old as the sport's ancient Polynesian roots (Chap. 2), and today, surfers continue to seek out new, more challenging and preferably uncrowded waves, often in culturally unique and geographically remote locations. For the purposes of this chapter, I follow the lead of Martin (2022) who, in turn, followed the lead of the International Surfing Association (ISA) that defines "surfing" to the act of waveriding on a shortboard, longboard, bodyboard, or stand-up paddle board and also includes the sub-disciplines of big-wave and tow-in surfing.

In defining surf tourism, Towner (2016, p. 63) stated simply that "Surfing becomes tourism as soon as surfers travel away from their local surf break, with riding waves as the primary purpose for travel." Dolnicar and Fluker (2003, p. 187) went a bit further, explaining that surf tourists travel to:

> … either domestic locations for a period of time not exceeding six months, or international locations for a period of time not exceeding 12 months, who stay at least one night, and where the surfer relies on the power of the wave for forward momentum as the primary motivation for destination selection.

By setting temporal boundaries and establishing surfing as the primary motivation for travel, this definition excludes people whose travel primarily relates to issues like employment, family, or attending an event. So, to be clear, we are talking about surfers who travel expressly to surf, not travelers who might choose to surf or do a surfing lesson while on a holiday or in town for work, an event, or family

D. O'Brien (✉)
Bond University, Robina, QLD, Australia
e-mail: daobrien@bond.edu.au

© The Author(s), under exclusive license to Springer Nature
Switzerland AG 2025
D. M. Kennedy (ed.), *The Science and Culture of Surfing*,
https://doi.org/10.1007/978-3-031-80979-8_10

reasons. Buckley (2002, p. 414) put it this way: "A surf tourist is a surfer first and a tourist second." We should also delineate between domestic and international surf tourism—both key aspects of the surf tourism economy that we explore later in this chapter. I should point out here that this chapter is not a literature review on surf tourism research. The aim is more general than that, to simply discuss the nature of modern surf tourism and then to look down the line and discuss what might be some future challenges and opportunities for surf tourism stakeholders.

Martin and Assenov (2012) and Martin (2022) systematically reviewed the surf tourism literature from 1997 to 2011 and 2011 to 2020, respectively, and demonstrated the inherently interdisciplinary nature of surf tourism. In the latter work, Martin reviewed 96 articles published across dozens of academic journals:

> ... with 12 journals focusing on management, planning and policy, eight journals in the environmental field, seven journals in the area of sport, six journals in the area of history, and four journals in the area of economics and marketing. Other areas of interest include four geography journals and two in anthropology. (Martin, 2022, p. 131)

This interdisciplinarity is no surprise when one considers surfers' predilection for quality, uncrowded surf combined with unique cultural experiences. Surfers' very specific travel appetite has them exploring remote corners of often developing countries in search of undiscovered, quality breaks and the elusive surfing nirvana. Historically, surfers have been responsible for inadvertently opening up and bringing to global attention formerly lesser known places like Margaret River, Byron Bay, and Bells Beach in Australia; Bali and the Mentawai Islands in Indonesia; Teahupo'o in Tahiti; Mundaka in Spain; Nazare in Portugal; Jeffrey's Bay in South Africa; and the Maldives—and the list goes on (Fig. 10.1).

For some destinations, the global attention brings welcome economic development. However, on the whole, the track record of surf tourism destinations has not been particularly positive (Towner & Davies, 2019). Indeed, surfers' tendency for venturing off the beaten track into isolated regions prompted Ponting et al. (2005) to describe surf tourism as "a colonizing activity," which has nudged unprepared

Fig. 10.1 Surfing in the Telo Islands, Indonesia. Waves here are for the intermediate to more advanced surfer. (Photos: Danny O'Brien)

destinations "down the slippery slope to large scale industrialized tourism and its related issues" (p. 152). And alarmingly, going completely against the grain of the surfing subculture's mantra of "sharing the stoke," in some cases surf tourism operators have discouraged or actively excluded host community members from even learning the sport, lest they crowd the lineup for visiting surf tourists (Mach, 2019; Towner, 2015). O'Brien and Ponting (2013) explain that:

> ... surf tourism has traditionally been characterized by unregulated, free-market development with unrestricted growth and little or no consultation with host communities. Far from sport acting as a lever for development, this approach led to overcrowding and deleterious economic, social and environmental impacts on host communities. (p. 160)

The delicate balancing act in managing the nexus of surf tourism, development, and sustainability goes some way to explaining the high levels of interdisciplinary interest in the field.

Fueling the growing popularity of surf tourism is the fact that surfing's global participant base continues to grow and diversify as technological advancement in surfboard design and construction enables previously marginalized demographic groups such as women, ethnic minorities, older people, children, and people with a disability more equitable access to the sport. Add to this the advent of surf parks, accurate and accessible surf forecasting (Chap. 3), the inclusion of surfing on the Olympic program, and the unrelenting commodification of the surfing subculture [see Chap. 9), and we see consistent growth in the numbers of people taking up the sport. Indeed, Statista Research Department (2024) reports that in the USA alone, surfing participation grew 8% from 2022 to reach a figure of 4,000,000 participants in 2023. Meanwhile, in Australia during the pandemic, surfing participation grew by 196,000 people, 60% of whom were women and girls (AusPlay, 2022). And despite continued perceptions of surfing as a youth sport, AusPlay (2022) reported that peak participation in surfing is actually in the 45–54-year-old age category. This democratization in both the number and diversity of surfing presents uncertainties, opportunities, and challenges, but fueled by more accessible travel, one thing is certain: more people are making plans for their next surf trip, making surf tourism a significant niche within the wider adventure tourism sector (O'Brien & Ponting, 2013, 2018; Ponting, 2009). In this chapter, I explore some of the key considerations related to surf tourism and then peer down the line to discuss what types of issues might be worth considering for the future.

10.2 The Nature of Surf Tourism

So, you want to take a surf trip? Well, you have a few things to consider: What's your skill level? Do you want to stay in-country or travel overseas? Will you travel with family members, with a friend group, or solo? Are you willing to handle the travel arrangements yourself, or will you use one of the many specialist surf tourism travel agencies? What's your budget? Would you like to be land-based in a surf camp/resort or ocean-based aboard a specialty surf charter boat, or perhaps some combination of both? It would be naive to presume that I could cover *all* surf tourism options here, but these are some of the main considerations that I elaborate upon below.

Perhaps the first consideration in organizing a surf trip is to make honest judgements around the surfing ability of the individual surf tourist/s. Finding a happy medium between those who want rolling, soft waves and those keen on heavier, more challenging waves is essential because established surf destinations are typically known for the types of waves they offer at different times of year. So, a little bit of up-front research will heighten the chances of a safe and enjoyable surf trip. For example, a place like Weligama in Sri Lanka is perfect for novice or older surfers who might seek easy, rolling beach-break waves (Fig. 10.2). But this same destination might be considered boring by those seeking heavier, more challenging waves such as those found at Kanduis in Indonesia's Mentawai Islands that, like many other breaks in the region, offers long barreling waves breaking over shallow reef that in-season (May to September) can handle waves up to triple overhead.

Alongside these considerations is deciding on overseas versus domestic surf travel. For most surfers, domestic surf trips take the form of overnight or weekend getaways, sometimes multiple times a year, and/or surf vacations around annual holiday periods, perhaps camping or renting accommodation and often with family and/or friend groups. Ponte et al. (2021) reported that 96.2% of adventure tourism participants prefer to travel with at least one companion. Surf tourism would appear to fit here—Manero et al. (2024) found that 73% of Australian surfers enjoy surfing

Fig. 10.2 A calmer break in Sri Lanka is more suited to the novice surfer. (Photos: Jane Dyson)

with friends and acquaintances, while 60% of respondents surfed with their children and/or spouses. Further, Porter and Usher (2018) noted that 19% of surfers travel with non-surfing children, partners, or spouses, all of which amplifies both the positive and negative impacts associated with surf tourism. Interestingly, Mach and Ponting (2021) identified domestic surf tourism as "a large, understudied market of unknown value" (p. 6). Of the scant research available, Wagner et al.'s (2011) socio-economic and recreational profile of US surfers established that the average American surfer travels at least 10 miles to surf around 100 times per year, spending around $40 each time, while Mach et al. (2018) found that 58% of American surfers drive upward of 40 km at least once annually in response to favorable surf forecasts. More recently, Manero et al.'s (2024) survey of Australian surfers (n = 569) found that 81% of respondents took at least 1 domestic surf trip per year (mean: 4.8 trips), staying an average of 7.2 nights away and spending an average of AU$1861 per year (Chap. 9).

In contrast to domestic surf tourism, much more is known about international surf tourism. Researchers from San Diego State University's Center for Surf Research (CSR) surveyed surfers across a range of issues related to surf travel and sustainability and presented their results at the inaugural Sustainable Stoke Conference at San Diego State University in September of 2015 (O'Brien & Ponting, 2018). Ultimately, 2994 viable responses were yielded from 68 countries of origin. Respondents revealed that, on average, they visited 3–5 different countries specifically to surf between 1 and 5 times over the previous 5 years and averaged a US$2500 spend on trips of 8–14 nights' duration. Sixteen percent of the sample had taken more than 6 international surf trips in the previous 5 years, and 2% made more than 20 international surf trips in the same period.

Hritz and Franzidis (2018) found that surfers normally travel in groups of two to four people and are often repeat visitors to the destination. Unsurprisingly, they also found that more experienced surfers tend to travel more frequently and spend more money per day. More recent figures are equally interesting and highlight the potential multiplier effects of surf tourism. Statista Research Department (2022a) reported the global surfing market (apparel and hardware such as surfboards, fins, leashes, tail-pads, wetsuits, etc.) was worth US$4 billion in 2020 and forecast to grow to US$4.8 billion by 2027. Meanwhile, Statista Research Department (2022b) valued the global surf tourism industry in 2022 at US$9.5 billion, expected to grow to US$17.1 billion by 2032. The second figure of US$9.5 billion (Statista Research Department, 2022b) is purely for surf tourism, and at more than double their figure of US$4 billion for the surfing market suggests considerable economic multipliers at work. Mach and Ponting's (2021) work suggests these figures may be conservative, estimating global surf tourism expenditure to be somewhere between US$31.5 and US$64.9 billion annually. While the figures reported above vary significantly, one thing is clear: international surf tourism is, "… a significant, yet overlooked, tourism niche" (Mach & Ponting, 2021, p. 6).

Once the decision is made on an international surf trip, the next decision is whether to handle the travel logistics yourself or to use one of the numerous travel agencies that specialize in surf tourism. Although modern surf tourism began in the late 1960s with mostly young male, self-guided independent travelers who roamed the world by whatever means necessary and slept rough or stayed in village homestays, the modern industry consists of thousands of small surf camps and resorts, live-aboard surf charter boats, wholesalers, retailers, and logistics handlers operating throughout the world (Barbieri & Sotomayor, 2013) (Chap. 13). Unlike "normal" tourism, at the core of surf tourism are very small owner-operated enterprises that typically only accommodate a maximum of 20 guests at a time, usually far less (O'Brien & Ponting, 2018). As surf tourists are comparatively cash rich but time poor (O'Brien & Ponting, 2013), surf travel agencies save them the hassle of gathering travel information from a plethora of small operators in remote locations, acting as conduits that connect these disparate components of the industry. Barbieri and Sotomayor (2013) summed it up this way:

> Although surf tourism started as a self-guided adventure driven by the quality of the surfing experience in other regions or climates (e.g., wave height and period, swell direction, tide), the majority of current surf travellers are no longer backpackers with plenty of free time, but travellers relying on surf tour operators to help them coordinate their travel arrangements and find the perfect wave. (p. 112)

Surf travel agencies operate differently in different countries, and this is dependent on regional market characteristics. For example, in mature surf tourism markets like Australia and the USA, well-established agencies like WorldSurfaris, Surf the Earth, Perfect Wave Travel, Soul Surf Travel, Wavehunters Global Surf Travel, and Water Ways Surf Adventures deal specifically in surf tourism and cater to a knowledgeable market of mostly experienced and proficient surfers. Meanwhile, in Europe, the tendency is for agencies to cater to a less core surf-specific market with many surf tourism packages including learn-to-surf lessons and board hire; some also couch surf tourism with other adventure sports and "mass" tourism products. For example, Luex offers both surf and snow tourism experiences; Zocotravel deals in surf, kitesurf, and windsurfing tourism; and 360 Degree Sport Travel Agency offers surf, ski, and snowboard tourism. These are but a selection of the companies servicing this niche market, but what they essentially all do is simplify the logistics of surf travel by offering a varied suite of bundled accommodation, transport, and insurance options at various price points for individuals and groups.

At the core of the commercial surf tourism experience are the camps/resorts and charter boats that surf tourists select for their surf tourism experience. So, why would a surf tourist choose a particular surf destination? And more specifically, why choose a land camp or resort over a surf charter, or vice versa? There are many answers here. Reis et al. (2022) found that the overall image of a surf destination, and particularly the strength of local surf culture, has a positive influence on surf tourists' loyalty and repeat visitation. But the core attraction will always come down to preferences based on wave quality and variety, the type of experience sought, and price point.

Interestingly, to understand the attributes that surf tourists base their destination choices on, Barbieri and Sotomayor (2013) found that "Abundance of Good Waves," "Variety of Wave Types," and "Area Never Crowded" were rated as Important or

Very Important by 96.5, 78.6, and 76.9% of respondents, respectively. These figures far overshadowed other destination attributes such as infrastructure. So, getting back to the question posed in the previous paragraph—why would a surf tourist choose a land camp over a surf charter, or vice versa—Barbieri and Sotomayer's work provides empirical evidence to suggest that above all other variables, surfers seek a variety of quality, uncrowded waves. Private surf charter boats are typically more expensive than land camps but offer greater mobility to venture further afield in search of a variety of wave types to give the surf tourist a much more diverse surfing experience. So, for surfers with a higher budget and who are more serious about the surfing rather than cultural aspects of their trip, surf charters are probably the way to go. Meanwhile, land camps offer a more immersive experience in the one location, enabling the surf tourist added opportunities to interact and become familiar with not only local surf breaks but also the host community and their culture. Demonstrating the evolution of surf tourism as a sector, within both surf charter and land-based options, there now exists an array of choice from budget-conscious, backpacker-style accommodation through to high-end luxury options.

10.3 Current and Future Considerations in Surf Tourism

The surf tourism sector is now far removed from its humble 1960s beginnings of nomadic young (mostly) men eking out a feral existence in rented fishing boats and jungle tree houses. So, in this section, we explore how some of the current changes both in surfing itself and society at large might influence the future of the surf tourism sector. To that end, we discuss changing business models and scalability issues in commercial surf tourism, the rise of "soft" surf tourism, public sector interest in surf tourism, and ongoing issues related to growth and sustainability.

10.3.1 Commercial Surf Tourism and Scalability

A prime example of how far the surf tourism sector has evolved from its humble beginnings is encapsulated in Resort Latitude Zero, a luxury surf resort in the Telo Islands, Indonesia. Owner Matt Cruden achieved legend status as one of the first surf charter operators in the Mentawai Islands back in the early 1990s with his boat, the *Mangalui Ndulu* ("wave searcher" in the local Nias dialect) (Fig. 10.3). By the mid-2000s, however, with a growing family, Matt and his wife Jen decided it was time to switch to life on land, and in 2009, they accepted their first guests at Resort Latitude Zero in the Telo Islands off the coast of North Sumatra, Indonesia (M. Cruden, personal communication, July 22, 2024). More recently, they have also opened a second Resort Latitude Zero on the island of Sumbawa, in Indonesia's Nusa Tenggara region. With over 30 years of accumulated local knowledge, they

Fig. 10.3 Resort Latitude Zero in Indonesia is a luxury tourism destination for surfers in the Telo Islands. (Photos (clockwise): Simon Williams, Mike Egan, Mick Curley, Mike Egan)

offer guests the option of land-based surf trips at either of their high-end luxury resorts or purely ocean-based surf trips on the "Manga."

This trend toward integrated, high-end luxury surf tourism reflects the fact that surfing is no longer a sport purely for low-income young males but also attracts a more discerning, older and diverse demographic with considerable spending power and who often travel with family groups. And as the example of the Crudens illustrates, this trend has not gone unrecognized by surf tourism operators. Indeed, Siloina Surf Travel is a new entrant to Indonesia's Mentawai Islands market in 2024, with a luxury surf resort on their own private island and two equally luxurious surf charter yachts. So, while the majority of surf camps/resorts and charter boats remain comparatively small owner-operators, this status quo may be shifting with a trend toward a more corporate model with multiple strategic business units.

There is also evidence that a type of branded conglomerate model of surf tourism is starting to emerge. For example, Pegasus Lodges has two branded surf resorts in the Telo Islands, Indonesia, plus one each in Samoa, Kiribati, Canada, and Portugal, all targeted at a high-end surf tourist. Meanwhile, operating at the more price-sensitive end of the market are operations like Rapture Surf Camps with branded operations in Bali, Costa Rica, Morocco, Nicaragua, and Portugal, all operating hostel-style options aimed more at the younger surf enthusiast. Similarly, Lapoint Surf Camps has two branded surf camps in Portugal and one each in Sri Lanka, Costa Rica, Bali, Norway, Spain, El Salvador, and Nicaragua and a surf charter boat in the Maldives; and The Stoke Travel Co. operates multiple budget hostel-style outlets in Spain and then one each in Germany, the UK, Andorra, Croatia, Australia, France, Ireland, the Netherlands, and Morocco.

What we are seeing here are two distinct approaches at seemingly polar ends of the market. That is, in the case of Resort Latitude Zero, Siloina Surf Travel, and Pegasus Lodges, the approach is one of upmarket pampered luxury, while for Rapture, Lapoint, and Stoke Travel Co., the target is more on the youth, budget, novice surfer market with hostel-style accommodation and inclusive of ancillary experiences like surf lessons, trekking, and music festivals. Whether the brands own each individual operation outright or run leasing, franchise-type models and whether these new (to surf tourism) corporate models—the budget approach versus the luxury approach—are truly scalable remain to be seen.

Intuitively, it would seem the niche element of surfing means operators must capture committed surfers consistently and repeatedly, but when the unique selling point is the opposite of volume—that is, no crowds—the equation becomes extremely difficult. Add in the high operating costs that come with remote locations, seasonality, depreciation of expensive equipment (boats, tenders, information technology equipment, etc.), and scalability becomes even more challenging, particularly for the luxury market. So, it may be that the branded approach to surf tourism is more scalable at the budget end of the market, where surf tourists are more "enthusiasts" than core surfers and may be less concerned about surf quality and variety and have a higher tolerance for crowds and hostel-style accommodations.

10.3.2 "Soft" Surf Tourism

Another emerging trend, which I would argue is also a response to the "democratization" of surf tourism discussed in the previous section, is the appearance of what might be described as "soft surf tourism." It seems logical to assume that an influx of new participants to the sport will skew the average proficiency of surfers toward the novice to intermediate level. For surf tourism operators, this means they are increasingly accommodating less skilled surfers who lack the proficiency for the more challenging reef breaks traditionally associated with remote or exotic surf locations. The worst-case scenario here is a slew of unnecessary deaths and gruesome injuries. Jed Smith, in his article, "1000 Ways to Die in Indonesia," discussed this very issue:

> With dozens of boats and land camps operating in the region, generating millions of dollars annually, it's safe to say Indonesian surf travel is as popular as ever and about as safe as it's gonna get for the time being. Amidst this glut of humans descending on its waves, is a whole bunch who shouldn't be out there. Kooks, in a word. "It took us a few years to get our head around it," says veteran captain, John McGroder, adding, "They wanna tick the boxes and get the waves under their belt, and some guys challenge themselves and fear for their life and they're actually stoked with that. And some guys hit the reef and they're stoked with that too." (Smith, 2024, ¶50–51).

Therefore, with an emerging core of novice-to-intermediate surfers, surf tourism operators can avoid the carnage described above by finding ways to provide more soft surf tourism experiences. We are already seeing evidence of this playing out in surf parks. Most surf parks have the capacity to create waves that are technically challenging even for advanced surfers, yet most only offer these settings rarely, with the bulk of their days dedicated to novice and intermediate settings. In arguing the logic for hotels and resorts to include a surf park in their property developments, Wilson (2023) cited Surf Park Central's, 2023 Consumer Trends Report that identified that surfers visit surf parks with an average of 2.9 people who stay 3–3.5 days and spend between US$220 and 380 daily, excluding accommodation (Surf Park Central, 2023). Therefore, with dozens of surf parks currently in operation and at various stages of development globally, and an influx of new participants, it is likely that surf parks will generate their own brand of soft surf tourism (Wilson, 2023).

Another manifestation of soft surf tourism is the emerging trend where operators are targeting female surfers, both individuals and groups, with curated, bespoke experiences that combine surfing with yoga, well-being, and retreat-style approaches. Bonanza Collective in France is a good example, describing themselves as:

> … an intimate and exclusive surf retreat for women in Biarritz, France. It's a merge of all the things we love: Waves, community, movement, food, wine, nature, BONANZA! But more than anything, it's a space to connect and welcome like minded people from all over the world into our community. Our authentic concept creates a unique experience where you get to immerse yourself in the Bonanza lifestyle. It's a romance of us sharing all the things that we love while making life-long connections and learning or improving how to surf. (Bonanza Collective, 2024, ¶1–3).

Interestingly, in another departure from the traditional surf tourism model, Bonanza Collective is not a 24/7 operation but offers what they call "editions" that accommodate a maximum of 12 women for 1-week stays in a Biarritz beach villa. At the time of writing, Bonanza is selling its 17th edition for late September 2024, with every previous edition completely sold out. The imagery and language used on Bonanza Collective's social media platforms encourage women of all ability levels and highlight the cultural and community-building aspects of the experience:

> No previous surf or yoga experience needed. Most girls have never surfed before …. It's the perfect place to travel solo to. On previous editions 95% came on their own – and now they're meeting up for reunions all over the world. The most beautiful thing about Bonanza is the community you'll become a part of. It's so special. (Bonanza Collective, 2024, ¶5).

A variation on the Bonanza Collective approach is Summer Sama that also targets women surfers of all levels with a surf and wellness-based approach. Based on the Indonesian island of Lombok, 1-week Summer Sama retreats include instructor-led surf lessons, surf skate sessions, and yoga and Pilates classes for a maximum of nine guests. Yet another player in this market is Soul Retreats that incorporates surfing as part of a fitness retreat that features "… yoga, breath work, meditation and other activities in our stunning locations to give you a holiday that nurtures both your body and mind" (Soul Retreats, 2024, ¶1). Like Bonanza Collective and Summer Sama, Soul Retreats utilize surfing as part of a suite of activities focused on wellness and community-building, particularly among women (Fig. 10.4). But, whereas Bonanza Collective and Summer Sama base themselves at the one location in Biarritz and Lombok, respectively, Soul Retreats operates weekend and week-long retreats in New Zealand, Fiji, Bali, Sri Lanka, and Australia. Perhaps negotiating the scalability issues related to the corporate branded model discussed in the previous section, Soul Retreats simply books out existing accommodation providers like Kommune Resort in Keramas, Bali, and Funky Fish Resort in Fiji, as venues for their retreats.

Emphasizing the "soft" aspect of this type of surf tourism, in each of the three cases discussed, the imagery on their respective social media platforms all heavily feature the "fun" aspect of surfing with smiling women riding predominantly longboards in small, rolling waves (Fig. 10.4). Clearly, for soft surf tourism, community-building is at the core of the product offering. And indeed, empirical research supports this approach, as Su et al. (2022) found that the "companion effect" is particularly prevalent among female adventure tourists where subjective well-being and satisfaction are derived from adventure travel experiences in the presence of a companion/s (friend/s, family member/s, tour member/s), particularly ones of comparable or greater ability.

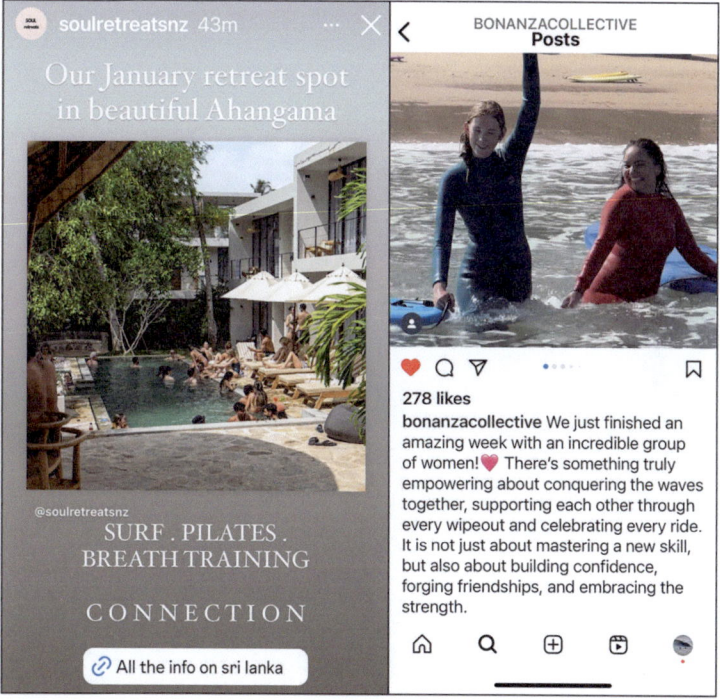

Fig. 10.4 Soul retreats and the Bonanza collective are part of a new wave of surf tourism reaching to a new generation and demographic of surfers. (Photos: Soul Retreats and Bonanza Collective)

10.3.3 Increasing Public Sector Interest in Surfing

While it is difficult to prove causation of the increased global participation in surfing from factors like Olympic inclusion or the proliferation of surf parks, one thing is certain, surfing has had a higher public profile in recent years. One outcome of this is the unprecedented involvement of non-endemic corporate partners in World Surf League (WSL) events. In 2023, the World Championship Tour and the Challenger Series, the top 2 tiers of professional competition, had 6 endemic surf industry corporate partners but 15 non-endemic partners spread across at least 7 industries. Significantly, two corporate naming rights partners on the Championship Tour were public sector tourism promotion authorities—the West Australian State Government for the Margaret River Pro and the El Salvador Ministry for Tourism for the Surf City Pro, El Salvador. Both leveraged their respective event investments with extensive surf destination brand building and advertising both on-site and during the global event broadcasts. Interestingly, the WSL recently announced a 3-year partnership with the Fijian Government. The iconic Fijian break of Cloudbreak (Fig. 10.5), a tour venue for the 2024 season, was announced as the site for the 2025 WSL Finals competition which determines the men's and women's World Champions, with the partnership going through until 2026. Brent Hill, CEO of

Fig. 10.5 Resort Island of Namotu located next to Cloudbreak in Fiji. (Photo: David M. Kennedy)

Tourism Fiji, highlighted the surf tourism motivation underpinning the Fijian Government's involvement.

> We are stoked that the World Surf League has chosen Cloudbreak, Fiji, as the venue for the 2025 WSL Finals …. Our waves and warm hospitality await surf enthusiasts from around the world. We look forward to showcasing Fiji as a world-class surfing destination. This event boosts our global visibility as well as uplifts communities and inspires our local surfers. Vinaka vakalevu, WSL, for recognizing Fiji as the ultimate destination for this event. (World Surf League, 2024, ¶6)

Public sector actors at more regional government levels throughout the world are also beginning to recognize value in supporting surf tourism. For example, in San Sebastian-Donostia, Spain, regional government supports a surf industry cluster called "Surf City Donostia" to assist with business initiatives and technological innovation and to globally position San Sebastian-Donostia as a surf tourism destination (WSCN, 2024). Similarly, in the city of Gold Coast, Australia, local government actively supports the surf industry and seeks to build on the city's reputation as a global surfing destination. Clearly, in destinations where surfing is practiced in numbers, it becomes economically and culturally significant and, therefore, politically relevant. For example, the California State Legislature adopted surfing as the official state sport in 2018 (California.com, 2024), and Australia's Gold Coast is one of few cities globally that has a government-backed surf management plan that was "… developed and implemented through an ongoing consultative process which enables surfing representatives to have oversight on the implementation of the plan" (Ware et al., 2017, p. 121). Gold Coast Mayor, Tom Tate, emphasized the importance of surfing to the city's destination brand:

> We recognise and appreciate the fact that our city's enviable worldwide reputation is largely driven by the appeal of our surfing beaches. So, … we're fully committed to protecting our beaches, supporting our surf industry and further developing our surf culture. (Tate, quoted in City of Gold Coast, 2015, p. 2).

Clearly, surf tourism is of interest to public sector actors because of its ability to contribute to reputation and repeat visitation and the economic benefits it brings to host communities. However, in Papua New Guinea (PNG), through the careful planning of the Surfing Association of Papua New Guinea (SAPNG), surf tourism has also contributed to social outcomes like vocational training; cross-cultural understanding; village-level sport development; poverty alleviation; and financial contributions to local infrastructure like aid posts, water reticulation projects, and school buildings and equipment (O'Brien & Ponting, 2013; Abel & O'Brien, 2015). More recently, since 2017, the SAPNG has been running the Pink Nose Revolution—a community-level program aimed at initiating discussions and raising awareness around gender-based violence. In March 2023, the US Government chose to involve itself in the SAPNG's Pink Nose Revolution program. Through the US State Department's "Sports Envoy Program," the US Government funded three high profile surfers from Hawaii to visit PNG and tour communities for 3 weeks running surf clinics and hosting discussions about gender-based violence and other community-level issues like equitable access to education and health. After visiting multiple communities, the three returned to the US Embassy in Port Moresby to present their findings to a roundtable discussion attended by diplomats from PNG, the USA, the UK, and the European Union.

Surfing and surf tourism have always played somewhat of a diplomatic role in that different nations' respective surfing reputations generate a certain level of soft power, which refers to a nation's positive reputation that results in a sense of aspiration and goodwill from other nations (Nye Jr., 2004). This initiative by the US State Department is significant because, although surfing has been generating soft power and *informally* acting as a diplomatic tool for decades, this initiative marks (as far as this author is aware) the first time that surfing has been incorporated into a federal government's formalized diplomatic agenda. While diplomacy itself is nothing new, the use of sport to achieve diplomatic ends *is* comparatively new. Sport diplomacy is a growing field and refers to the practice of "using sport as a tool for overcoming and mediating separation between people, non-state actors, and states" (Murray, 2020, p. 1). Murray is at pains to point out that, contrary to popular assumptions, diplomacy is not purely the domain of state actors. He notes that "Diplomacy is a word, as well as a profession. A person does not have to work for a ministry of foreign affairs to be considered a diplomat. The state does not have a monopoly on the 'business of peace,' nor should it" (p. 3). In this case, the US State Department used diplomats in boardshorts (the three Hawaiians) to help shore up US relations with PNG, presumably to combat an increasingly assertive China in the Indo-Pacific region. In explaining its use of sport diplomacy, the US State Department explains:

> When leveraged thoughtfully and strategically, we know that sports can be a platform to champion foreign policy priorities—inclusion, youth empowerment, gender equality, health & wellness, conflict resolution, and entrepreneurism. (United States Department of State, 2024, ¶2).

As global participation in surfing continues to grow and given the fact that much of surf tourism takes place in geopolitical hot spots like the Indo-Pacific region, this use of surf tourism as a tool for soft power and sport diplomacy may become increasingly desirable to public sector actors.

10.3.4 Surf Tourism and Sustainability

There is an uncomfortable paradox in surfing. On the one hand, as surfers, we are the veritable "canaries in the coal mine" who are among the first in society to notice changes in the marine and coastal environment. This has led to formation of Surfrider Foundation, Surfers Against Sewage, Save the Waves, and other nongovernment organizations and community-based action groups founded to engage surfing and coastal communities to protect their marine and coastal environments. However, the reality is that the toxic petrochemicals that go into the construction of our surfboards and other surfing hardware, not to mention the carbon footprint of our surf travel, directly contribute to the very same ecological problems we are being recruited to combat. Essentially, we are a big part of the problems we are trying to alleviate. Worse still, Mach and Ponting (2021) empirically demonstrated that there exists an attitude-behavior gap in surf tourism. They found that:

> While surf tourists appear to care deeply about sustainability issues in surf tourism, scores reflecting personal sustainability behaviours are significantly lower. At the same time, surf tourists have an expectation that surf tourism providers will be actively engaged in sustainable tourism initiatives. (Mach & Ponting, 2021, p. 7).

Essentially, Mach and Ponting's findings suggest that surfers prefer to leave the heavy lifting of acting sustainably to industry practitioners, "with many preferring to delegate responsibility to their surf tourism providers" (2021, p. 7).

Nonetheless, while the spirit of adventure and discovery that fueled the very first surf tourists continues, there is growing recognition that the ecological and social costs of this adventure and discovery have invariably been high and borne mostly by host communities. Mach and Ponting (2021) identified these costs as including "ecological issues related to coral reef damage, untreated effluent in waterways, unscrupulous coastal development, and, social issues including gentrification, loss of cultural values, drug and alcohol abuse, and prostitution" (p. 2). In response, researchers have applied a sustainability lens to empirical surf tourism research, leading to a subset of literature termed sustainable surf tourism (SST). However, Ruttenberg (2023) has criticized the SST literature as being too growth-based, income-oriented, and capitalocentric. She states that:

> … the field of SST is thus critiqued by decolonial surf scholarship for aligning itself with the persistently dominant discourse of sustainable development that continues to inform the neoliberal international development agenda … in which SST runs the risk of reproducing the same colonial-capitalist logics and practices it seeks to remedy in Global South surfing destinations. (Ruttenberg, 2023, p. 1085).

Ruttenberg argues the need to move away from the sustainable development paradigm and more toward Foucauldian post-structuralist, post-development approaches complemented by Marxian perspectives on the conventional neoliberal economic-growth-for-development paradigm. However, given that surf tourism is already a well-established and growing industry that generates hundreds of millions of dollars globally (Mach & Ponting, 2021), downplaying sustainable development approaches seems, at best, unwise.

Recent work by Manero et al. (2024) offers a more pragmatic way forward by positioning surfing as a cultural ecosystem service (CES). Manero and her colleagues' approach does not ignore or partition off crucial economic impacts but integrates them with social impacts to build a platform to advocate for supportive public sector policy frameworks that protect the marine and coastal environments from which these impacts derive. They employ an ecosystem service approach to explain and interpret how ecosystems contribute to market economies and social well-being. In doing so, they conceive surf breaks and their surrounding coastal environments as a CES: "the spaces and interactions between multiple components of coastal environments where surfing takes place, including elements like waves, reefs, currents, sediment, flora, fauna and humans" (Manero, 2023, p. 1). The authors posit that verified socioeconomic impacts should be the evidence base upon which to argue the case for effective public sector policies that protect surfing cultural ecosystems and encourage positive impacts while building capabilities to recognize and minimize the negative.

10.4 Conclusions

As stated up front, the aim of this chapter was not to provide an exhaustive literature review but more to discuss the general nature of surf tourism and how we currently experience it and to consider some of the industry's emerging trends and issues. Of course, word limits prevent me from covering *every* topic and including *all* published work on the subject. But what I have tried to do is in equal parts incorporate my reflections as a scholarly researcher with my observations as a lifelong surfer. Indeed, I had to ask the editor for a deadline extension on the writing of this chapter to join friends on an 11-night surf charter in the waters off North Sumatra, Indonesia, but I figured this gave me pause to reflect on what I was writing about—collecting field observations if you will!

After an introduction and explaining the general nature of surf tourism, I discussed how we are now seeing significant change and evolution in the business models of commercial surf tourism and explained how scalability issues may ultimately determine the longevity of these models. The second topic for consideration—the rise of "soft" surf tourism—refers to the trend where surf tourism operators are increasingly targeting previously marginalized groups like women and novice surfers with bundled packages that integrate surfing, typically in less challenging conditions, with wellness retreat-style activities. Interestingly, soft surf tourism operators seem to negotiate scalability issues by staying light and nimble with rented rather than owned resorts and running short, time-bound retreats of around 1 week, predominately targeting upmarket, health-conscious females with a focus on learning, friendship, and community-building.

The third topic for consideration dealt with the public sector's increasing interest in surf tourism, particularly the economic benefits that flow from a region's reputation as a surf destination. But these reputation benefits also bestow soft power on a destination and, as such, can be leveraged as a vehicle for international sport diplomacy—a fact recognized by the US State Department that recently deployed diplomats in boardshorts to build friendly relations with Papua New Guinea with an eye toward the Indo-Pacific region generally. The public sector also featured prominently in the fourth and final topic for consideration in that we discussed the ongoing sustainability challenges that surf tourism presents its stakeholders. It was highlighted that an attitude-behavior gap persists as an uncomfortable truth in surf tourism where surfers expect surf tourism providers to actively engage in sustainable tourism initiatives but are less enthusiastic about changing their own behavior (Mach & Ponting, 2021). Much of our current understanding about surf tourism and sustainability issues stems from the work of researchers such as Ralf Buckley, Nick

Towner, Jess Ponting, and others over the last couple of decades who collectively have built a body of sustainable surf tourism (SST) literature. Ruttenberg (2023) recently criticized the alignment of SST literature with the sustainable development paradigm, arguing that "SST runs the risk of reproducing the same colonial-capitalist logics and practices it seeks to remedy …" (p. 1085). Recent work by Manero et al. (2024) that positions surfing as part of a cultural ecosystem service was highlighted as a pragmatic way forward. This approach recognizes the importance of monitoring and measuring the social and economic contributions of surfing and surf tourism to a community and then using this evidence base as a platform to lobby public sector stakeholders on the need for robust policy frameworks that protect the coastal and marine environments within which surfing takes place.

It is significant that all four of the future considerations discussed essentially stem from the same source—the fact that all over the world, more people, and more diverse people, are going surfing. Surfing is, albeit not uniformly and perhaps too slowly, becoming democratized. As a result, the utility of surf tourism as a mechanism for creating change in communities is growing, and with that recognition comes political relevance. The extent to which surfing communities can find ways to leverage their newfound political relevance to advocate for protection of coastal and marine environments—that is, protect the goose that laid the golden egg—will largely shape the future of surfing and, with that, surf tourism.

References

Abel, A., & O'Brien, D. (2015). Negotiating communities—Sustainable cultural surf tourism. In J. Ponting & G. Borne (Eds.), *Sustainable stoke – Transitions to sustainability in the surfing world* (pp. 154–165). University of Plymouth Press.

AusPlay, (2022). Surfing report. Australian Sports Commission. https://app.powerbi.com/view?-r=eyJrIjoiNzQyYzYzOTYtYjA4MC00NzZiLTliMzItMzMyOWE3YjZlYTZjIiwidCI6IjhkMmUwZjRjLTU1ZjItNGNiMS04ZWU3LWRhNWRkM2ZmMzYwMCJ9

Barbieri, C., & Sotomayor, S. (2013). Surf travel behavior and destination preferences: An application of the serious leisure inventory and measure. *Tourism Management, 35*, 111–121. https://doi.org/10.1016/j.tourman.2012.06.005

Bonanza Collective. (2024). *Bonanza collective*. https://www.bonanzacollective.com/

Buckley, R. (2002). Surf tourism and sustainable development in Indo-Pacific islands: I. The industry and the islands. *Journal of Sustainable Tourism, 10*, 405–424. https://doi.org/10.1080/09669580208667176

California.com. (2024). *Explaining California surf culture: The state's official sport*. https://www.california.com/explaining-california-surf-culture-the-states-official-sport/#:~:text=It's%20a%20culture%20shaped%20by,has%20helped%20define%20its%20identity

City of Gold Coast. (2015). *Gold Coast Surf Management Plan*. https://www.goldcoast.qld.gov.au/files/sharedassets/public/v/1/pdfs/policies-plans-amp-strategies/surf-management-plan.pdf

Dolnicar, S., & Fluker, M. (2003). Behavioural market segments among surf tourists: Investigating past destination choice. *Journal of Sport & Tourism, 8*(3), 186–119. https://doi.org/10.1080/14775080310001690503

Hritz, N., & Franzidis, A. F. (2018). Exploring the economic significance of the surf tourism market by experience level. *Journal of Destination Marketing & Management, 7*, 164–169. https://doi.org/10.1016/j.jdmm.2016.09.009

Mach, L. (2019). Surf-for-development: An exploration of program recipient perspectives in Lobitos, Peru. *Journal of Sport and Social Issues, 43*(6), 438–461. https://journals.sagepub.com/doi/pdf/10.1177/0193723519850875

Mach, L., & Ponting, J. (2021). Establishing a pre-COVID-19 baseline for surf tourism: Trip expenditure and attitudes, behaviors and willingness to pay for sustainability. *Annals of Tourism Research Empirical Insights, 2*(1), 100011. https://doi.org/10.1016/j.annale.2021.100011

Mach, L., Ponting, J., Brown, J., & Savage, J. (2018). Riding waves of intra-seasonal demand in surf tourism: Analysing the nexus of seasonality and 21st century surf forecasting technology. *Annals of Leisure Research, 23*(2), 184–202. https://doi.org/10.1080/11745398.2018.1491801

Manero, A. (2023). A case for protecting the value of 'surfing ecosystems'. *npj Ocean Sustainability, 2*(6). https://doi.org/10.1038/s44183-023-00014-w

Manero, A., Yusoff, A., Lane, M., & Verreydt, K. (2024). A national assessment of the economic and Well-being impacts of recreational surfing in Australia. *Marine Policy, 167*, 106267. https://doi.org/10.1016/j.marpol.2024.106267

Martin, S. A. (2022). From shades of grey to web of science: A systematic review of surf tourism research in international journals (2011–2020). *Journal of Sport & Tourism, 26*(2), 125–146. https://doi.org/10.1080/14775085.2022.2037453

Martin, S. A., & Assenov, I. (2012). The genesis of a new body of sport tourism literature: A systematic review of surf tourism research (1997–2011). *Journal of Sport and Tourism, 17*(4), 257–287. https://doi.org/10.1080/14775085.2013.766528

Murray, S. (2020). Sports diplomacy: History, theory and practice. In *Oxford research Encyclopedia of international studies*. Oxford University Press. https://doi.org/10.1093/acrefore/9780190846626.013.542

Nye, J. S., Jr. (2004). *Soft power: The means to success in world politic* (4th ed.). Public Affairs.

O'Brien, D., & Ponting, J. (2013). Sustainable surf tourism: A community centred approach in Papua New Guinea. *Journal of Sport Management, 27*, 158–172.

O'Brien, D., & Ponting, J. (2018). STOKE certified: Initiating sustainability certification in surf tourism. In B. McCullough & T. Kellison (Eds.), *Handbook on sport, sustainability, and the environment* (pp. 301–316). Routledge.

Ponte, J., Couto, G., Sousa, A., Pimentel, P., & Oliveira, A. (2021). Idealizing adventure tourism experiences: Tourists' self-assessment and expectations. *Journal of Outdoor Recreation and Tourism, 35*. https://doi.org/10.1016/j.jort.2021.100379

Ponting, J. (2009). Projecting paradise: The surf media and the hermeneutic circle in surfing tourism. *Tourism Analysis, 14*(2), 175–185.

Ponting, J., McDonald, M. G., & Wearing, S. L. (2005). Deconstructing wonderland: Surfing tourism in the Mentawai Islands, Indonesia. *Loisir et Societe. Society and Leisure, 28*(1), 141–162.

Porter, B. A., & Usher, L. E. (2018). Sole surfers?: Exploring family status and travel behaviour among surf travellers. *Annals of Leisure Research, 22*(4), 424–443. https://doi.org/10.1080/11745398.2018.1484782

Reis, P., Caldeira, A., & Carneiro, M. J. (2022). Can surf culture foster loyalty towards surf destinations? *Journal of Sport & Tourism, 26*(4), 387–407. https://doi.org/10.1080/14775085.2022.2105387

Ruttenberg, T. (2023). Alternatives to development in surfing tourism: A diverse economies approach. *Tourism Planning & Development, 20*(6), 1082–1103. https://doi.org/10.1080/21568316.2022.2077420

Smith, J. (2024). 1000 ways to die in Indonesia. Stab Premium. https://stabmag.com/premium/long-read-a-thousand-ways-to-die-in-indonesia/

Soul Retreats. (2024). *Fitness retreats around the world*. https://www.soulretreats.co.nz/

Statista Research Department. (2022a). *Global surfing industry market size 2022–2027*. https://www.statista.com/statistics/1327319/surfing-market-size-worldwide/

Statista Research Department. (2022b). *Size of the global surfing tourism market 2022–2032*. https://www.statista.com/statistics/1327449/surfing-tourism-market-size-worldwide/

Statista Research Department. (2024). *Surfing participation in the U.S. 2010–2023.* https://www.statista.com/statistics/191328/participants-in-surfing-in-the-us-since-2006/

Su, L., Cheng, J., & Swanson, S. (2022). The companion effect on adventure tourists' satisfaction and subjective well-being: The moderating role of gender. *Tourism Review, 77*(3), 897–912. https://doi.org/10.1108/TR-02-2021-0063

Surf Park Central. (2023, January). *Consumer trends report.* https://surfparkcentral.com/2023-surf-park-consumer-trends-report/

Towner, N. (2015). Surf tourism and sustainable community development in the Mentawai islands, Indonesia: A multiple stakeholder perspective. *European Journal of Tourism Research, 11,* 166–170.

Towner, N. & Davies, S. (2019). Surfing tourism and community in Indonesia. *Journal of Tourism and Cultural Change, 17*(5), 642–661. https://doi.org/10.1080/14766825.2018.1457036.

Towner, N. (2016). Searching for the perfect wave: Profiling surf tourists who visit the Mentawai Islands. *Journal of Hospitality and Tourism Research, 26,* 63–71. https://doi.org/10.1016/j.jhtm.2015.11.003

United States Department of State, Bureau of Educational and Cultural Affairs. (2024). *Why does the U.S. Department of State support sports diplomacy?* https://eca.state.gov/sports-diplomacy

Wagner, S., Nelson, C., & Walker, M. (2011). *A socioeconomic and recreational profile of surfers in the United States.* Surf-First & The Surfrider Foundation. https://www.surfrider.org/news/a-socioeconomic-and-recreational-profile-of-surfers-in-the-united-states

Ware, D., Lazarow, N., & Hales, R. (2017). Surfing voices in coastal management: Gold Coast Surf Management Plan—A case study. In G. Borne & J. Ponting (Eds.), *Surfing and sustainability* (pp. 107–124). Routledge.

Warshaw, M. (2017). *A brief history of surfing.* Chronicle Books.

Wilson, T. (2023). 3 reasons why hotels and resorts should build a surf lagoon. *Endless Surf.* https://endlesssurf.com/2023/05/30/3-reasons-why-hotels-resorts-should-build-a-surf-lagoon/

World Surf League. (2024). World surf league announces Cloudbreak, Fiji as 2025 WSL finals location: Cloudbreak to determine 2025 world champions in one-day, winner-take-all season finale. World Surf League. https://www.worldsurfleague.com/posts/531693/world-surf-league-announces-cloudbreak-fiji-as-2025-wsl-finals-location

WSCN. (2024). San Sebastian-Donostia: Our interest. *World Surf Cities Network.* https://www.worldsurfcitiesnetwork.com/en/cities/san-sebastian/our-interest

Chapter 11
The Surf Industry

Craig Sims and Danny O'Brien

11.1 Introduction

Although records show that surfing dates back thousands of years (Chap. 2), the modern subcultural phenomenon of surfing had its genesis in America, particularly California in the late 1950s. Since then, surfing's influence upon mainstream society has been felt well beyond its size (Jarratt, 2010; McGregor, 1966; Sims, 2022; Stranger, 2010; Stratton, 1985). Surfing attracts the interest of the broader non-surfing public because some of its characteristics fortuitously combine to meet an aspirational lifestyle proposition. The notion of gliding across the final moments of an oceanic swell and the hedonistic and rebellious undertones expressed in surfing's subcultural values have a broad-based aspirational appeal that has spawned a US$4 billion global surf industry (Statista, 2022). This chapter explains the key underlying drivers of the surfing industry (also see Chap. 9) through the lens of the social theories of identity, authenticity, subculture, and hegemony. It proposes a new theoretical concept called *subcultural power*, and it highlights structural changes in the surfing subculture that could bring both opportunities and challenges.

11.2 The Surf Industry

The surf industry is a broad term that refers to a collective of commercial enterprises that benefit from the commercialization or commodification of the surfing subculture. The surf industry can be further delineated into two categories. The first is the endemic surf industry which operates from within the subculture and comprises

---

C. Sims (✉) · D. O'Brien
Bond University, Gold Coast, QLD, Australia
e-mail: csims@bond.edu.au; daobrien@bond.edu.au

several sectors that produce the core products that enable the "thrilling, embodied practice of waveriding itself" (Ford & Brown, 2006, p. 68) such as surfboards, fins, surf wax, leashes, and wetsuits. Also included in this category are sectors that are peripheral to the act of riding waves but still considered central to the lifestyle such as surf clothing, media, tourism, and wave-riding instruction. The second category is the non-endemic surf industry which "operates externally from the subculture and is engaged in the production of surf culture products by mainstream companies for consumption in the mainstream market" (Stranger, 2011, p. 188). This category includes the fashion sector but extends far and wide to homewares, stationary, entertainment, cars, alcoholic beverages, mobile services, etc. These two categories represent two extremes on a continuum, so the boundary between them is blurry if not entirely indistinct. The midrange typically include sectors such as sunglasses, footwear, and watches; however, some of the brands within the midrange can be positioned closer to the extremes and vice versa.

The endemic surf industry has its genesis in the manufacture of surfboards, which still, for the most part, consists of small-scale localized artisan operations (Fig. 11.1). Over time, the surf industry expanded to include ancillary equipment—like fins, leashes, surf wax—that service the functional needs of the surfing community and are commonly categorized as the surfing hard goods sector. A pioneer product of the surfing soft goods sector was the surfing boardshort. Stranger (2011) describes how the surf fashion industry emerged from a functional need:

> Boardshorts were designed in the US to withstand the punishment that conventional bathing costumes could not (Doyle, 1994). They were longer in the leg, which helped to prevent chaffing on the inside of surfers' legs that occurred as a result of straddling the board while waiting their turn in the line-up for waves. The shorts became a medium for the display of surfing style (through colour, prints, and cut). (p. 192 parentheses in original quote)

From these humble beginnings the surf industry grew, first as an emerging body of "small, flexible, innovative and trend setting manufacturers" (Booth, 1996, p. 24), but over time, and through a process of mergers, acquisitions, and takeovers, some of the original "backyard" brands, such as Rip Curl or Billabong, have become multinational, multi-sector conglomerates. The burgeoning growth of the sector and some of the businesses within it has sparked debate about authenticity. Indeed, it will become apparent later in this chapter that how factors such as scale and distribution have significant implications for the perceived authenticity of surf industry brands.

The original channel of distribution for the surf industry was through direct sales out of the back of founders' cars in the car parks of their local surf breaks (Jarratt, 2010). As business grew, these budding entrepreneurs moved into surf shops, described by Stranger (2011) as being "cult enclaves, intimidating to the outside world" (p. 192). Surf shops, usually owned and run by local surfers, existed in most coastal towns where there were surfable waves. But as surfing became more popular and the lifestyle more attractive to the non-surfing public, brands that started on the back of home sewing machines like Billabong, Rip Curl, and Quiksilver became multinational corporations. Distribution moved from the car boot to the local surf

Fig. 11.1 Examples of end members of the surfing industry found in Victoria, Australia. (**a**) A mall-type franchise store where surfing is one of many adventure-sport offerings and (**b**) a backyard boutique shop where boards are made and designed on-site (**c** and **d**). (Photos: David M. Kennedy)

shop to franchised surf shops that began to emerge in major cities around the world, many of which were nowhere near the ocean (Fig. 11.1). A classic example of this is the Desert Wave Surf Shop located almost 1000 km away from any beach in the town of Alice Springs, Australia. Today, surf shops range along a similar positioning continuum between no-frills purveyors of surfing hard goods to surf fashion boutiques located in strip malls that target the more lucrative mainstream market with soft goods (Fig. 11.1). Indeed, some "surf shops" today do not actually sell surfing hard goods!

11.3 Surfing's Aspirational Appeal

That there are surf shops doing a roaring trade in landlocked countries is evidence of the aspirational appeal of the surfing subculture. There is some debate about whether surfing is a sport, artform, or lifestyle (Brennan, 2021), but less disputed is the sense that it is a subculture (Ford & Brown, 2006). A sport subculture is defined by Jary and Jary (2006) as "any system of beliefs, values and norms which is shared and actively participated in by an appreciable minority of people within a particular culture" (in Ford & Brown, 2006, p. 59). Stranger (2010) explains that the surfing subculture is grounded by the act of riding waves—which he describes as the "pursuit of an ecstatic communion with nature" (p. 118). Various factors conspire to afford only a fortunate few the experience of riding the ocean's waves. Access to the coast, a degree of athleticism and swimming ability, and the means to buy the required surfing equipment are among the barriers to entry that contribute to its sense of exclusivity and mystery. Market economists posit that when you withhold supply, value increases. In a similar way, it is perhaps surfing's inaccessibility that enhances its cache or value. The value of scarcity in surfing was succinctly articulated by surf industry legend, the late Mike Tomson, with this sage warning to the surfing industry in the late 1990s: "Size is the enemy of cool" (Nettle, 2020, 6). This statement came at a time when surf brands expanded in both scope and scale, and the ubiquity of the surfer "look," he contended, diminished the value of its cool status.

Surfing's "cool status" (Ford & Brown, 2006), however, does not only arise from the exclusive act of surfing. Since the 1950s when surfing first became popularized as a sport, surfers have been on the fringes of society, viewed with a combination of disdain and intrigue. In the early 1950s, during post-World War II austerity, surfers were seen as lazy and irresponsible citizens indulging in an unproductive pastime. In the 1960s, surfers were the Vietnam draft dodgers, and in the 1970s, they were the hippies and "stoners." By the 1980s, surfers were at the vanguard of the action sport movement that brashly disrupted the institution of team sport. Surfing, therefore, has its roots in resistance or rebellion, and this is why Taylor (2007) characterizes the subculture as having an anti-bourgeois ethos and countercultural roots.

Around the turn of the twenty-first century, however, surfing found itself in the slipstream of a worldwide social trend as people looked to regain a sense of life balance and responded by introducing more leisure into their lifestyle. While surfing fits the description of a youth leisure lifestyle, Wheaton (2019) describes how "shifting definitions of ageing" (p. 387) have seen increasing numbers of older people engaging in youthful leisure lifestyles and, in so doing, are "creating new and meaningful identities via immersion in forms of serious leisure associated with youth" (p. 391). The growth of so-called post-youth leisure lifestyles reveals that youthful thinking and behaviors are not limited by age or a temporal definition of youth. Youthfulness, as a state-of-mind, is yet another factor that contributes to surfing's broad-based aspirational appeal.

In summary, surfing's ability to attract a consumer base that is far larger than its participant base stems from a blend of factors, which are all founded upon the exclusive act of riding the ocean's waves, which supports an adventurous, healthy, youthful lifestyle that has grown from rebellious and, sadly, misogynistic antisocial roots to being a widely accepted and popular leisure pastime. We pick up on the misogynistic theme later in this chapter, but combined, these factors both underpin and fuel a dynamic subculture that itself underpins the commercialization of surfing and is integral to the success of the surfing industry.

11.4 Personal and Group Identity

To understand how the aspirational appeal of surf culture extends to the mass market, it is necessary to understand the core tenets of individual and group identity. Elliott and Wattanasuwan (2015) postulate a person's identity formation is a lifelong affair that is actively created and constantly validated. Thompson (1995) describes this construction of the self as a symbolic project, during which a consumer seeks out symbolic materials, which are weaved into "a narrative of self-identity" (p. 120). Elliott and Wattanasuwan argue that "all voluntary consumption carries, either consciously or unconsciously, symbolic meanings; if the consumer has choices to consume, he or she will consume things that hold particular symbolic meanings" (p. 134). Consumption choices, therefore, are based not only on a product's utility but also on the contribution its symbolic meanings make to the construction of the consumer's identity. The "enormous appeal and commercial success" (Ford & Brown, 2006, p. 67) of the surfing industry, therefore, was made possible because surfing's cultural values and symbolic meanings were successfully articulated through products which are targeted to the wider non-surfing society.

Group or social identity, as explained by Tajfel's (1978) social identity theory, postulates that identity can be influenced by group affiliations, which, when emotionally embraced, become a source of individual pride and self-esteem. Anderson and Stone (1981) identified that sport can be a powerful catalyst for group bonding. They argue that participation in sport facilitates socialization into that sport's subculture, allowing participants to learn and adopt the attitudes, outlooks, and values common to the sport, which then become part of their identity. Thus, an affiliation with a sport includes a sense of belonging and an attachment to the culture of a larger social structure (Wann & Branscombe, 1991).

Surf culture, therefore, caters to the identity needs of a wide range of people whose affiliation can be expressed on both an individual and group level. The effect of these expressions can impact the perceived strength of a person's cultural affiliation by others within the subculture. To understand this, it is necessary to delve into systems of distinction within cultural groupings and the idea of subcultural capital (Thornton, 1995).

11.5 Social, Cultural, and Subcultural Capital

From an anthropological or sociological perspective, culture can be defined as "a set of attitudes, practices and beliefs that are ... expressed in a particular society's values and customs, which evolve over time as they are transmitted from one generation to another" (Throsby, 1995, p. 202). Therefore, it is possible to claim that something has cultural value if it contributes to the shared elements that comprise a culture. Thus, the concept of culture "carries with it a concomitant notion of cultural value" (Throsby, 1999, p. 6).

A measure of cultural value is "cultural capital," a concept popularized by Bourdieu (1986) in his analysis of society's high-status culture. According to Bourdieu, cultural capital is the knowledge that one accumulates to gain status and opportunity in cultural settings. Cultural capital can also exist in cultural objects (such as art or media) and intuitional qualifications (such as certificates and awards); however, understanding their value requires a level of insider knowledge in the first instance, so knowledge accumulation is central to the concept of cultural capital (Throsby, 1999).

Bourdieu explains that cultural capital is closely linked to "social capital." Social capital can be understood as the accumulation of a network of relationships which serves as a socially instituted credential (Bourdieu, 1986) providing status and opportunity in social settings. Social capital can also be passed on through group membership such as a family name, the alumni of a school, a club, a tribe, etc., where each individual member enjoys the benefits of the collective's social capital. When relationships between individuals and groups in society are being examined, the idea of cultural capital is often entwined with that of social capital (Throsby, 1999). Indeed, Bourdieu argues that the two are "never completely independent of each other" (p. 21).

Drawing on the work of Bourdieu 10 years later, Thornton (1995) introduced the term "subcultural capital" while researching cultural distinctions in the British music scene. She argued little attention has been paid to the systems of social and cultural distinction that divide and demarcate contemporary youth culture. Thornton's focus was on the role of media as a facilitator of the hierarchies within music subcultures. Her concept of subcultural capital can be defined as an accumulation of insider knowledge, style, and artifacts that contribute to a perception of subcultural affiliation (Thornton, 1995). An individual's subcultural capital is a demonstration of their authenticity, and as such, subcultural capital is a vehicle for the establishment of an individual's subcultural hierarchy and status (Smith-Lahrman, 1996; Thornton, 1995). Green and Chalip (1998) noted how members of sport subcultures like to demonstrate and celebrate their subcultural affiliation, particularly through conspicuous attire, or what Force (2009) referred to as "sartorial adornment." Indeed, surf clothing is a commonly used method of displaying cultural affiliation in surfing, and individuals will be willing to pay "for the embodied cultural content" of a product by offering a price higher than that which they would offer for the physical entity alone (Throsby, 1999, p. 8). Herein lies the foundation of the commercialization of surf culture. In a study by Sims (2022), respondent Anna (aged 17) admits, "the clothes and swimmers I have from surf brands would probably identify me as a surfer because, you know, I buy certain bikinis to wear in the surf, and that makes me feel more like a surfer." Anna's comment shows that some brands are perceived to be more authentically aligned with the surfing subculture, and the demonstration of this insider knowledge enhances her credibility, or subcultural capital, within that subculture. To better understand this link, it is necessary to explore the fundamentals of authenticity theory.

11.6 Authenticity and Subcultures

In the philosophical literature, definitions of authenticity are fraught with controversy because the term can be used in so many contexts. Plesa (2023) suggests authenticity is linked to the concept of sincerity, "which constitutes a moral approach to dealing honestly with others" (p. 556). Taking the same perspective, Rings (2017) describes authenticity as being "true to oneself," while Varga and Guignon (2020) describe it as being "faithful to an original" or a "reliable or accurate representation" or being of "undisputed origin or authorship."

In the sociological literature, authenticity plays a key role in developing and displaying identity within a subculture. In this context, authenticity is referred to as "a set of qualities that people in a particular time and place have come to agree represent an ideal or exemplar" (Vannini & Williams, 2020, p. 3). Thus, the word "authentic" has no inherent meaning because its meaning is based on how we perceive it as representative of a subcultural ideal. Moreover, the notion of a subcultural ideal is constantly evolving. Subcultural norms and values are negotiated daily through the interactions of their participants (Olive, 2015), so, in a sociological sense, authenticity is a created construct that is constantly changing.

McLeod (1999) argues that when a subculture enters the mainstream, the authenticity of that subculture becomes threatened, and this can impact the subculture as whole. In this context, authenticity aligns with Varga and Guignon's (2020) description of it as being "faithful to an original" or a "reliable or accurate representation." Mike Tomson's cautionary warning to the surfing industry through his declaration that "size is the enemy of cool" is an example of how the authenticity of a subculture can diminish when a subculture enters the mainstream. Tomson is warning that as surfing expands, it risks losing the original qualities that gave it the power to attract a customer base greater than the size of its participant base. We propose this power that surfing has to attract a customer base beyond the size of its participant base be referred to as its "subcultural power." So, subcultural power can be defined as the relative value of a subculture's aspirational appeal that affords it the ability to attract its share of mainstream societal interest.

11.7 Subcultural Power: A New Construct

Lifestyle sports involve dynamic interaction between participants and physical geographical features as well as the dynamic forces that produce these features (Booth, 2024). Lifestyle sports are distinguishable from traditional sports by their:

> 'pacing and feel' (Krein 2014, 193), sense of adventure (Van Bottenburg and Salome 2010), fewer formal rules (Humberstone 2011), higher hedonistic content (Thorpe and Rinehart 2010), and less competitive relationships (Midol and Broyer 1995). (Booth, 2024, p. 193)

Many lifestyle sports, such as skateboarding, snowboarding, mountain bike riding, and rock climbing, have the capacity to capture the hearts and minds of the mainstream public through their own respective subcultural values and symbolisms. In the same way that subcultural *capital* is used to characterize personal membership and status within a subculture, we propose a new concept called subcultural *power* to characterize a subculture's hierarchy and status relative to other subcultures. In this manner, the surfing subculture competes with the subcultures of other lifestyle sports for aspirational appeal.

Since the surf industry relies on surfing's broad-based aspirational appeal for its commercial success, surfing's subcultural power is an important contributor to the surfing economy. It is, therefore, in the collective interest of each and every lifestyle sport to be attentive to subcultural power. Indeed, actors should nurture and maintain the strength of their sport's subcultural power in order to maximize their share of mainstream appeal. From a business perspective, this is not so much of a coordinated collective action, but more about each brand's individual interest in demonstrating an authentic connection to the subculture. By doing this individually, the collective result of strong subcultural power is achieved.

11.8 Authenticity and the Surf Industry

In the surf industry, actions of authenticity come in various forms with commensurately varying results but tend to include sponsorship of influential professional surfers, hosting or alignment with cultural events (such as the Noosa Festival of Surfing in Queensland, Australia), commercial partnerships with surfing events, and engaging in advertising or content partnerships with the surfing media. Authenticity is also expressed through the goods and services produced for the core surfing market. The term "core" is critical to authenticity here; it refers to the committed surfer who approaches the sport from a serious leisure perspective, as opposed to the casual enthusiast who surfs occasionally. If these goods and services are designed for and meet the needs of the active or core surfer, then that affords a brand the appearance of authenticity. This alignment to the core surfing experience is a key marketing objective and is captured in the brand slogans of legacy surf brands like Rip Curl, "Made by surfers for surfers"; Rusty, "Good boards come first"; and Billabong, "Only a surfer knows the feeling." Equally, with surf travel a long-established component of the surfing subculture, surf brands have sought alignment with marketing campaigns called "The Search" (Rip Curl) or "The Crossing" (Quiksilver). More recently, brands such as Patagonia have become the harbingers of core authenticity in surfing by celebrating and aligning with values around the environment and sustainability (Fig. 11.2).

The consistent demonstration and articulation of authenticity by individual brands is how surfing has managed to maintain its subcultural power despite its expansion in both scope and scale (Fig. 11.3). This is, however, a fine line. When individual brands fail to demonstrate their authenticity, the collective result is potentially diminishing value in surfing's subcultural power, and this results in a loss of surfing's share of mainstream appeal and, ultimately, consumer spend. This is succinctly articulated by Jarratt (2010) who observed that "Surfing's biggest brands (were able to) cross the billion-dollar threshold by thinking big and staying cool …

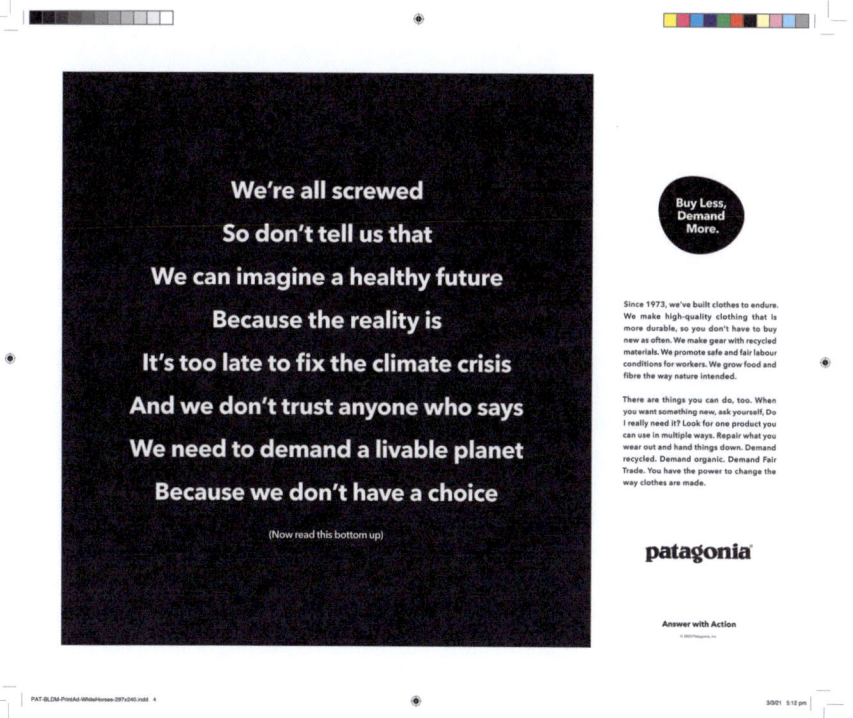

Fig. 11.2 This advertisement from Patagonia has become the harbingers of core authenticity in surfing that attempt to align the brand with core surfing identify. (Reproduced with permission from Patagonia)

Fig. 11.3 Olympian and two-time World Champion surfer John Florence's brand partnership with the North Shore Lifeguards Association is prime example of authenticity in the surf industry. (Photos: FLORENCE/Arto Saari)

and that's a hell of a balancing act" (p. 10). Interestingly, and demonstrating the dynamic nature of a market founded upon the vagaries of subcultural alignment, those same brands such as Quiksilver, Rip Curl, and Billabong that Jarratt described as "thinking big and staying cool" have since fallen into administration and/or been gobbled up in corporate raids by much larger private equity firms. And all, to varying degrees, have lost their foothold on the precipice of surfing authenticity. Indeed, the eponymously named Authentic Brands Group (ABG), an American brand management company headquartered in New York City that counts Shaquille O'Neal and David Beckham among its owners, sent shockwaves through the surf industry in September 2023 with its US$1.25 billion takeover of Boardriders, the parent company of surf brands like Quiksilver, Billabong, Roxy, RVCA, DC Shoes, Element, Von Zipper, and Honolua (LaFrenz, 2023).

As surfing continues to grow in both participation and mainstream acceptance, the balancing act of maintaining its subcultural power will become more difficult. The ABG deal referred to above now has some of surfing's most iconic legacy brands under the same corporate umbrella as Juicy Couture, Van Heusen, Nine West, and 60-odd other brands, most of which are completely unaffiliated to surfing. To what extent will this dilute surfing's subcultural power? This is one of the biggest challenges facing the surf industry today.

11.9 Subcultural Capital and Subcultural Power: The Case of Surf Clothing

Surf clothing brands derive commercial success from two arenas. First, they pursue customers who relate and aspire to the surfing subculture. These customers buy surf clothing to showcase the extent of their subcultural affiliation and/or to bolster their subcultural capital. To win these customers, surf brands seek to present themselves as authentically connected to the subculture and, in so doing, compete with each other to deliver the most subcultural capital for the customer. Second, surf clothing brands also rely on the aspirational appeal of the subculture for their commercial success. If the subculture fluctuates in mainstream appeal, its subcultural power is impacted as mainstream consumer interest shifts to and from likeminded and competing subcultures. So, while individual surf brands compete among themselves to deliver maximum subcultural capital to their customers, they also compete as a collective to maximize their subcultural power in a broader competition among subcultures for mainstream aspirational appeal.

11.10 The Existential Challenge of an Aging Surfing Demographic

Another challenge the surf industry faces is demographic change in surfing. Older people are engaging in youthful leisure lifestyles, and this is altering the demographic composition of the affected lifestyle niches (Bennett & Hodkinson, 2020; Featherstone & Hepworth, 1995; Wheaton, 2019). Surfing has not been excluded from this trend. The average age of surfers has been creeping up for decades (Jarratt, 2010). Indeed, two decades ago, Warshaw (2005) observed that in the 1970s, the average age of a surfer was around 18 years of age, but it "is now somewhere near 30 today" (p. 54). Recent research shows a clear and present trend of an aging surfing demographic in Australia (AusPlay 2021, 2022). With peak surfing participation now in the 45–54 age category for men and in the 35–44 age category for women, surfing can no longer be legitimately positioned as a youth sport tied to youth culture (Sims, 2022). A more accurate descriptor of surfing's social position is that it is "a youthful sport and surfers of all ages form part of a youthful subculture" (Sims, 2022, p. 97). The distinction between youth and youth*ful* is important because while the average surfing participant no longer falls into the traditional definition of "youth" (14–24 years of age), the surfing lifestyle remains youthful by nature. The youthful appeal of the surfing life is, therefore, an aspirational positionality, and so "consumers of surfing related goods and services, should not be defined by age but rather their attraction to a youthful aspirational proposition" (Sims, 2022, p. 98).

This understanding of surfing's positionality alters a range of previously held assumptions about the nature and characteristics of the archetypal surfer. For example, older surfers have different media uses and content preferences to younger surfers (Sims, 2022). Relatedly, Wheaton (2019) found that as surfers age, their interest moves toward broader surfing experiences such as "the natural blue space, intergenerational friendships and the sense of excitement, achievement and wellbeing" (Wheaton, 2019, p. 404). Much of these experiences are satisfied by surf travel, a content category that Sims (2022) found to rank highest among surfers who are members of Generations X and Y and lower in the younger Generation Z cohort. In a pragmatic sense, surf brands might also need to adjust sizing, styling, and pricing because the average surfer today is more mature and has more disposable income than the average surfer of 10–15 years ago. Equally, brand messaging will also need to change as surf brands shift from communicating to youth, to communicating a youthful image to a more diverse age group.

The phenomenon of an aging surf population is especially relevant in Australia because it is combined with an aging general population (ABS, 2020). The effect of an aging trend in surf participation will therefore occur more rapidly than other countries where the population is aging at a slower pace. This will challenge surf producers who have traded off surfing's youthful image for decades (Pearson, 1982; Ford & Brown, 2006). Jarratt (2010) concurs, writing that:

> Every season, surfwear manufacturers from California to Cornwall, Biarritz to Bali, Sydney to Sao Paulo, ask themselves the same question: How much longer will kids continue to wear the clothes their fathers wear and how can we convince them they're not? (p. 8)

11.10.1 Gender Power Relations and Hegemonic Masculinity

For the sake of clarity, discussion of gender power relations in surfing should be prefaced by the separation of the terms sex and gender. Sexual distinctions of men and women describe anatomical or physical differences, whereas gender distinctions describe the relational dynamic between men and women. Hall (1996) explains gender distinctions are socially constructed and contested phenomena and as such are dynamic. So, the way gender power relations are understood is determined by the way masculinity and femininity are expressed in the social world. When these expressions become accepted and entrenched in a social setting, they can be described as having a certain hegemonic status. A hegemonic status describes the origins and influences of the accepted relational norms within a social setting (Gramsci, 1971).

It has been widely published that the surfing subculture has a masculine hegemonic social order (Ford & Brown, 2006; Henderson, 2001; Heywood, 2008; Olive 2015; Stedman, 1997; Wheaton & Thorpe, 2018). The way surfing is practiced (the performative norms), the way the lineup is ordered and regulated (the operational norms), and the way surfers interact with each other (the social norms) have been determined by a masculine world view. Ford and Brown (2006) point out that female participation in surfing has, historically, been an unproblematic and uncontested aspect of this "male sphere of activity" (p. 83). This is because female participants have, for the most part, conformed to the performative, operational, and social norms of the masculine hegemony. Demetriou (2001) argues that their compliance makes them complicit actors which further entrenches the status quo. An example of complicity is evident in an analysis of surfing's performative norms. Women's competitive surfing uses institutionalized (masculine hegemonic) judging criteria to determine the winner. Shields (2004) contends that surfing's hegemonic masculinity has resulted in an aggressive, heroic, and technically competent approach to riding waves. Because the performative norms of surfing have been defined by a masculine ideal, surfing has struggled to incorporate femininity into performance, and thus there exists a disconnect between the corporate image of the surfer-girl and what the sport considers good surfing.

In recent years, however, women have taken to surfing in an ever-increasing proportion to men. Of the 196,000 people who took up surfing in Australia since the start of the COVID-19 pandemic, 118,000 or 60% were female (AusPlay, 2022). Moreover, the AusPlay report (2022) indicates the adult participation gender split is 72:28 (men/women), while for under 15s, it is 54:46 (boys/girls), which suggests a much more gender-neutral surfing future, at least in Australia. Increasing female participation is an interesting trend to watch because it could alter surfing's male-dominated cultural norms (Ford & Brown, 2006; Olive, 2016; Shields, 2004). Brennan (2021) takes a contrary view, arguing that an increase in female participation will "reinforce the oppressive structures of the sport rather than diminish them" (p. 158).

Certainly, the accepted gender power relations in surfing are being challenged through the excellent written work of Booth (2008) and Ford and Brown (2006) and, more recently, Wheaton (2019) and Olive et al. (2015). Olive (2016) argues that within subcultures like surfing, understandings and behaviors are learned through the day-to-day cultural and subjective relationships that individuals have with people and the contexts in which they occur. Thus, as we move toward a gender-neutral lineup, the increasing presence of women will inevitably change power relations and, ultimately, cultural norms, which up until now have been shaped by a masculine hegemony.

As feminine perspectives become inculcated into the cultural melting pot, social, operational, and performative norms will change. Social norms will evolve as surfers adapt to a more inclusive lineup with more tolerance and respect demonstrated in the interactions between male and female surfers. Operational norms—the way waves are allocated—are likely to accommodate female interpretations of surf etiquette as the fittest, strongest, and best performing surfers will no longer feel entitled to the most waves. Finally, performative norms, currently centered on power and aggression, could evolve to more holistically accommodate style, flow, and more feminine interpretations of what constitutes good wave riding.

The industry implications of these shifts could be profound. Since surf culture is an ever-evolving phenomenon, media and brand representations of the subculture must always align with the status quo, or they risk being dismissed as inauthentic. As already discussed, when a brand fails to demonstrate authenticity, loss of credibility and market share typically follows. And when this occurs at scale, the collective result is a decline in the value of surfing's subcultural power. Women filling senior management roles in the surf industry is moving slowly, with only two as far as we are aware: Brooke Farris, Rip Curl CEO, and and Jennifer Vandekreeke, CEO of URBNSURF (and not forgetting Sophie Goldschmidt, former CEO of the WSL). However, perhaps we are seeing a portent of the future with smaller surf brand start-ups like Seea, Carve Designs, Left on Friday, and Kassia Surf all being founded by females and established solely to cater to female surfers (Roberts, 2023).

The World Surf League made a significant move toward gender equity in 2021 by committing to equal pay in the Men's and Women's divisions. In 2023, Surfing Australia followed suit, making equal pay part of their rules of governance. In 2024, the Australian Government legislated equal pay for men and women in sport. This sequence of events provides support for Stranger's (2010) assertion that surfing is a postmodern culture existing within a broader postmodernizing environment and suggests, therefore, that "surfing exerts a postmodernising influence on its parent culture" (Stranger, 2010, p. 14). Since most sports display a hypermasculine characteristic (Hall, 1996; Wearing, 1996), surfing could be, as Stranger suggests, ahead of the curve by advancing out of its condition of hegemonic masculinity before most other sports.

11.11 Conclusions

In this chapter, we have attempted to explain the underlying drivers of the surfing industry through the lens of social theory. Understanding the psychological and sociological drivers of personal identity and the deep and often enduring connections that form in group identities allows us to understand the aspirational appeal of the surfing subculture. We argued that the surf industry is founded upon the commercialization of the surfing subculture. Therefore, awareness of how appealing the surfing subculture is in comparison with other similar sport subcultures, what we termed its "subcultural power," is essential. Subcultural power helps us understand that the surfing industry exists in a competitive ecosystem of aspirational subcultures. This exposes the importance of surfing brands demonstrating their authentic links to the subculture, both for their own commercial success and for the wider benefit of maintaining surfing's subcultural power. As authenticity is a created construct that is constantly changing, it highlights the need for members of the surfing industry to be especially mindful of cultural shifts. We proposed a looming cultural shift that stems from recent changes in both the age and gender composition of surfing participants. These changes will inevitably destabilize industry-specific drivers of demand that have underpinned the surfing economy for decades. It is therefore essential for surf industry actors to embrace these changes and help drive surfing's new cultural zeitgeist into the future.

References

Anderson, D. F., & Stone, G. P. (1981). Sport: A search for community. In S. L. Greendorfer & A. Yiannakis (Eds.), *Sociology of sport: Diverse perspectives* (pp. 164–172). Leisure Press.

Australian Bureau of Statistics. (2020, June). Australian National Accounts: *National Income, Expenditure and Product*. ABS Website. Accessed 11 Dec 2020.

Australian Sports Commission. (2021). AusPlay State of Play Report: *Surfing: State of Play Report*. Australian Government. https://app.powerbi.com/view?r=eyJrIjoiNzQyYzYzO-TYtYjA4MC00NzZiLTliMzItMzMyOWE3YjZlYTZjIiwidCI6IjhkMmUwZjRjLTU1ZjItNG NiMS04ZWU3LWRhNWRkM2ZmMzYwMCJ9

Australian Sports Commission. (2022). AusPlay State of Play Report: *Surfing: State of Play Report*. Australian Government. https://app.powerbi.com/view?r=eyJrIjoiNzQyYzYzO-TYtYjA4MC00NzZiLTliMzItMzMyOWE3YjZlYTZjIiwidCI6IjhkMmUwZjRjLTU1ZjItNG NiMS04ZWU3LWRhNWRkM2ZmMzYwMCJ9

Bennett and Hodkinson (2020). Ageing and youth cultures: music, style and identity. Routledge, Taylor & Francis Group.

Booth, D. (1996). Surfing films and videos: Adolescent fun, alternative lifestyle, adventure industry. *Journal of Sport History, 23*(3), 313–327.

Booth, D. (2008). (Re)reading the surfers' Bible: The affects of tracks. *Continuum (Mount Lawley, W.A.), 22*(1), 17–35.

Booth. (2024). The power of nature (sports)? From anthropocentrism to ecocentrism. *Journal of the Philosophy of Sport, 51*(2), 191–207. https://doi.org/10.1080/00948705.2024.2332907

Bourdieu, P. (1986). The forms of capital. In J. Richardson (Ed.), *Handbook of theory and research for the sociology of education* (pp. 241–258). Greenwood.

Brennan, D. (2021). *Surfing and the philosophy of sport*. Lexington Books.

Demetriou, D. Z. (2001). Connell's concept of hegemonic masculinity: A critique. *Theory and Society, 30*(3), 337–361. https://doi.org/10.1023/A:1017596718715

Elliott, R., & Wattanasuwan, K. (2015). Brands as symbolic resources for the construction of identity. *International Journal of Advertising, 17*(2), 131–144.

Featherstone M, Hepworth M (1995) Images of positive aging. In: Featherstone M, Wernick A (eds) Images of Aging: Cultural Representations of Later Life. *London: Routledge*, pp. 29–47.

Force, W. R. (2009). consumption styles and the fluid complexity of punk authenticity. *Symbolic Interaction, 32*(4), 289–309. https://doi.org/10.1525/si.2009.32.4.289

Ford, N., & Brown, D. (2006). *Surfing and social theory experience, embodiment and narrative of the dream glide*. Routledge.

Gramsci, A. (1971). *Selections from the prison notebooks of Antonio Gramsci*. Lawrence & Wishart Limited.

Green, B. C., & Chalip, L. (1998). Sport tourism as the celebration of subculture. *Annals of Tourism Research, 25*, 275–291. https://doi.org/10.1016/S0160-7383(97)00073-X

Hall, M. (1996). *Feminism and sporting bodies: Essays on theory and practice*. Human Kinetics.

Henderson, M. (2001). A Shifting Line Up: Men, women, and Tracks surfing magazine. *Continuum (Mount Lawley, W.A.), 15*(3), 319–332.

Heywood, L. (2008). Third-wave feminism, the global economy, and women's surfing: sport as stealth feminism in girl's surf culture. In A. Harris (Ed.), *Next wave cultures: Feminism, subcultures, activism* (pp. 63–82). Routledge.

Jarratt, P. (2010). *Salts and suits*. Hardie Grant Publishing.

Jary, D., & Jary, J. (2006). *Collins Web-linked dictionary of sociology* (1st Collins U.S. ed.). Collins.

LaFrenz, C. (2023, April 3). Big names back US group buying Billabong and Quiksilver parent. *Australian Financial Review*. https://www.afr.com/companies/retail/billabong-quiksilver-parent-to-be-sold-to-group-with-star-backers-20230403-p5cxl4

McGregor, C. (1966). *Profile of Australia*. Penguin.

McLeod, K. (1999). Authenticity within hip-hop and other cultures threatened with assimilation. *Journal of Communication, 49*(4), 134–150.

Nettle, S. (2020, October 9). *Michael Tomson dead at 66*. Swellnet. https://www.swellnet.com/news/swellnet-dispatch/2020/10/09/michael-tomson-dead-66

Olive, R. (2015). Reframing surfing: Physical culture in online spaces. *Media International Australia, 155*(1), 99–107.

Olive, R. (2016). Women Who Surf: Female difference, intersecting subjectivities and cultural pedagogies. In A. T. Hickey (Ed.), *The pedagogies of cultural studies* (Vol. 85). Routledge.

Olive, R., McCuaig, L., & Phillips, M. G. (2015). Women's recreational surfing: A patronising experience. *Sport, Education and Society, 20*(2), 258–276.

Pearson, K. (1982). Conflict, stereotypes and masculinity in Australian and New Zealand surfing. *Journal of Sociology, 18*, 117–135.

Plesa, P. (2023). Authenticization: Consuming commodified authenticity to become "authentic" subjects. *Theory & Psychology, 33*(4), 555–576. https://doi.org/10.1177/09593543231174030

Rings, M. (2017). Authenticity, Self-fulfillment, and Self-acknowledgment. *The Journal of Value Inquiry, 51*(3), 475–489. https://doi.org/10.1007/s10790-017-9589-6

Roberts, O. (2023). 5 female-founded women's surf brands we love. *American Surf Magazine*. https://www.americansurfmagazine.com/article/women-surf-brands

Shields, R. (2004). Surfing: Global space or dwelling the waves? In M. Sheller & J. Urry (Eds.), *Tourism mobilities: Places to play, places in play* (pp. 44–51). Routledge.

Sims, C. (2022). *The nature of Gen-Z's influence on the future of printed surf magazines*. Bond University.

Smith-Lahrman, M. B. (1996). Club cultures: Music, media and subcultural capital. Sarah Thornton [Review of *Club Cultures: Music, Media and Subcultural Capital. Sarah Thornton*]. *American Journal of Sociology, 102*(3), 927–928. University of Chicago Press. https://doi.org/10.1086/231032

Statista (2022). *Global surfing industry market size 2022–2027*. https://www.statista.com/statistics/1327319/surfing-market-size-worldwide/

Stedman, L. (1997). From gidget to gonad man: Surfers, feminists and postmodernisation. *Australian and New Zealand Journal of Sociology, 33*(1), 75–90.

Stranger, M. (2010). Surface and substructure: Beneath surfing's commodified surface. *Sport in Society, 13*(7-8), 1117–1134.

Stranger, M. (2011). *Surfing life: Surface, substructure and the commodification of the sublime*. Routledge.

Stratton, J. (1985). Youth subcultures and their cultural contexts. *Journal of Sociology, 21*(2), 194–218.

Tajfel, H. (1978). *Differentiation between social groups: Studies in the social psychology of intergroup relations*. Academic.

Taylor. (2007). Surfing into spirituality and a new, aquatic nature religion. *Journal of the American Academy of Religion, 75*(4), 923–951.

Thompson, J. B. (1995). *The media and modernity: A social theory of the media*. Polity Press.

Thornton, S. (1995). *Club cultures: Music, media and subcultural capital*. Polity Press.

Throsby, D. (1995). Culture, economics and sustainability. *Journal of Cultural Economics, 19*(3), 199–206.

Throsby, D. (1999). Cultural capital. *Journal of Cultural Economics, 23*, 3–12.

Vannini, P., & Williams, J. P. (2020). *Authenticity in culture, self, and society*. Routledge, Taylor & Francis Group.

Varga, S., & Guignon, C. (2020). Authenticity. In E. N. Zalta (Ed.), *The Stanford encyclopedia of philosophy* (Spring 2020 ed.). Stanford University. https://plato.stanford.edu/archives/spr2020/entries/authenticity/

Wann, D. L., & Branscombe, N. R. (1991). The positive social and self-concept consequences of sports team identification. *Journal of Sport and Social Issues, 15*, 115–127.

Warshaw, M. (2005). Winterland. Fred Van Dyke and the dynamics of the aging surfer. *The Surfer's Journal, 14*(4), 46–59.

Wearing, B. (1996). *Gender: The pain and pleasure of difference*. Longman Australia.

Wheaton. (2019). Staying "stoked": Surfing, ageing and post-youth identities. *International Review for the Sociology of Sport, 54*(4), 387–409.

Wheaton, B., & Thorpe, H. (2018). Action sports, the Olympic games, and the opportunities and challenges for gender equity: The cases of surfing and skateboarding. *Journal of Sport and Social Issues, 42*(5), 315–342. https://doi.org/10.1177/0193723518781230

Chapter 12
The Surf Media

Craig Sims

12.1 Introduction

While the traditional purpose of media is to entertain, inform, and hold those in power to account, the surf media also serves to communicate and validate surfing's subcultural values, norms, and symbolisms. This extra role of being a key source of reference for what is acceptable, desirable, and special in surfing is important to the functioning of the surfing economy, the success of which depends on the allure of the surfing lifestyle as it is presented to the mainstream consumer. This chapter examines the evolution of the surf media and explains their contribution to the surfing ecosystem. The growing body of surfing research is then discussed with particular attention paid to two theoretical paradigms apposite to the nature and pace of media change. Through these theoretical lenses, the future of the printed surf magazine is considered. Finally, a short case study introduces a process framework to assist surf media owners, marketers, and brand owners navigate media change, today and into the future.

12.2 A Brief History of the Surf Media

During the early 1950s, when surfers had very little contact with each other beyond their home beach, the surf film was the medium that facilitated an exchange of ideas between surfers (Jarratt, 2010) (Fig. 12.1). Films were few and far between and most came out of Southern California, where surfing's popularity was germinating. Bud Browne, a schoolteacher studying film part-time, produced a 45-minute edit of

C. Sims (✉)
Bond University, Gold Coast, QLD, Australia
e-mail: csims@bond.edu.au

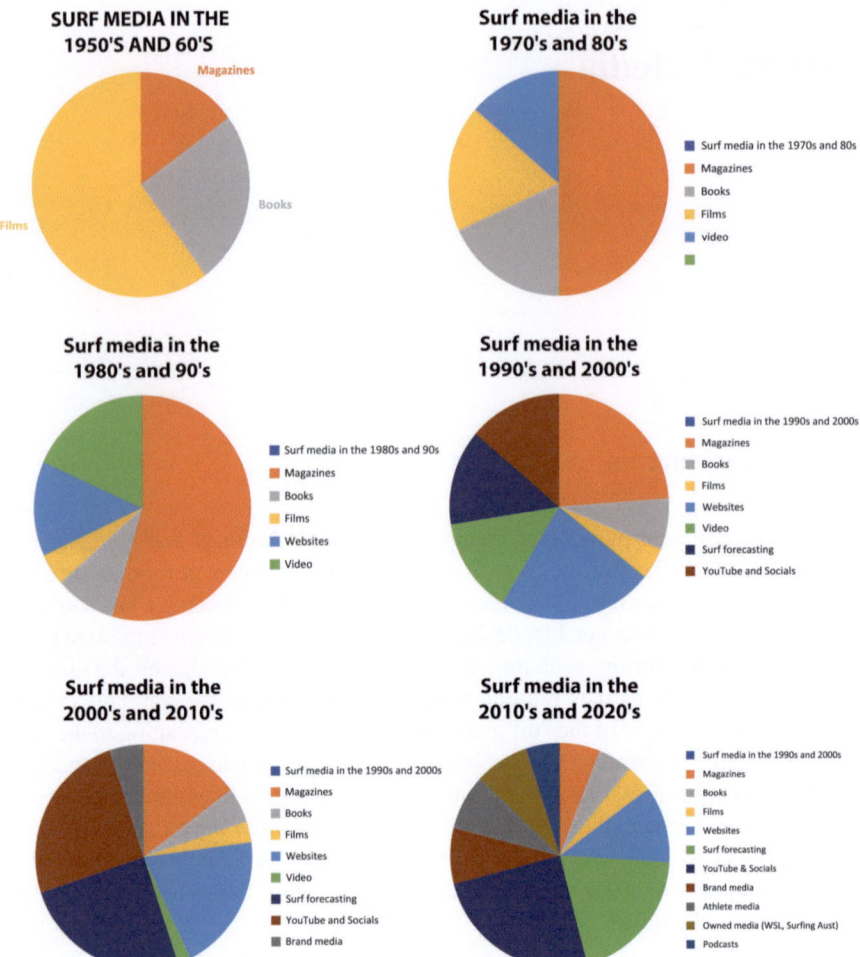

Fig. 12.1 The changing nature of surf media from the 1950s to the present has seen major shifts in the forms of communication used to capture surfing culture

16-mm film footage taken on vacation in Hawaii. He called the film *Hawaiian Surfing Movies* (1953), and it was shown to a packed audience in the auditorium of the John Adams Junior High School in Santa Monica, USA. This was the genesis of the commercial surf film (Jarratt, 2010). The following year, Browne had more success with another film he named *Hawaiian Holiday* (1954), showing it to audiences from Santa Monica to San Francisco. Meanwhile, famous big wave surfer Greg Noll produced a series of movies, all called *Search for Surf* (1957), featuring exotic locations like Australia, Mexico, and the Hawaiian island of Molokai, where surfing was still relatively unknown to Californian surfers. One of the surfers featured in Noll's films was Bruce Brown, a talented filmmaker who later would take the surf travelogue genre to a whole new level, with the release of the film *The Endless Summer*

(Brown, 1966), which popularized surfing and surf culture globally, inspiring many to take up the sport and seek out new surfing destinations (Chap. 13). The most influential film of this era, however, was not one that emerged from the surfing community. *Gidget* (1959) was a Hollywood movie made from a book of the same name written by Frederick Kohner in 1957 about his 15-year-old daughter Kathy, who spent the summer of 1956 fraternizing with surfers on Surfrider Beach in Malibu. The story reflected the surfing lifestyle in all its "wine-swilling, wave-ripping, sex-crazed" glory, accurately reflecting what was happening up and down the Californian coast (Jarratt, 2010, p. 64). *Gidget* was distributed worldwide showcasing the surfing lifestyle to a global audience activating dormant markets for surf cultures in places like Europe and South Africa and accelerating the growth in existing markets like North America, Hawaii, and Australia. The Australian public's fascination with surf culture, post *Gidget*, is supported by McGregor (1967, as cited in Stranger, 2011), who described the extent to which surf culture pervaded Australian mainstream consciousness at the time:

> All the mass media and channels of publicity have thrown their weight behind the surfies: the Sunday newspapers carry surfing supplements, disc jockeys plug surf music remorselessly, the advertising agencies flatter and glamorize the beach life. They know what the coming thing is. (Stranger, 2011, p. 188)

A few months after *Gidget* the movie was first released, a young surfer and talented artist named John Severson, who dabbled in surf movies while doing military service in Hawaii, produced a 36-page black and white magazine to promote his surf movie tour which featured two of his movies, *Surf Safari* and *Surf Fever* (Severson, 1960). He named the magazine *The Surfer* and sold it at his movie showings. He printed 5000 copies and it sold out before the movie tour had completed. In 1962, *Surfer* was launched as a quarterly magazine, becoming surfing's first dedicated magazine (Jarratt, 2010) which would remain a pillar of surf culture for another six decades.

The surf media industry boomed during the late 1960s and 1970s, with books, magazines (Fig. 12.2), and film productions driving the diffusion of surf culture around the world (Fig. 12.1). The surf media reflected the countercultural spirit of surfing, an image that attracted more surfers and non-surfers and which laid the foundation for the fledgling surfing industry. Through the 1980s and 1990s, both the surf media and the surf industry were engaged in a productive symbiosis with surf brands providing advertising support to the magazines that fueled the subculture they profited from (Fig. 12.1). Competition among advertisers for premium space was fierce as brands sought to express their importance in the subculture through their advertising placements. Surf magazine editors were gatekeeping powerbrokers whose editorial decisions had the capacity to profoundly affect the credibility of the surf brands and the sponsorship value of their surfers.

By the late 1980s, the surf video had lowered the barrier to entry for budding surf filmmakers, and video productions featuring high tempo surfing accompanied by equally high tempo music become the standard genre. Surf brands embraced the technology, expanding their promotion through the release of video productions

Fig. 12.2 A selection of classic surf magazine covers. (With permission from Keith Grisman— Private Collection)

featuring their sponsored team riders, which often become DVD cover mounts on surf magazines (Booth, 1996). Indeed, Booth (1996) declared the surf industry dominated video production to such an extent that that they ultimately "squeezed out the independent producers" (p. 19). Stranger (2001) summed up the pre-Internet surf media landscape as follows:

> Collections of surfing magazines are pored over countless times and favourite images adorn the walls of surfers' bedrooms and beyond. Surf videos are watched repeatedly, either as whole programs or in fragmented snippets as viewers scan for their favourite sequences or watch short segments as the whim and opportunity coincide. (p. 188)

12.2.1 Digital Disruption

By the mid-1990s, the arrival of the Internet attracted copious investment along with the predicted death of any media without a domain name (Fig. 12.1). In 1999, swell. com launched with surprising aggression hiring the best surf editors, photographers, and filmers to form a global surf media empire by locking down exclusive access to the world's finest content creators. But a year later, it all fell apart when the dot-com bubble burst as the initial wave of Internet investment collapsed. Meanwhile, surf magazines scrambled to respond to this new existential threat, hastily building websites as extensions of their print products. A flood of online-only surf media followed, and within a short period of time, the surf media landscape became fragmented and overcrowded.

Around the turn of the century, the Internet pivoted toward Web 2.0, offering the interactivity that heralded the arrival of social media and resulted in further fragmentation of the surf media. By now, the low barriers to entry enabled anyone, and indeed everyone, to be a publisher of surf content. Surf magazines, once the gatekeepers to the culture—the *influencers* of the pre-Internet era—were under threat. Both advertising and copy sales plummeted as advertisers and readers migrated to the flood of website and social media accounts delivering surf content more frequently and in a shorter timeframe. With declining revenues and tighter production budgets, magazine staff found themselves spread thin trying to curate their magazines while attempting to populate their own digital media channels on an ever-shortening news cycle. The result was that magazine content diminished in depth and substance and was no longer distinguishable from what was available for free via the Internet. As revenues continued their downward spiral, the cacophony of claims about the death of print grew louder.

Amidst the confusion and reactionism of digital disruption, Surfline, a surf forecasting service that started in 1985, offering twice-daily pay-per-call telephone surf reports in Southern California, saw a gap that surf magazine publishers missed. While the surf magazines were preoccupied with restructuring their operations to keep up with their new online competitors, Surfline understood that surf conditions, not surf news, were what surfers cared most about. So, during the turmoil of disruption, in 1995 Surfline quietly migrated their service to the Internet drawing on wind and wave data provided by the National Weather Service to provide daily surf reports and surf forecasting with real time updates. A year later, they launched their first live camera feed overlooking the waves at Huntington Beach. Today, Surfline is an international surf media juggernaut providing accurate long-range forecasts and live camera feeds to most of the world's surfable breaks, as well as surf news, feature articles, and a YouTube channel. Because they hold the key to answering the first question all surfers want to know—what is the surf like?—Surfline is the first port of call for most surfers and has the largest audience of the formal surf media sector.

The World Surf League (WSL), the governing body of professional surfing, has also capitalized on the rapid development of digital communication technologies. During the 1980s, 1990s, and 2000s, the surf league relied on the media to broadcast event-related stories, photos, and video footage to surf fans around the world. Over this period, mainstream media and the endemic surf media were invited to major surfing events where plush media hubs were created, food and drink was provided, and life was made very easy for content producers and media executives. Today, the WSL restrict the capture of any media content within the confines of their events and have much less contact with the media. This is because the means of capturing and broadcasting the stories, photos, and video footage is now owned and tightly controlled by the WSL due to the availability of high-speed broadband, live streaming technology, and the low cost of filming and editing equipment. The mainstream media and the endemic surf media are no longer an essential component of the WSL's business model, and as a result, the WSL's relationship with the endemic surf media is distant and sometimes hostile.

Social media influencers have brought a new dimension to the complex and fragmented surf media landscape. Initially, surfers who were famous for their competitive surfing prowess were sponsored by surf brands that promoted them through their advertising and promotional campaigns. Their high-profile status drove followers to their personal social media platforms in large numbers. Some surfers now have a larger social media following than the sponsors who initially made them famous, and as a result, they are now brands in their own right. For example, when John Florence left his sponsor, Hurley, in January 2020, his personal Instagram account had as many followers as his sponsor (1.5 million followers). The transition to his own surf clothing brand, Florence Marine, was enabled by his vast social media following, which was enabled, at least in part, by Hurley who promoted John extensively through the media over the 6 years of their sponsorship deal. The surfing-influencer sector has now expanded beyond successful competitive surfers to include specialists in all spheres of the sport and lifestyle. Jamie O'Brien, for example, is a video blogger specializing in risky surfing antics. He has over 1.2 million subscribers to his YouTube channel, more than the WSL, and another 1.3 million followers of his Instagram account. Indeed, O'Brien has become a media empire unto himself, selling merchandise to fans and advertising opportunities to brands. These surfing-influencers, along with a growing legion of podcasters, represent a new, emerging informal sector of the surf media.

The surf film genre has also undergone transformation due to the widespread availability of digital production technologies and access to distribution via streaming platforms (Boyd, 2024). Independent video producers, previously squeezed out by the big brands, returned with edgy, grassroots productions that proved popular at a local or tribal level (Stranger, 2011). Today, the surf film genre includes a broad spread of film productions screened via a network of surf and adventure film festivals and "a thriving online film culture via video hosting sites such as YouTube, Vimeo and Garage Entertainment" (Boyd, 2024, p. 946). Boyd (2024) argues that "in an era when surf magazines have lost their role as a primary communicator of surf culture, surf films (and online videos) have an enhanced role within the subculture" (p. 946).

12.3 The Influence of the Surf Media

Despite the Internet and a slew of digital media options fragmenting the surf media landscape that was once dominated by printed magazines and films, the surf media remains a cornerstone of the surfing subculture. Olive et al. (2015) explain that a subculture is an always evolving phenomenon that is negotiated daily by the people and contexts within it and suggest that the surf media can be "as productive of a culture as they are reflective of it" (p. 262). Booth (2008) suggests the nature of the surf media's influence is more direct, explaining that they "frame the cultural precepts of young surfers, telling them how to think and act" (p. 17). Whether it be through film/video, text, or still images, the surf media reflects a sense of and an insight into the surfing lifestyle, allowing surfers and non-surfers to experience the subculture they relate and aspire to. Indeed, Stranger (2011) asserts the surf media play a crucial role as an arbiter and disseminator of surfing's cultural norms and values. This is important to the businesses that comprise the surf industry that, according to Ford and Brown (2006), have developed "lucrative niche markets" due to the successful articulation of surfing's cultural "values and symbolisms ... to the wider non-surfing society" (p. 57). (Chap. 9). Since the surf media are so central to the culture and industry, it is appropriate to turn our attention to the study of surf media, starting with a brief overview of the scholarly research followed by an investigation into the most popular and relevant theoretical paradigms for media research.

12.4 Surf Media Research

There is a growing body of scholarly work exploring and explaining the many facets of surfing. Within most of this research, media plays a central role. For example, Ormrod (2005) looked at how the surf movie *The Endless Summer* (Brown, 1966) instigated a cultural fascination with surfing in America during the 1960s (Chap. 13); Ponting et al. (2005; Ponting and McDonald 2013) investigated the surf media's role in the growth of surf tourism in the Mentawai islands; and Evers and Phoenix (2022) studied how emotional labor is practiced by male free surfers who are paid to live an aspirational lifestyle and communicate this through their digital media work. However, very little research has been conducted into surf media itself, and much of what has been studied concerns surfing's masculine hegemony and its disturbing track record of sexism and misogyny showing the surf media to be "an ideological apparatus of gender that frames readers' thoughts and actions" (Booth, 2008, p. 30). For example, Stedman (1997) and Henderson (2001) conducted a systematic review of *tracks* magazine's representation of women over decades of publication (Fig. 12.2); Booth (2008) also conducted an investigation into *tracks*, focusing more on the affective characteristics that connect that magazine with young male surfers; and Olive (2015) investigated how media representations by women on Instagram are challenging and reinforcing the sexualization and differentiation of women in surf culture. Other studies that focused on surf media directly include—but are not limited to—Booth (1996) who categorized nearly five decades of the surf film and video industry into distinctive genres, Sims (2022) who considered the nature of Generation Z's influence on the future of the printed surf magazine, and Boyd (2024) who analyzed the effect of Irish surf films on the commercialization of Irish surf culture. The theoretical scope of studies into surf media is as broad and varied as the fields of inquiry; however, uses and gratifications (U&G) theory and media substitution theory (MST) are effective theoretical approaches that can help us to understand responses to the abundance of choice and ongoing disruption in the current media landscape.

U&G theory has been a reliable theoretical approach in the initial stages of developing understanding of each new mass communications medium over the last two decades (Lin, 1993; Ruggiero, 2000; McQuail, 2010). U&G theory recognizes that individuals actively make media choices for the gratification of their needs and that they are able to discern the reasons for making such choices (Katz et al., 1973). This contrasts with one of the largest fields of communications study, media effect theory, which views people as passive consumers of media with predictable responses to them (Vorderer et al., 2020). So, while media effect theory focuses on what the media do to audiences, U&G theory focuses on what audiences do to the media through the choices they make about the media they consume. The U&G approach is most useful when there is a range of media options to choose from, and choice becomes a function of personal, social, and psychological needs (Kilian et al., 2012). By examining the motives behind the consumption of a particular form of media, researchers can understand the reasons for that medium's popularity, or failure, and the roles that the medium fills in society.

McQuail (1983) advanced four common reasons for media use: information, personal identity, integration and social interaction, and entertainment. Personal identity and integration and social interaction are especially relevant to surfers, whose acceptance into the surfing subculture is largely dependent on their demonstration of insider knowledge (Langseth, 2011; Stranger, 2010). The entertainment motive also has relevance to the surfing subculture, which has been described as a hedonistic lifestyle that offers a diversion from the rigors of everyday life (Humphreys, 1997). In recent years, habit and trust have emerged more frequently in media studies concerning the pervasive use of the Internet, smartphones, and social media, especially among youth where their use is prolific (Valkenburg & Piotrowski, 2017). For this reason, Sims (2022) added Habit and Trust to McQuail's (1983) typology in his study of the future of the printed surf magazine.

A suitable approach for examining the effect of digital disruption is media substitution theory (MST). This theory asserts that when a new media technology is introduced, audiences redistribute the allocation of their time among available media options and, as a result, new patterns of media consumption emerge (Kaye & Johnson, 2003). Extending MST, Adoni and Nossek (2001) advanced three possible outcomes relating to the extent to which one media option is substituted for another:

1. Displacement: as a result of the new medium being "functionally equivalent," creating conditions that cause the incumbent to become either displaced or obsolete
2. Coexistence: from a process of "functional differentiation" whereby the incumbent makes adjustments that facilitate a functional point of difference and reestablish its uniqueness
3. Convergence: which is a "functional synthesis" of the two media sources resulting in a new medium that simultaneously utilizes the functionality of both

MST focuses on time allocation as a key mechanism of media substitution. The assumption is that time is finite and, consequently, increased time spent on one medium will result in decreased time spent on another. This concept of media use as a zero-sum gain can be challenged. Billings et al. (2015) describe how, in the digital age, the existence of multiple platforms has allowed for "simultaneous and intermittent selections to jointly aid the quest of the optimal media gratification" (p. 3). This information was advanced to explain the existence and use of second and third screen viewing of sport and to suggest that when a particular need is not being adequately fulfilled by one source, audiences will satisfy that need via other sources simultaneously. Nevertheless, MST is still an effective theoretical framework to explain and understand responses to disruption.

As an example of how these theories can apply, Sims (2022) drew from a participant group of Australian surfers aged between 15 and 60 years ($n = 1039$), to identify the most preferred media source from an adapted set of motives for media choice originally developed by McQuail (1983). He found websites enjoy a strong association with Information, Entertainment, and Habit; magazines enjoy a strong association with Trust and Identity Development; and social media had a strong association with Interaction and Habit. Interestingly, it was found that the participant group engaged in high use of social media but had low trust in them and, paradoxically, they engaged in low use of magazines but had high trust in them. Taking an MST approach, Sims (2022) then argued that in the face of extreme disruption, if surf magazines are going to survive in the digital age, they should focus on the MST option of Coexistence. This would require surf magazines to functionally differentiate by capitalizing on their strength, which is the authority that comes with trustworthiness and the influence that this authority has on identity development. He also suggested that printed surf magazines should focus on providing high quality, well-researched content to maintain the trust that has been attributed to the medium. To this end, Sims (2022) argued that surf magazines could functionally differentiate themselves from their digital disruptors by concentrating on content about surf travel and technique. This was informed by his finding that certain media sources were better suited to the delivery of certain content types as indicated in Table 12.1.

Another theoretical construct that is relevant to responses to the ever-changing media environment is creative destruction, a concept introduced by Joseph Schumpeter (1943) who, in explaining why he refutes Karl Marx's view that economies grow as a result of competition between companies, argued that it is competition between companies over new innovations that drives economies. Here Schumpeter suggests that companies with an innate propensity to innovate will not only gain a substantial and lasting competitive edge but collectively will drive economies forward. He also suggested that new innovations would destroy established companies that fail to innovate or disrupt themselves, a process he called *creative destruction* (Storsul & Krumsvik, 2013). Indeed, creative destruction aptly describes what happened to the printed surf magazine with the arrival of the Internet. Surf

Table 12.1 Content suitability by media source

Media source	Content type
Magazines	Travel and technique
Social media	Competition and interviews
Websites	Conditions and travel
Television	Competition and conditions

magazines, once the primary communicators of surf culture (Boyd, 2024), missed the opportunity to maintain their connection to the surfing public through the delivery of surf forecasts and then later surf cameras. This failure to disrupt their production process is a prime example of creative destruction within the surf magazine sector—an ironic outcome for such a creative medium. Despite claims about the imminent death of print in the broad media context, surf magazines have somehow displayed remarkable resilience and remain an important, albeit significantly reduced, contributor to the maintenance and dissemination of surfing's cultural norms and values. This anomaly warrants further analysis.

12.5 The Printed Surf Magazine in the Digital Age

Applying their MST model of interactions, Nossek et al. (2015) found that displacement of print media by their digital equivalents was not occurring; instead, they identified that sophisticated audiences who are aware of their "idiosyncratic psychosocial needs" are finding ways to "use various media simultaneously to fulfil those needs in the best possible way" (p. 381). Nossek and colleagues suggest this is because readers favor specific features of a medium that go beyond their functional utilitarian value, "such as the smell and touch of paper and various book formats" (p. 380).

The qualities of print beyond their functional utilitarian value have been well documented: as a tactile object, with material substance and permanence (Kitch, 2009), as a signifier of higher social status (Bourdieu, 1984), and as a collectable material artifact that people keep to enhance identity or remember major or significant events (Kitch, 2009). Bonner and Roberts (2017), Webb and Fulton (2019), and Haniff (2012) found aesthetics, which includes overall look and feel, design, and photography, to also be a unique distinguishing feature of magazines. These result in an emotional connection, which Randle (2003) posits is the affective characteristic of magazines. The affective attachment to magazines is not easily articulated and is often expressed in a colorful, if not idiosyncratic, description. Sax (2016), for example, expressed this through a quote from Monocle magazine's Editor, Andrew Tuck, who observed that:

> There is no romance in the world of digital. In a gentle way, there is romance about the print product. It is tactile, beautiful, and you can smell the ambition on the page. You can't smell ambition when you are on a website. (p. 113)

The ability of magazines to elicit an emotional connection is explained by Rosenblatt's (1986) transactional theory, which analyzes the reciprocal interactions between the reader and text. Rosenblatt's aesthetic stance recognizes that the reader derives an emotional, aesthetic, and intellectual experience when consuming texts (such as books and magazines). The aesthetic stance recognizes that the reader is attentive not only to the content but also to the feelings evoked. These feelings are experienced when the reader is open to the "experiential aura" (p. 125) that the text offers. This experiential aura can also originate from the aesthetics of the text, such as color, design, and photographic images (p. 127). Photographic images are especially relevant to surfing because surfing is such a visual activity (Ford & Brown, 2006). Indeed, Ford and Brown state that "Although the words of surfing magazines are important, what really drives the purchase of such material is their rich and evocative visual images" (Ford & Brown, 2006, p. 33).

12.5.1 The Primacy of the Surfing Photograph

According to Stranger (2011), a central focus of surf magazines is photographic representations of the "pursuit of an ecstatic communion with nature" (p. 118). Ford and Brown (2006) contend that the act of surfing requires a presence of mind that "denies that momentary reflection which allows the possibility of imprinting an image upon the mind's eye" (p. 33). They argue that the observer of surfing sees the act of surfing from a similar perspective—a flowing series of idiosyncratic responses by the surfer to the changing form of the wave, which is over in a matter of seconds. The act of surfing on a wave is, therefore, an ephemeral event—an activity bound by space and time that leaves no trace of its existence (Ford & Brown, 2006). A surf photo captures a moment in this fleeting interaction between the surfer and the wave, allowing the viewer to take in all aspects frozen within the frame—the sunlight, texture of the water, shape of the wave, surfer's posture, position of the board, and shards of spray emanating from it, as well as features of the background and the foreground. The richness of detail provided in the frozen moment illuminates the surfing aesthetic.

Ford and Brown (2006) suggest images of surfing trigger a desire that stems from the experience of surfing: "an experientially sedimented stimulus" (p. 142). We identify detail that aligns with our level of ability, sense of place, and kinship with the culture, so the memories and feelings it evokes are unique to each of us. Barrett (1985) describes this as the "interpretive ambiguity" of photographs. Where a still image freezes the moment allowing the viewer to take the time to study and interpret every aspect in detail, surf video captures movement and the dynamic interplay between the surfer and the wave. According to Stranger (2011), video is a closed narrative with no mystery and no fantasization, while photos, he contends, offer an open-ended narrative to take forward as your experience allows. Thus, the still surfing photo represents not just the capture of reality but also the creation of it (Stedman, 1997) as exemplified in the excerpt below:

> In the early Summer of 1966, twelve-year-old Robert Buchner McKnight Jr cut a photo out of Surfer magazine and tacked it to the wall beside his bed. The photo showed 17-year-old Jeff Hackman, who had just won the Duke Kahanamoku Invitational crouched on the nose of his board, driving hard through the inside section of a Sunset Beach wave that seemed five times taller than him. Bob wanted to ride waves like that, and as he lay in bed, waiting for sleep, he imagined himself doing so on his new Harbour Surfboards Bannana Model that stood in a corner of the room where he could watch over it. (Jarratt, 2010, p. 130)

Ford and Brown (2006) also argue that rich and evocative images tap into a repository of memories that "engage the body and senses" (p. 41), eliciting "a powerful aesthetic, contemplative experience" (p. 42) and creating "fantasies of possible lives (p. 52). As such, still surf photos can be "a stronger stimulant to action than intellectual lines of argument" (p. 42). This is why most surf advertisements preference photos over words.

So, to conclude, I propose that the printed photograph is a powerful and evocative medium for recording surfing. And through their superiority for showcasing surf photography, surf magazines are both eloquent and resonant in their depiction of the surfing experience, which is the foundational layer that underpins both the subculture and the industry. This, I believe, is a key reason why surf magazines have displayed such resilience in the face of digital disruption.

All surf media face disruptive pressure because media is continuously evolving, and creative destruction is an ever-present threat. Drawing on my experience as a surf media publisher in South Africa, New Zealand, and Australia, I have developed a seven-step model to build resilience in the face of a rapidly evolving media landscape. This model provides a response framework that is as relevant to surf media owners as it is to surf marketers and surf brand owners. The model is best presented through a case study.

12.6 A Case Study: Facing Media Change

Between 2008 and 2017, I worked as the Publisher of seven action sports magazine brands at Morrison Media, a publishing company based in Queensland, Australia. In April 2010, we invested a great deal of time and money into what looked like salvation for publishers, the Apple Newsstand. This was a digital platform for magazine publishers to sell digital editions of their print publications. It provided a desperately needed way to monetize digital content, something that had eluded them for several years. At Morrison Media, we were quick to respond to the invite by Apple to be on the beta testing team and get the early adopter (Rogers, 2003) advantage over our competitors. We restructured production teams, committed ourselves to weeks of R&D, and purchased the required software (Adobe Digital Publishing Suite) and the accompanying prepaid download packages. We were "all in"! We had to be because we had committed substantial upfront investment in the software and download packages. But the results were lower than expected. Despite this, we persisted justifying our strategy as a clear demonstration to our advertisers that our magazine brands had an expanding digital footprint. However, in August 2015, with the release of iOS9, Apple discontinued the Newsstand without warning and announced the launch of Apple News, along with a string of monetizable options to attract content creators. In response, Adobe changed (and renamed) their Digital Publishing Suite software to enable new functionality and abandoned their download packages with many publishers still having hundreds of prepaid downloads in credit. Overnight, publishers lost their digital edition distribution platform.

Technology had made digital editions redundant before they even gained market traction. To many small magazine publishers, this was the last straw, and they threw in the towel.

12.6.1 A Process Framework for Resilience Against Media Change

Since the fifteenth century when Johannes Gutenberg invented the printing press, technology has been the driver of media change. The lesson this case study offers is that the ongoing interaction between technology and media creates an unstable media environment that is constantly changing. Under these conditions, it is dangerous to set a digital strategy and commit to it wholeheartedly. A more prudent digital strategy would be to get involved early so as not to lose the early adopter advantage but start small and progress iteratively while being hawkishly vigilant and prepared to adjust on demand. In simpler terms, under these conditions, it is best to move fast but take small bites and evaluate continuously as you go. This can be further distilled into a convenient mantra: "It's not what you know, it's how you respond to the unknown." The rationale is in the rapidly evolving media landscape—what you know today will be redundant tomorrow, so the focus should be on how you respond to the inevitability of change. In addition to the mantra mentioned above, I propose a seven-step process framework to assist decision-makers to successfully navigate change in a rapidly evolving media environment, today and into the future. The process framework below is followed by a brief explanation of each step.

1. Be curious.
2. Identify the issues.
3. Analyze the context.
4. Interpret the issue.
5. Strategize a response.
6. Implement in stages.
7. Repeat.

Be curious
While shifts in consumer behaviors ultimately reveal themselves in market data, it is usually after the fact. Real time detection is ideal, and this requires a sensitivity that comes with being curious. Similarly, shifts in technology are not always publicly announced, as it was in this case study. New technologies and innovations often diffuse into the market slowly (Rogers, 2003), and cultivating a curiosity about new technologies will help to remain in touch with innovations in the early stages of their release. "Be curious" is a call to question everything and remain alert.

Identify the shift
Shifts in media uses and/or audience gratifications can also be insidious. There is a tendency to label a shift as an anomaly that will correct next week, month, or

quarter. This delays taking appropriate steps to address the shift. For this reason, identifying change and labeling it correctly—as a shift to be addressed or an anomaly to be ignored—are important steps in being responsive.

Analyze the context

Systems thinking is the process of recognizing the existence, influence, and interdependence of the individual parts of a larger system (Senge, 1992). This reminds us that shifts always occur against the backdrop of something greater than the problem area itself. An example could be changes in technology, audience behavior, or competitor activity. Analyzing the context of change allows us to understand the broader underlying causes and interpret the issue more accurately.

Interpret the issue

The application of the first three steps in this framework lays a sturdy foundation for the application of this, the fourth. Interpret the issue is a call to understand the cause and effects of the shift. As an example, taking a U&G perspective will assist in understanding the shift as a potential response to the fulfillment of certain psychosocial needs, while analyzing it from an MST perspective will assist in understanding the shift as a potential consequence of the allocation of time and attention given to a medium due to the arrival of a new alternative.

Strategize a response

Having established that a shift is indeed underway and the context and nature of it is understood, you are well placed to respond. It is not uncommon to get to this point and not proceed because confidence in the strategic response is low. Indeed, in their proposed model for managing organizational change, Robbins and Barnwell (1998) declare that "successful change requires unfreezing the status quo" (p. 339) as step 1 in the change process. The key to this step is to decide whether the response requires a system or a process change. A system change might entail the adoption or invention of a new media platform or, as in this case study, the adoption of a new distribution platform. A process change might entail an adjustment to any of the steps taken in order to make a system work more efficiently. Using this case study as an example, a process change would be to populate our digital editions with more hyperlinks to make them more engaging and current.

Implement in stages

This step advocates for an iterative approach to the rollout of any response. Taking small steps and monitoring them to ensure a desired outcome before proceeding with the next step have multiple advantages. The first is to allow for strategic adjustments in the event that the shift originally identified evolves. With the pace of change in media today, this is a realistic possibility. The second advantage of an iterative rollout is the opportunity to make adjustments if the agreed strategy is not achieving the desired result, and finally, an iterative rollout alleviates the wasted investment of a fully committed wholesale strategic change. All of the scenarios used to support an iterative approach to implementation occurred in this case study. Had an iterative approach been applied, it's likely that Apple's shock closure of their digital newsstand would have had a less catastrophic effect on the many small publishers who bought into the idea that this was the panacea for the digital disruption of the printed magazine.

Repeat
The final stage advocates for the adoption of a culture of continuous learning and improvement. Even if the desired strategic response to the identified shift is effective, ongoing curiosity will ensure that the inevitability of future shifts will be identified, analyzed, and interpreted, laying a sturdy foundation for strategizing and implementing a suitable response. While this framework appears to draw on the five stages of the design thinking process, which is a nonlinear iterative process, my framework advocates a linear process that is circular ensuring it is continually repeated. This is an important cultural precondition to counteracting the possibility of falling victim to creative destruction.

12.7 Conclusion

In this chapter, we examined the state of the surf media by situating the present in the context of its historical continuum. In an environment as fluid as media, understanding its evolution from the past greatly enhances our understanding of the present and, indeed, our predictions for the future. It was never the intention to document a complete and comprehensive summary of the evolution of the surf media as this would be better suited to a book than a chapter. The focus instead was on the remarkable resilience of the printed surf magazine because while magazines have diminished in both number and influence in most other interest categories, it seems to be occurring to a lesser extent in surfing. This makes the surf media and indeed the surfing subculture somewhat unique. To understand the unusual composition of the surf media sector, it was first necessary to explain the surf media's influence on the culture and business of surfing. More depth about this is provided in Chap. 11.

The surf media's value to the subculture is highlighted by the fact that the surf media features prominently in most scholarly analyses of surf culture. There are many theoretical frameworks through which media phenomena are analyzed, but this chapter proposes two media theories that are relevant to the current surf media landscape. The subsequent application of these theories as a lens through which the printed surf magazine can achieve functional differentiation demonstrates how theory can inform practice.

The act of riding the ocean's waves is a fundamental underpinning feature of the surfing ecosystem (Stranger, 2011), and while video might seem like the best medium through which this act is showcased, it has been widely argued that the printed photograph is both powerful and evocative in its representation of surfing (e.g., Stranger, 2011; Ford & Brown, 2006; Sims, 2022). For this reason, we investigated the special characteristics of print, and then the surf photograph, to explain why the humble surf magazine demonstrates such resilience in the digital age. Finally, to assist decision-makers to successfully navigate change in a rapidly evolving media environment, a seven-step process framework is advanced and explained through the lived experience of this author.

References

Adoni, H., & Nossek, H. (2001). The new media consumers: Media convergence and the displacement effect. *Communications. The European Journal of Communication Research, 26*(1), 59–83.

Barrett, T. (1985). Photographs and contexts. *The Journal of Aesthetic Education, 19*(3), 51–64. https://doi.org/10.2307/3332643

Billings, A. C., Qiao, F., Conlin, L., & Nie, T. (2015). Permanently desiring the temporary? Snapchat, social media, and the shifting motivations of sports fans. *Communication & Sport, 5*(1), 10–26.

Bonner, E., & Roberts, C. (2017). Millennials and the future of magazines: How the generation of digital natives will determine whether print magazines survive. *Journal of Magazine & New Media Research, 17*(2), 1–13.

Booth, D. (1996). Surfing films and videos: Adolescent fun, alternative lifestyle, adventure industry. *Journal of Sport History, 23*(3), 313–327.

Booth, D. (2008). (Re)reading the surfers' bible: The affects of tracks. *Continuum (Mount Lawley, W.A.), 22*(1), 17–35.

Bourdieu, P. (1984). *Distinction: A social critique of the judgment of taste*. Harvard University Press.

Boyd, S. (2024). Beyond the noise: The cultural (or subcultural) politics of Irish surf films. *Sport in Society, 27*(6), 946–964. https://doi.org/10.1080/17430437.2024.2334599

Brown, B. (Director). (1966). *The endless summer* [Film]. Cinema V distribution.

Evers, C., & Phoenix, C. (2022). Relationships between recreation and pollution when striving for wellbeing in blue spaces. *International Journal of Environmental Research and Public Health, 19*(7), 4170. https://doi.org/10.3390/ijerph19074170

Ford, N., & Brown, D. (2006). *Surfing and social theory experience, embodiment and narrative of the dream glide*. Routledge.

Haniff, Z. (2012). *Niche theory in new media: Is digital overtaking the print magazine industry?* [Masters thesis, University of Navada]. ProQuest Dissertations Publishing.

Henderson, M. (2001). A shifting line up: Men, women, and tracks surfing magazine. *Continuum (Mount Lawley, W.A.), 15*(3), 319–332. https://doi.org/10.1080/10304310120086803

Humphreys, D. (1997). 'Shredheads go mainstream?' Snowboarding and alternative youth. *International Review for the Sociology of Sport, 32*(2), 147–160.

Jarratt, P. (2010). *Salts and suits*. Hardie Grant Publishing.

Katz, E., Blumler, J. G., & Gurevitch, M. (1973). Uses and gratifications research. *Public Opinion Quarterly, 37*(4), 509–523.

Kaye, & Johnson, T. J. (2003). From here to obscurity?: Media substitution theory and traditional media in an on-line world. *Journal of the American Society for Information Science and Technology, 51*(3), 260–273.

Kilian, T., Hennigs, N., & Langner, S. (2012). Do Millennials read books or blogs? Introducing a media usage typology of the Internet generation. *Journal of Consumer Marketing, 29*(2), 114–124.

Kitch, C. (2009). The afterlife of print. *Journalism, 10*(3), 340–342.

Langseth, T. (2011). Liquid ice surfers-the construction of surfer identities in Norway. *Journal of Adventure Education and Outdoor Learning, 12*(1), 3–23.

Lin, C. A. (1993). Exploring the role of VCR use in the emerging home entertainment culture. *Journalism Quarterly, 70*(4), 833–842.

McQuail, D. (1983). *Mass communication theory: An introduction*. Sage.

McQuail, D. (2010). *McQuail's mass communication theory* (6th ed.). SAGE.
Nossek, Adoni, H., & Nimrod, G. (2015). Is print really dying? The state of print media use in Europe. *International Journal of Communication (Online)*, 365–385. https://ijoc.org/index.php/ijoc/article/view/3549
Nyre, L. (2013). Editors: Tanja Storsul, Arne H. Krumsvik: Media innovations. A multidisciplinary study of change. *Norsk Medietidsskrift, 20*, 283–287. https://doi.org/10.18261/ISSN0805-9535-2013-03-11
Olive, R. (2015). Reframing surfing: Physical culture in online spaces. *Media International Australia, 155*(1), 99–107. https://doi.org/10.1177/1329878x1515500112
Olive, R., McCuaig, L., & Phillips, M. G. (2015). Women's recreational surfing: A patronising experience. *Sport, Education and Society, 20*(2), 258–276.
Ormrod, J. (2005). Endless summer (1964): Consuming waves and surfing the frontier. *Film & History, 35*(1), 39–51. https://doi.org/10.1353/flm.2005.0022
Ponting, J., & McDonald, M. G. (2013). Performance, agency and change in surfing tourist space. *Annals of Tourism Research, 43*, 415–434. https://doi.org/10.1016/j.annals.2013.06.006
Ponting, J., McDonald, M., & Wearing, S. (2005). De-constructing wonderland: Surfing tourism in the Mentawai Islands, Indonesia. *Loisir et Société, 28*(1), 141–162. https://doi.org/10.1080/07053436.2005.10707674
Randle, Q. (2003). Gratification niches of monthly print magazines and the world wide web among a group of special-interest magazine subscribers. *Journal of Computer-Mediated Communication, 8*(4), 00.
Robbins, S. P., & Barnwell, N. (1998). Organisation theory: concepts and cases (Third edition.). Prentice Hall.
Rogers, E. (2003). *Diffusion of innovations* (5th ed.). Free Press.
Rosenblatt, L. (1986). The aesthetic transaction. *The Journal of Aesthetic Education, 20*(4), 122–128.
Ruggiero, T. E. (2000). Uses and gratifications theory in the 21st century. *Mass Communication & Society, 3*(1), 3–37.
Sax, D. (2016). *The revenge of analog: Real things and why they matter*. PublicAffairs.
Schumpeter, J. A. (1943). *Capitalism, socialism, and democracy*. Allen & Unwin.
Senge, P. M. (1992). *The fifth discipline: The art and practice of the learning organization*. Random House Australia.
Severson, J. (Director). (1960). *Surf fever and surf safari*. John Severson.
Sims, C. (2022). *The nature of Gen-Z's influence on the future of printed surf magazines*. Bond University.
Stedman, L. (1997). From gidget to gonad man: Surfers, feminists and postmodernisation. *Australian and New Zealand Journal of Sociology, 33*(1), 75–90.
Stranger, M. (2001). *Risk-taking and postmodernity: Commodification and the ecstatic in leisure lifestyles – The case of surfing* [Unpublished Phd Thesis]. University of Tasmania.
Stranger, M. (2010). Surface and substructure: beneath surfing's commodified surface. *Sport in Society, 13*(7–8), 1117–1134.
Stranger, M. (2011). *Surfing life: Surface, substructure and the commodification of the sublime*. Routledge.
Valkenburg, P., & Piotrowski, J. (2017). *Plugged in: How media attract and affect youth*. Yale University Press.
Vorderer, P., Park, D. W., & Lutz, S. (2020). *A history of media effects research traditions*. Routledge, Taylor & Francis Group.
Webb, S., & Fulton, J. (2019). "I want to read it in my hands": The aesthetic attraction of independent women's magazines. *Australian Journalism Review, 41*(2), 273–287.

Chapter 13
Sonic Waves and Acid Screens: Surf Culture and the Long 1970s

Sean Lowry, Danny Butt, and Jason Beech

13.1 Introduction

Surfing is at once a physical activity without inherent meaning and a performative act richly infused with social and cultural significance. For many, the experience of surfing is deeply personal and meaningful. Some describe deep, humbling, and awe-inspiring connections with the ocean and its power. Others point to the sublime challenge of balancing fear and pleasure offering a way to live in the moment. Sharing these experiences and their representations can create strong senses of community and belonging that in turn shape personal and societal priorities and values. Surfing has also long reflected the priorities and values of the times, and this was particularly apparent in the late 1960s through 1970s. Di Donato (2020) describes the "long 1970s" as a period in the West where the social transitions of the mid-1960s continued to echo through economic shocks into the neoliberal era ascending in the early 1980s. This period saw immense changes in political and economic changes in Western culture and social ferment characterized by post-Vietnam war disillusionment and civil rights activism. Significantly, these were movements in which young people were prominent, as was the case with the rapidly expanding surfing subculture. In this chapter, we consider three epochal surf films produced across a dynamic 11-year period: *The Endless Summer* (Brown, 1966), *Morning of the Earth* (Falzon & Elfick, 1972), and *Free Ride* (Delaney, 1977). Each film reflects significant moments in the evolution of surfing culture and the associated aesthetics of surf music and film during the "long 1970s."

The post-World War II era marked a significant transformation in "youth cultures." As noted by Birmingham's Centre for Contemporary Cultural Studies

---

S. Lowry (✉) · D. Butt · J. Beech
The University of Melbourne, Southbank, VIC, Australia
e-mail: sean.lowry@unimelb.edu.au; danny.butt@unimelb.edu.au; jason.beech@unimelb.edu.au

© The Author(s), under exclusive license to Springer Nature Switzerland AG 2025
D. M. Kennedy (ed.), *The Science and Culture of Surfing*,
https://doi.org/10.1007/978-3-031-80979-8_13

(CCCS), the emergence of new social dynamics and subcultures in the 1960s unleashed a crisis in ruling class conceptions of social organization in the West (Clarke et al., 2006). The growth of the "teenage consumer" and an accompanying expansion of cultural production oriented around leisure underscored the formation of new subcultures and generational changes concerned with "style" and aesthetics in music, fashion, and film. The late 1960s saw the codification and commodification of place-based working-class leisure or "hanging out," alongside a middle-class countercultural "attempt to explore 'alternative institutions' to the central institutions of the dominant culture: new patterns of living, of family-life, of work or even 'un-careers'"(Clarke et al., 2006, p. 48). Through this period, countercultures expanded various lived critiques of "dominant culture from a privileged position inside it" and, accordingly, began to "inhabit, embody and express many of the contradictions of the system itself" (Clarke et al., 2006, p. 55). As Booth (1995) suggests, in transforming the "work-leisure dichotomy into a work-is-play philosophy," they rejected "high consumption, materialism and competition" to express a "'fraternal' individualism" that "extolled creativity and self-expression within a cooperative milieu" (p.195). At the same time, however, the experimental fraternity within surfing counterculture, with its emphasis on aggressive innovation, also led to new gender dynamics that invariably marginalized women in surf culture.

Surfing has long functioned as a meeting point and catalyst for social development. For coastal youth with newfound recreational time on their hands, surf culture coincided with the emergence of magazines, film, and other new pedagogical forms of youth culture documentation. Although surfing lessons are widely available today, surfing was a largely an autodidactic experience of learning to surf by watching others surf in the 1960s and 1970s. Consequently, surf films constituted pedagogic devices that not only instructed in surfing performance but also helped to shape surf culture and the identity of surfers, particularly in the Anglosphere. Further, surfing tracked the emergence of a participatory culture, marked by novel valorizations of labor and leisure that could just as easily incorporate a psychedelic, hippie, or even punk ethos. Through surf music and surf film, we can sense the remarkable power for surf culture to function in an ambiguous negative dialectic, critiquing mass cultural forms with a countercultural, "dropout" lifestyle that also in turn complicitly reproduced the dominant social and political dynamics of its time.

The three films we consider in this chapter all interpellated young surfers with tropes such as travel, adventure, freedom, pleasure, unspoiled exotic cultures, and nature, though with little sensitivity to either the politics of the time nor the intrinsic coloniality of the cultural encounters represented. Over time, these messages were packaged quite differently across each of the three films, with different emphases and perspectives reflecting changing values. Where *The Endless Summer* uses a didactic pedagogy of surf travel and adventure as an essential element in surf culture, *Morning of the Earth* promotes a pedagogy of the simple life from a more constructivist approach in which viewers/learners are expected to develop their own vision. Meanwhile, *Free Ride* implicitly promotes a tutelage of performance and competition underscored by the development of radical new surfing techniques and accompanying innovations in board design.

The films both reflect and promote cultural and economic ideologies. Where *Endless Summer* both helped to establish the utopian allure of eternal travel and fuel the sometimes extractive rise of surf tourism, *Morning of the Earth* offered a counternarrative that romanticized the era's emerging eco-conscious lifestyle. By the time of *Free Ride*, a burgeoning professional competitive surfing scene starts to link a branded high-performance culture of excellence to a lifestyle economy that prefigures the growth of creative industries. While surfing often exalts a mythic pre-corporate past, the progressive and potential of surf culture as an agent of social change is nevertheless inextricably connected to its incorporation into mass culture through this period. At this time, what would later grow into more pronounced social debates on colonization, equitable access to resources, and the development of women's surfing were at best only occasional glimpses in what is largely a patriarchal settler colonial story.

Surf music would also play a significant concurrent role in both reinforcing and critiquing cultural conditions and the aesthetic dimensions of surfing. Echoing broader cultural and social changes, surf music—a reverb-soaked guitar driven rock genre emerging in Southern California in the early 1960s—would undergo substantial lateral development through the late 1960s and 1970s. Across this period, surf music would diversify significantly, reflecting and influencing broader cultural shifts both within surf culture and beyond. Variously encapsulating the ethos of lifestyle experimentation, grassroots authenticity, rebellion, performance-centered brashness, and corporatization, this adaptability was underscored by a voracious integration of musical styles, from psychedelic rock to reggae to folk and punk, all the while promoting a distinctive celebration of freedom and adventure. Notable surf bands and solo artists that grew out of the broader popular cultural zeitgeist of the 1960s include The Beach Boys, Dick Dale, The Surfaris, Jan and Dean, The Chantays, The Ventures, and The Sandals.

In oscillating between reinforcing and critiquing established cultural tropes, surf films and surf music alike would mirror broader societal shifts. All the while, a recurring motif is the aesthetic dimension of surfing as a performative act, which is in turn continually redefined through representations and reperformance. Considered in succession, each of these three films and accompanying soundtracks offer useful case studies for examining the evolution of surf culture, from iconic sun soaked silhouettes and simpler melodic tunes that capture the essence of a carefree lifestyle through to a growing interest in philosophical and environmental consciousness and more complex musical compositions and, finally, the beginning of the transformation of surf culture from niche interest to a performance-centered global corporate phenomenon.

13.2 The Endless Summer (1966)

Bruce Brown's 1966 film *The Endless Summer* represents something of the beginning of a new phase in experimental surf culture. Although Brown had been filming surfers in a documentary form since the late 1950s, their distribution was essentially limited to films by surfers for surfers. By the 1960s, however, surfing was well established in US popular consciousness. As Rutsky (1999) notes, US "beach party" films in the early 1960s functioned more or less as nationalist morality plays, with groups of white men and women reflecting suburban norms and evading the threats posed by ethnic others and degenerate lifestyles. Although the emerging political crises of the 1960s remained well out of the frame in these films, the world was nevertheless changing rapidly. Accordingly, *The Endless Summer* foreshadowed a cultural shift in youth subculture from an organized group activities described in the beach party films to an ascendant search for individual expression and meaning (Booth, 1995, p.192). Consequently, surfing as a way of life soon involved less of the hometown gang and pursued travel, adventure, and mingling with distant locals while riding waves for pleasure. Significantly, the film's showcasing of Cape St. Francis in South Africa as a "perfect wave" not only placed this once obscure location on the map but also the trope of exploration and discovery—a decidedly Romantic and Orientalist vision (Fig. 13.1).

In the late 1960s, the social mobility and affluence was represented in West Coast US popular culture that had not yet reached white surfers in Australia and Hawai'i. Accordingly, it was, at least in the first instance, Californian surfers that created "the search" for perfect and challenging waves while taking advantage of increasingly

Fig. 13.1 Brown, B. (Director). (1966, June 15). *The Endless Summer* (still from publicly available promotional trailer). Retrieved from https://www.youtube.com/watch?v=yZsuQXKkPdw (accessed June 25, 2024)

affordable international travel and, furthermore, refracting the experiences of many men who, like Brown, had previously traveled overseas as part of US military expansion (Booth, 1995, p. 192). Significantly, jet planes and the growth of international commercial flights had opened the door for surfers to follow in the footsteps of Mike Hynson and Robert August, the two main Californian characters in *The Endless Summer*, in their search for the perfect wave through Australia, New Zealand, Tahiti, Hawaii, Senegal, Ghana, Nigeria, and South Africa. The accompanying road trips and extended walks carrying what were then heavy surfboards are also central to the film. In time, these adventures would underscore a significant symbolic addition to the romanticization of the surfing lifestyle—as was later so successfully exploited by Rip Curl with their motto "The Search"—as emblematic of the pioneering spirit of adventure. Importantly, despite representing limited examples of cultural exchange, *The Endless Summer* also invariably raised questions about the sustainability of travel-centric lifestyles that seek pristine, untouched places yet ultimately lead to their subsequent transformation and commercialization.

The Endless Summer is didactic in its pedagogic style, assuming a joking distance that provisionally invited the non-surfer in the narrative but nevertheless held emerging societal transformation at bay. On the one hand, the film is aimed at explaining what surfing is about for non-surfers, but on the other, it instructs surfers as to how an idealized surfing life should unfold. Without allowing for much space for the viewers to develop their own interpretations, the narrator maintains a constant prescription as to what the camera is portraying and how it might be best understood. This is particularly evident when the travelers encounter and navigate racial and cultural difference. A passing reference to "porpoises and sharks not integrating too well in South Africa," for example, reveals an implicit critical reference to the enduring background conditions of Apartheid while a group of white surfers are out enjoying the waves. Beyond this comment, however, the film does not explicitly engage with the politics of Apartheid. South Africa is instead largely represented through beautiful scenery, waves, and friendly largely white locals. Indeed, very few black people appear in the South African section of the film, and when they do—as rickshaw or taxi drivers—they are offered as lighthearted objects of ridicule. The two main characters of the film do however mingle with black Africans in their previous stop in Ghana, where the narrator casually uses colonialist tropes such as "no one saw this wave before" or repeatedly uses the word "primitive" to describe the locals and their culture (Brown, 1966, 18:20). Poverty is also romanticized, with locals essentialized as living a happy life despite evident material deprivations. A racialized gaze is also explicitly evident when the film moves to Tahiti, where Irving, who is playing in the shallows, is described as having "the shiniest skin on the block" (Brown, 1966, 1:22:55).

Notwithstanding blind spots such as these—which are certainly more conspicuous to a contemporary eye—*The Endless Summer* is much more than an epoch-marking surf documentary. It was a cultural phenomenon, with its widely known soundtrack capturing a slice of the still burgeoning exportability of postwar American popular culture and, by extension, global popular culture. The film's soundtrack, which was composed and performed by The Sandals (originally known as The Sandells), contributed significantly to its mood and call to a nomadic spirit of adventure. Indeed, The Sandals' original music is a compelling example of cinematic and musical elements synergizing to indivisibly carry the vehicular essence of cultural and historical moment. The jazz and rock-influenced soundtrack not only complemented the visual experience but actively extended the film's influence beyond the cinematic to a musical style that would carry the ethos of surfing across far broader cultural landscapes through extensive radio play and other forms of media. This highly influential blend of music styles not only enhances the visual journey across diverse landscapes but also embedded a nostalgic and iconic sound that continues to resonate as emblematic of surf culture today.

Clearly, as both film and soundtrack, *The Endless Summer* helped romanticize the surfer as lone explorer, a motif that tapped into and extended deeper American and broader colonial narratives of frontier exploration and manifest destiny. This is particularly apparent in a moment in which we hear "frontier-type, television Western music as they drive to the beach in a rented car [...] behind Brown's narration and film of Robert and Mike surfing a long stretch of Manu Bay" (Rinehart, 2015, p. 547). The soundtrack's role in reinforcing these themes through music evokes a sense of freedom and boundless possibility, thus helping to sustain romantic one-dimensional understandings of other peoples and cultures as mere backdrops for personal discovery and adventure. Herein lies the dual nature of surf culture across this era, being at once a powerful vehicle for cultural dissemination and catalyst for cultural oversimplification and environmental extraction.

The Sandals' soundtrack is itself a fusion of rock 'n' roll, albeit with more relaxed, reverb-heavy sound that mimics oceanic rhythms. The music is largely upbeat and imbued with a light eclecticism that implicitly mirrors the film's adventurous global pursuit of the perfect wave. No doubt for many viewers at the time, this laid-back sensibility contrasted sharply with the structured, often oppressive societal norms of the time. From livelier tracks with an upbeat tempo, stronger guitar presence, and energetic riffs, such as "Scrambler," through to slower paced pairings with melancholic moments of waiting, missed opportunities, and sunsets, such as "Trailing," and, finally, the more mysterious vibes exemplified in "Jet Black" or the dreamy guitar lines of the title track "Endless Summer," the soundtrack borrows heavily from other musical traditions. Like much surf rock, the soundtrack draws heavily from African American musical traditions, particularly rock 'n' roll and R&B, which were themselves deeply rooted in the blues. Jazz, albeit in lighter gentrified "cocktail" variations, also constitutes a core ingredient in the soundtrack. This appropriation, to be sure, and certainly in the context of a film which focuses on the largely carefree experiences of white surfers traveling in Africa and elsewhere, simply forms part of a much larger continuum that romanticizes and privileges individual freedom while overlooking deeper cultural narratives and histories with their own intrinsic values and complexities. Perhaps, on balance, the very notion of an "endless summer" essentializes a nomadic Western fantasy that overshadows and trivializes the lived realities of local populations and places, reducing them at best to exotic landscapes and novel recreational venues. In time, the utopian dream would come undone for one of the film's two main nomadic surf stars. For Mike Hynson, whose later life would publicly track through drug smuggling, addiction, and prison time, life is never quite as simple as the seductive prospect of an "endless summer." In 1994, Bruce Brown would however release a sequel titled *Endless Summer II*.

13.3 Morning of the Earth (1972)

Fast forward 6 years to the release of Alby Falzon and David Elfic's classic *Morning of the Earth*, the aesthetic qualities of the surf film and its accompanying soundtracks have shifted markedly. This groundbreaking surf film, like *The Endless Summer*, has also become a vibrant cultural artifact broadly understood to encapsulate the spirit of its era. Where *The Endless Summer's* anthropological bent offers a vehicular expression of American economic supremacy when it was released in 1966, by the early 1970s, the geopolitical axis of surfing had expanded substantially. As declared confidently in surfing magazines at the time, the growing currency of Australian competitive surfers represented "a new era," and as Booth has put it, "aggression, power and radical (creative) manoeuvres on short boards" were the hallmarks of this new approach (Booth, 1995, p.195). This insurgent group would eventually dominate international competition through this transitional period, and consequently, the style of surfing in *Morning of the Earth* more closely resembles contemporary surfing today than the documentation captured in *The Endless Summer* a mere 6 years earlier. Yet competition is antithetical to the spirit and ethos of *Morning of the Earth*. This was an altogether different take and representation of the utopian potential of surfing, a radical approach to the waves becoming a Romantic grasping for a holistic union between human and ocean.

Although, just like *The Endless* Summer, *Morning of the Earth* also promoted images of surfers as pleasure-seeking carefree travelers immersed in unspoiled nature and cultures in exotic lands, it presents a very different perspective of surfing and the good life. Pedagogically, it promotes the value of a simple life which is more connected with nature and less concerned with material consumption. Accordingly, the jet planes, cars, taxis, and fancy hotels central to *The Endless Summer* are not present. Instead, the viewer is presented with humble cabins in natural settings and surfers harvesting fruit trees or playing with chickens and dogs. They are also depicted building their own surfboards (albeit with the latest toxic resins and foam materials made available through the forces of industrial capitalism kept outside the frame). Although the ethos of travel and adventure is clearly still present across the various locations presented in the film—which include Australia's northeast coast, Bali, and Hawai'i—the idea of travel is implicit, and the trope of discovery is oriented toward self-discovery and connections with the land rather than "tourism." The opening title of the film makes its message evident by stating that it is a film about "A fantasy of surfers living in three unspoiled lands and playing in nature's oceans" (Falzon & Elfick, 1972, 4:33). Following this opening title, there is no further explanation or narration. Instead of being told what they are seeing and how they should interpret it, the viewer is left free to make low-frequency associative connections between the imagery, music, and lyrics. Significantly, although the Indonesian sections were filmed during Suharto's bloody dictatorship, locals are represented as living a simple, happy life with no signs of conflict. Meanwhile, the white visitors reveal their innocence by appearing naked on Indonesian beaches, evidently with little consideration of local cultural practices. The only inkling of blood and violence is the detailed portrayal of a cockfight. Here, at a stretch, if animals are again used as an implicit metaphor of dissent in local politics (such as was the case with the aforementioned "porpoises and sharks" in *The Endless Summer*), the reference in *Morning of the Earth* is much more subtle.

The film's soundtrack was produced by G. Wayne Thomas and included music and songs by Australian acts such as John J. Francis, Brian Cadd, G. Wayne Thomas, and Tamam Shud. It has long served as a sonic embodiment of the 1970s "soul surfing" experience and its close relationship with nature. The music and lyrics encapsulate themes of freedom, harmony with nature, and a return to essential life values. Surfing, in this context, is portrayed not merely as a sport but as a holistic interaction with the elemental forces of nature. The fluid, rhythmic movements of surfers riding waves mirror the acoustic and lyrical flows found in the soundtrack. "Open Up Your Heart" by G. Wayne Thomas, for example, invites an environmental connection and symbiosis between the surfer and the ocean. Here, the film's soundtrack lyrically and visually encapsulated the union exemplifying surfing as a performative act that transcends mere physical activity, positioning it as a meditative, almost spiritual practice. This utopian quality in both the film and soundtrack, however, also offers an opportunity to go a little deeper into the dual natures of surf culture. There is something, for example, about the rolling hypnotic bassline, wandering flute, and "I don't care" lyricism in Australian psychedelic surf rock band Tamam Shud's 1971 contribution "Sea the Swells" that at once represents the utopian idealism of a then emerging nomadic surf culture

and, perhaps in retrospect, its concurrent obliviousness to a certain complicity within the continuum of colonial mindsets.*Sea the swells out on the ocean*
All a-movin' in a rhythm
If I didn't have to be here
I would surely be-ee there with them

I don't care-are, I don't care-are
Walk around and breathe the air-air
Drink fresh water from a fountain
Watch the sun rise from a mountain

I don't care, I don't care, I don't care, I don't care
I don't care…are…are
I don't care…are…
I don't care…are
—Tamam Shud, "Sea the Swells" (excerpt) (Falzon & Elfick, 1972; 1:04:48)

These playfully homophonic lyrics invite us to connect with the majesties of the ocean and to the prospect of losing worldly cares in an imagined island paradise. Yet, like the iconic film itself, which portrays surfers living in spiritual harmony with nature as they traveled in search of perfect waves, the lyrics also testify to an essentially white male imposition upon colonized lands and waters. Here, the "I don't care" refrain potentially suggests a more problematic doubled sentiment: "I do not care" about what, exactly? Indeed, in appropriating a polytheistic grab bag of non-Western cultural and spiritual traditions, as was common in the late 1960s and 1970s "countercultures," Tamam Shud's vocalist and guitarist Lindsay Bjerre had taken the band's name from the Persian phrase "tamám shud" (roughly translated as "ended," "finished," or "the very end") in the closing words of the eleventh-century poetry collection *Rubáiyát of Omar Khayyám*. To be sure, this kind of cultural appropriation by white artists in the Anglosphere would likely be viewed a little differently with the vantage of time.

The early 1970s were marked by a surge of environmental consciousness and a generational zeitgeist of disillusionment with industrial progress, consumerism, and urban sprawl. Accordingly, the soundtrack of *Morning of the Earth* broadly reflects this turning away from materialism and an embrace of a simpler, more sustainable way of living. This call for simplicity is particularly evident in pastoral and serene tracks such as John J. Francis' "Simple Ben," which narrates a lifestyle deeply connected with the land and sea, free from the complexities of modern urban existence. Once again, from the vantage of a more intersectionally literate present, this idealization of a rustic lifestyle overlooked the challenges faced by less privileged communities for whom such a life is neither a choice nor a feasible escape. Moreover, while the soundtrack and the film's imagery champion a return to nature and simpler living on the North Coast of New South Wales, Australia, they do not critically address the implications of such a lifestyle on Indigenous or local communities and environments affected by increasing popularity of surf tourism and coastal habitation. Although the film presents liberating images of surfers gliding on dawn waves in primal rituals of renewal, it glosses over the question of who gets to participate in this idealized new dawn. The conspicuous absence of voices and themes actively connecting different genders, races, or socioeconomic statuses to the core romantic liberation of "soul surfing" clearly underscores limitations in the inclusivity of this cultural moment. Although presenting itself as part of an "evolution" in consciousness more broadly, women's participation in surfing had regressed through this period, and accordingly, women in the film appear bathing nude and carrying children, rather than riding waves. Although the film and its music present an idyllic world that is appealing, it is nevertheless unable to imagine broader societal structures that might limit access to such a life (Fig. 13.2).

The shaper/surfer as an artisanal worker invested only in their own self-development also found resonance with a hippie movement looking to establish alternative lifestyles, far away from competitive surfing. Albert Falzon, who shot *Morning of the Earth*, was also partly responsible for the magazine *Tracks*, which "brought together countercultural stories that included communal living, travel stories, meditation, and yoga, alongside the more traditional sports reportage"

Fig. 13.2 Falzon, A., & Elfick, D. (Directors). (1972, February 25). *Morning of the Earth* (still from publicly available promotional trailer). Retrieved from https://vimeo.com/ondemand/morningoftheearth (accessed June 25, 2024)

(Blackwood, 2022, p. 50). Interestingly, it also documented and promoted Indonesia as not only a site for perfect waves but a potential place of spiritual transformation through tourism, prefiguring the Western counterculture's role in gentrification and commodification of the good life, wherever it is to be found.

13.4 Free Ride (1977)

By the late 1970s, the surf culture had begun to shift again. Bill Delaney's 1977 surf film *Free Ride* would, together with a broad explosion in surf movies that followed, begin to articulate a passage from the soulful, exploratory surfing of the early 1970s through to the more competitive, high-performance world of the 1980s, 1990s, and beyond. Unlike *Morning of the Earth*, the crisp, polished, ultra-saturated color and still revolutionary slow motion water photography in *Free Ride* captured the next stage in a shortboard revolution that was now all about high-performance surfing. Meanwhile, the film's prog-rock, jazz, R&B, and country rock soundtrack, with narration by Jan Michael Vincent, again helped to articulate this next evolution in surfing culture, from a laid-back lifestyle to a more performance-oriented professional sport.

Free Ride assumed surfing performance and technique as its primary focus. Rather than offering a storyline or a message, this film marked the return of surfing to the genre of sport. Accordingly, the pedagogic motif focuses more directly on the power, beauty, and aesthetics of surfing, while the world beyond the shoreline becomes far less prominent. Although the protagonists of the movie are mostly men, some female surfers were depicted mastering Hawaiian waves, representing a modest but still relevant shift toward the growing participation of women in mainstream surf culture.

The images of surfing are filmed from the water in a tighter frame, allowing the (presumably surfing literate) viewer to analyze in detail the style of surfers such as Mark Richards and Shaun Thompson as they both glide and "thrash" waves in the more radical style that was only made possible because of new innovations in shortboard design. Here, surfing progression and high performance are seen as the path to liberation, with more critical maneuvers and new styles of boards to support this process. Even the fins of the surfboards feature in a scene in which Richards and Thompson discuss the newly invented twin fin set up in comparison with the traditional single fin. Meanwhile, new ways in which surfing is filmed are the result of new technical apparatuses and filmographic techniques, often involving swimming directly in path of surfers to capture very different perspectives. With the concurrent growing popularity of video cassette recorders, *Free Ride* would become the kind of movie that young surfers all over the world would watch over and over again, pausing and rewinding the scenes to study the technique of their idols in detail, ready to imitate the maneuvers themselves at their local break. Mirroring this forensic analysis of performance via moving image replay, a new generation of musicians were similarly obsessed with technical perfection, as is evidenced in some parts of the film's accompanying soundtrack.

There is one part of the soundtrack to *Free Ride* that particularly epitomizes this transition toward performance-oriented professionalism. Accompanying one of the most spectacular high-performance surfing sequences in the film (which is now the most viewed clip from *Free Ride* on YouTube) is the technically sophisticated prog-rock instrumentation of Pablo Cruise's 1976 track "Zero to Sixty in Five," which was also used as theme music in various high-performance sports television programs of that era (Delaney, 1977 1:18:00). Like the surfing itself, this soundtrack represents an emerging emphasis on an aspirational level of proficiency that would be unattainable for the amateur participant. Interestingly, to at least to contemporary ears, this track exudes a quintessentially 1970s vibe with its exuberant extended-play virtuosity while simultaneously hinting at the corporate "yacht rock" FM radio sound and musical palette of action sports television that would later help underscore 1980s sports consumerism. Certainly, at least in the context of *Free Ride*, this track perfectly prosecutes the transition from the casual ease of earlier surfing culture to an emerging emphasis on skill, performance, and mastery (Fig. 13.3).

With the shift from lifestyle to performance in the water, surfing is opened for a return to competition. Alongside surfing, Australia is presented as a context in which people are obsessed with competitive sports, and as the narrator of the surf contest at Bells Beach suggests in *Free Ride,* it is only through competition that surfers push

Fig. 13.3 Delaney, B. (Director). (1977). *Free Ride* (still from publicly available promotional trailer) Retrieved from https://www.freeridefilm.com (accessed June 25, 2024)

each other and the sport progresses. Other contests in Waimea and Pipeline in Hawaii are presented as stages for the best surfing. Clearly, the adventurous pleasure-seeking carefree soul-surfer of less than a half-generation earlier was beginning to be usurped by the ambitious professional competitor, guardians of their own brand, and, in the case of the Bronzed Aussies, a return to a decidedly nationalized tone. Accordingly, *Free Ride* shows the struggles and the austere lives of young men that were adamant to make a living out of surfing in this period of nascent commercialization. Nevertheless, what becomes evident in the pedagogy of *Free Ride* is that surfing was being transformed, at least in its mainstream incarnations, from a pastime based on pleasure and communion with nature to a competitive sport and global business.

13.5 Conclusion: The Dialectics of Surfing

The dialectical relationship between surfing and politics is largely suppressed in these films. In part, this suppression aligns with the spirit of Tamam Shud's gentle call to resistance, as embodied in the ethos of "I don't care," which implicitly advocates for dropping out and being accountable to no one. In many ways, the individual surfer's concerns appear to end at the water's edge. Indeed, when any broader context is depicted, it is invariably presented as unspoiled nature and culture, without attention to their locational particularities or the global geopolitical structures that enabled them to be there in the first place. Yet a dialectical tension has nevertheless existed ever since the (re)birth of (modern) surfing, for as Walker (2011) argues, surfing is insidiously connected to colonialism since its origins, as Westerners appropriated the sport from their traditional owners in Hawai'i, while at the same time surfing was rearticulated as a form of resistance for Hawai'ians. At the times in which these three films were made, some of the global seascapes depicted were proxy arenas in Cold War contests between the USA and USSR for the hearts and minds of the so-called Third World. The paradisiac representations offered in these films ignore the implications of economic or military interventions and ideological contestations in the Global South, including Western support of Suharto's dictatorship in Indonesia. In his critical account of surfing history, (Laderman, 2014, p. 14) argues that "those charged with crafting U.S. foreign policy gave the surfing lifestyle a big part in official Cold War cultural diplomacy."

The surfers portrayed in these movies were not only largely oblivious to the implications of these tragic conflicts; they did not seem to care too much for the evils of Apartheid in South Africa either. It was only in 1985 that Australian professional surfer and then world champion Tom Carroll decided to boycott South Africa in support of the anti-Apartheid movement. His bold move was highly criticized by many at the time, who argued that surfing should be kept clean from politics. Carroll was however then followed by other great surfers, such as Tom Curren and Martin Potter, and eventually many others (Laderman, 2014). This commitment of surfers to political causes has continued, and expanded, mostly in relation with environmental issues and the protection of oceans and coastlines but also in solidarity with other causes. Accordingly, surf films, videos, and music have more broadly become vehicles for a range of environmental concerns, gender politics, and Indigenous rights both within and beyond surfing communities.

So, what can we learn about surfing's imbrication in politics through these three films? Historiographically, surfing can be seen to move through eras or phases that emphasize its relationship to lifestyle practices, coastal management, technology, media and communications, and the dynamics of professionalization. As we have argued, these phases have all variously reflected shifting contemporary global political economies, irrespective of how much it is conspicuously held outside the frame. As in the 1970s, surfing is once again in a transitional mode between eras in which all these relationships are changing. In a recent essay, Jay Caspian Kang suggests that the availability of beginner friendly equipment, ubiquitous cameras, and surf reports at breaks and the burgeoning of short form online videos have all contributed to new audiences experiencing the pastime through the Internet's "filter of unreality"(Kang, 2024). At the same time, the global surfing industry lurches from crisis to crisis. In the mid-2010s, as Booth (2014) has noted, notwithstanding the "diversification of the major surfing corporations—Rip Curl, Quiksilver, Billabong, O'Neill—into fragrances, bedroom accessories and travel agencies" (p. 585), these brands have themselves become emptied of all power as they are bought and sold by venture capital firms such as Oaktree Capital Management. The World Surf League itself has been bought by venture investor Dirk Ziff, whose hiring of Joe Carr, an executive who steered the UFC's eventual sale into an integrated capital firm, epitomizes the broader subsumption of all cultural activities into oligarchal networks of global finance with no particular attachment to the sport (Doherty, 2013).

With a surf culture now firmly in the mainstream, and the 1970s surf industry players largely cashed out, a new generation of content creators are again reworking the innovations of the previous era. Consequently, the dynamics of authorship of surfing content have changed, with Instagram and YouTube fostering surfer-led profiles and a constant stream of low-budget short form documentary content for the consumers of the surfing lifestyle. Surfers such as Torren Martyn, Nathan Florence, or Nick von Rupp integrate the independent "slice of life" content in *Morning of the Earth* in a branded digital content platform with flexible monetization possibilities. Meanwhile, Dane Reynolds left the tour to focus on Chap. 11 tv, a daily documentary of Ventura-area surfing whose hallmark is a self-conscious use of ironic post-punk soundtracks through a recursion of the 1990s slacker culture that seems aware that their 1970s precursors were imbricated in bad politics. And the ubiquity of video has produced new genres such as "Surfers of Bali" that push the limits of how much content can be produced out of a particular place's surfing, relying on diegetic sound as a kind of anti-soundtrack (Surfers of Bali, 2024). They all in their different ways strive for a glimpse of the all-encompassing rhythmic experience of surfing, its unique performative qualities, and the rich aesthetic world within which these three films and accompanying soundtracks of the 1970s have set the tone.

This Romantic spirit and the political complexities of surfing were reflected in the surf films and soundtrack of the "long 1970s" in several ways. They emphasized the value of relationships with the beauty and power of nature, a central theme of counterculture at the time. They also celebrated freedom and individuality, which are at once markers of both outsider rebellion and corporate capitalism. And herein lies the paradox: while surfing cultures have historically embraced and promoted the countercultural values of freedom and individuality, they also became commodified symbols of the very corporate capitalism that the counterculture sought to resist. This duality has underscored a complex interplay between rebellion and assimilation, demonstrating how popularized countercultural expressions are invariably co-opted and commercialized as they challenge the status quo.

References

Blackwood, G. (2022). "Forever Bali": Surf tourism and morning of the earth (1972). In *Screen tourism and affective landscapes* (1st ed., pp. 46–63). Routledge.
Booth, D. (1995). Ambiguities in pleasure and discipline: The development of competitive surfing. *Journal of Sport History, 22*(3), 189–206.
Booth, D. (2014). Invitation to historians: The historiographical turn of a practicing (sport) historian. *Rethinking History, 18*(4), 583–598.

Brown, B. (Director). (1966, June 15). *The endless summer* [Surf documentary]. Cinema V; Apple TV. https://tv.apple.com/us/movie/the-endless-summer/umc.cmc.2b4rfovo6qwmoebwjwadxj1ya

Clarke, J., Hall, S., Jefferson, T., & Roberts, B. (2006). Subcultures, cultures and class. In S. Hall & T. Jefferson (Eds.), *Resistance through rituals* (pp. 3–60). Taylor & Francis Group.

Delaney, B. (Director). (1977). *Free ride* [Surf documentary]. Vimeo. https://vimeo.com/ondemand/freeride

Di Donato, M. (2020). Landslides, shocks, and new global rules: The US and Western Europe in the new international history of the 1970s. *Journal of Contemporary History, 55*(1), 182–205.

Doherty, S. (2013, July 1). *The new ASP*. Surfer. https://www.surfer.com/features/the-new-asp

Falzon, A., & Elfick, D. (Directors). (1972, February 25). *Morning of the earth* [Surf film]. Youtube. https://youtu.be/3l40HyPuaFs?si=vzF2QnoqO_0hegpp

Kang, J. C. (2024). Arguing ourselves to death. *New Yorker*. https://www.newyorker.com/news/fault-lines/arguing-ourselves-to-death

Laderman, S. (2014). *Empire in waves: A political history of surfing*. University of California Press.

Rinehart, R. E. (2015). Surf film, then & now: The endless summer meets slow dance. *Journal of Sport and Social Issues, 39*(6), 545–561. https://doi.org/10.1177/0193723515594210

Rutsky, R. L. (1999). Surfing the other: ideology on the beach. Film Quarterly, 52(4):12–23. https://doi.org/10.2307/1213771

Surfers of Bali. (2024, June 13). *Surfers of Bali*. https://surfersofbali.com/

Walker I. H. (2011). Waves of resistance: surfing and history in twentieth-century hawai'i. Honolulu: University of Hawaii Press. https://doi.org/10.1515/9780824860912

Chapter 14
Conclusions and the Future of Surfing

David M. Kennedy

14.1 Away from the Ocean in Parks and Pools

Surfing is the act of riding a wave; therefore, where there are waves, people will surf. While this seems an obvious statement when considering the breaks of the worlds' coast, things get interesting when reflecting on other locations where waves can form. While the emphasis of this book is on waves formed by the wind, surfable waves can be found in a wide variety of settings where wind is not the driver of wave creation, such as tidal bores.

Waves in the ocean move; as energy is transferred through the water, the wave form shifts its position in time. A different type of wave form is known as a standing wave, which often marks the location where there is a transition between the flow characteristics of the water in a natural (e.g., at the entrance to an estuary) or artificial (e.g., a concrete drain) channel. Such waves can produce bores that are surfable. A classic example is the Eisbach River in Munich, Germany, where a standing wave 1 m high is found in city park (Fig. 14.1). This break ranks in the top 10 most unusual surf breaks in many blogs such as Surfer Magazine and Surfer Today. While the Munich wave was not intended as a surf destination, it is not unusual today for such breaks to be deliberately created for tourism as part of river engineering works. The key condition to create a standing wave is there needs to be a significant difference in the water elevation between the two ends of the channel, termed hydraulic head. In natural settings, such as when an estuary suddenly breaches its berm and rapidly drains, standing waves may occur for only a short period of time (Fig. 14.2). In artificial settings, so long as water is flowing in the channel, then the standing wave may persist for very extended periods.

D. M. Kennedy (✉)
The University of Melbourne, Parkville, VIC, Australia
e-mail: davidmk@unimelb.edu.au

© The Author(s), under exclusive license to Springer Nature Switzerland AG 2025
D. M. Kennedy (ed.), *The Science and Culture of Surfing*,
https://doi.org/10.1007/978-3-031-80979-8_14

Fig. 14.1 The Eisbach River in Munich, Germany, is a popular inland surfing location. The standing wave develops as the river exits the bridge producing rides of 3–30 seconds in duration. (Photo: clu on iStock)

Artificial waves, just for surfing, can also be created in dedicated parks that may be located hundreds of kilometers from the ocean (Fig. 14.3). Interest in creating artificial waves in parks has been around since the mid-nineteenth century when King Ludwig II of Bavaria electrified one of his private lakes to create waves, but it was not until 1969 that the first dedicated wave riding surf pool was built in Arizona (SurferToday, 2024a). Today, interest in surf parks is flourishing. The waves in these artificial settings are created using a range of techniques to displace water in order to create the break. These include giant plungers, moving foils, water drops, and paddles (see RWR (2024) for a summary). The advantage of these technologies is that they can be controlled. This allows the operator to create waves of different types and difficulties allowing surfers to challenge themselves in varying conditions and according to their skill level, all at the same location. Often it is the desire to experience the "perfect wave" that has driven surf park development, and this has been championed by former world champion Kelly Slater and his Kelly Slater Wave Company in central California (KSW, 2024). Such is the success of these parks in creating ridable waves that there now are critiques on whether such experiences detract from what is considered a traditional surfing lifestyle (Roberts & Ponting, 2020). These narratives are often similar to those of previous generations who opposed the commodification of surf culture (Roberts & Ponting, 2020), and the

Fig. 14.2 A small standing wave, decimeters in height, perfect for young beginners. The wave pictured formed during the opening of Spring Creek, Torquay, Australia, on May 31, 2019. (Photo: Sarah McSweeney)

debate shows how the simple act of surfing remains a focal point for questioning assumptions of contemporary society.

14.2 Technological Limitations Overcome

The design of boards and fins is an ever-evolving space. Countless books have been written on the subject, and as surfers chase more and more extreme environments and push the human body to its limit, design will continue to evolve. Probably the most significant advance in surfing technology has been the development of the fin. The original boards of Hawai'i were smooth planks which lacked the ability to turn to a great extent (Chap. 2). Fins have evolved rapidly in shape, design, and number placed on a board. The specific design and setup are often tailored to the wave conditions that will be encountered. Modern studies now use three-dimensional printing to subtly alter the shape of the fin to improve surfing performance (Forsyth et al., 2024) and even derive inspiration from nature (humpback whales' fin shape) (Shormann & In Het Panhuis, 2020). A more visually striking innovation is the use of hydrofoils, where under the right conditions, a wave can be surfed without the

Fig. 14.3 An artificial wave pool at UrbnSurf in Melbourne, Australia, located 70 km inland from the nearest open-ocean surf break. This park utilizes WaveGarden Cove technology to create the wave. (Photo: David M. Kennedy)

board even being in contact with the water. Foils almost made it onto Hollywood screen in the early 2000s. In the opening sequence of the James Bond film *Die Another Day,* three riders (Laird Hamilton, Dave Kalama, and Darrick Doerner) are seen surfing a big wave. In the documentary *Laird,* Hamilton recounts his original intention for this sequence was to use a foil but the producers declined for fear that the audience would think it was a special effect rather than real surfing. Riding these largest waves on Earth has also required significant technological innovations from tow-in jet skis to give surfers the momentum to get onto the wave face to self-inflating vests to overcome the hold-downs that occur when caught in the break.

14.3 A Humanized Environment

Our coast faces twin challenges of climate change and overdevelopment. Climate change and associated rising sea levels and varying storm patterns are causing changes in the wave energy field throughout the ocean basins, as already seen in the Southern Ocean (Young et al., 2020). The more energy that is inputted into the ocean, the greater the waves will be—this can be considered a positive in some respects. Rising sea levels will change breaks through gradually increasing water depths (see Chap. 4 for more detail); future generations will see different breaks than exist today. The biggest impact will however be the geomorphic response, namely, how sand is moved around the coast and therefore the forms of banks and bars. It is these breaks on soft foundations that will see the soonest, and likely most rapid, responses to higher sea levels.

A more immediate challenge however is the progressive hardening of the coast through artificial structures. While built as "protection," these features aim to prioritize a property or survey line rather than the natural environment. Such features are now commonly greenwashed as being "nature-based" when marine organisms can grow on their surface, but the principal aim remains—prioritization of human infrastructure over nature.

Offshore structures have unintentionally created surf breaks. This has led some companies to specialize in deliberately placing structures offshore in order to create artificial surf reefs (e.g., Narrowneck, Gold Coast, Australia, and Bournemouth, UK). The success of these structures is highly mixed with only a couple producing good ridable waves (SurferToday, 2024b), and when a suitable break is created, continued maintenance is required. An underlying problem with measurement of such artificial breaks is the published reviews are often conducted by those with a financial interest in their outcome. There is need for unbiased research in this space, including impacts outside of the surfing space such as downstream erosion and public amenity.

Finally, one last challenge that many surfers will have experienced is crowding. With the global human population ever growing and increased participation in surfing, the pressure on surf breaks only intensifies. This inevitably leads to tension and conflict between "locals" and "outsiders." This conflict can spread into unforeseen realms such as erosion monitoring and wave studies where cameras set up to measure the natural environment are viewed as also promoting a site, thereby increasing crowd numbers. While the vast majority of surfers respect the customs and "rules" of a break, such as no drop-ins, the thoughtless few create impact beyond their individual selfishness.

References

Forsyth, J. R., Barnsley, G., Amirghasemi, M., Barthelemy, J., Elshahomi, A., Kosasih, B., Perez, P., Beirne, S., Steele, J. R., & In Het Panhuis, M. (2024). Understanding the relationship between surfing performance and fin design. *Scientific Reports, 14*(1), 8734.

KSW. (2024). *Kelly Slater Wace Co: The science of stoke*. Retrieved August, from https://kswaveco.com/

Roberts, M., & Ponting, J. (2020). Waves of simulation: Arguing authenticity in an era of surfing the hyperreal. *International Review for the Sociology of Sport, 55*(2), 229–245.

RWR. (2024). *Raised water research: Wave pool technology comparison*. Retrieved August, from https://raisedwaterresearch.com/wave-pool-technology-comparison/

Shormann, D. E., & In Het Panhuis, M. (2020). Performance evaluation of humpback whale-inspired shortboard surfing fins based on ocean wave fieldwork. *PLoS One, 15*(4), e0232035.

SurferToday. (2024a). *A brief history of artificial wave pools*. Retrieved 2/9/24, from https://www.surfertoday.com/surfing/a-brief-history-of-artificial-wave-pools

SurferToday. (2024b). *The history of artifical surf reefs*. https://www.surfertoday.com/surfing/the-history-of-artificial-surf-reefs

Young, I. R., Fontaine, E., Liu, Q., & Babanin, A. V. (2020). The wave climate of the Southern Ocean. *Journal of Physical Oceanography, 50*(5), 1417–1433.

Index

A
Accounting frameworks, 197
Activism, 259
Aging, 166, 233–236
Anatomy, 126, 129, 136, 142, 146, 147
Aspiration, 216
Australian Beach Safety Management Programme (ABSAMP), 86–88, 90
Austronesian, 3, 9, 10, 13–17, 23, 24
Authenticity, 4, 223, 224, 228–232, 235, 236, 261

B
Backyard, 224, 225
Balance, 4, 44, 45, 124–126, 134, 135, 139, 141, 186, 193, 226, 264
Bathymetry, 45, 52, 56, 64, 65, 68, 81
Beach, 1, 4, 12, 29, 34, 41–43, 64, 68, 75–90, 93–96, 98–104, 110, 111, 159, 161, 167, 172, 174, 190, 191, 195, 215, 225, 241, 243, 264, 266
Beach models, 84
Blogs, 275
Boardrider, 105, 232
Bores, 3, 51, 76, 77, 79, 84, 129, 275
Break intensity, 52, 60, 63
Breaks, 1, 3, 10, 12, 27, 30, 31, 39–43, 45, 51–69, 75–77, 79, 81, 84–90, 94, 103, 120, 133, 161, 171–175, 179, 181–183, 187–197, 203, 204, 206, 209, 211, 214, 217, 224, 245, 269, 271, 275, 276, 278, 279
Bystanders, 90, 94, 102

C
Cardiopulmonary resuscitation (CPR), 94, 95, 97, 99, 102, 103, 105–107, 111, 197
Climate, 45–48, 54, 55, 68, 147, 148, 208
Climate changes, 28, 52, 54, 68, 183, 278
Colonisation, 163
Commercial, 5, 160, 208–211, 218, 223, 227, 230, 233, 236, 242, 263
Common pool, 192–197
Conservation, 41, 52, 69
Consumers, 159, 186, 190, 197, 227, 231, 233, 241, 248, 253, 260, 272
Cultures, 1–6, 10, 11, 13–16, 18–20, 22, 24, 93, 109, 157, 159, 161, 164, 168, 169, 171, 179, 188, 208, 209, 215, 224, 226–229, 233, 235, 236, 242, 243, 245–247, 250, 251, 255, 259–272, 276
Currents, 4, 11, 35, 37, 51, 54–56, 64, 76–85, 94, 95, 97, 98, 100–102, 105, 146, 147, 160, 175, 179, 208–218, 247, 254, 255

D
Dangers, 79, 90, 107, 132, 175, 193
Demographics, 98, 110, 205, 210, 213, 233–236
Digital, 244–255, 272
Diplomacy, 5, 216, 218, 271
Dispersion, 36–44
Diversity, 3, 164, 205

© The Editor(s) (if applicable) and The Author(s), under exclusive license to Springer Nature Switzerland AG 2025
D. M. Kennedy (ed.), *The Science and Culture of Surfing*,
https://doi.org/10.1007/978-3-031-80979-8

E

Earth, 5, 27, 32, 90, 147, 148, 259, 265, 268, 272, 278
Ecosystems, 52, 69, 179–197, 217–219, 236, 241, 255
Endless Summer, 19, 242, 247, 259–266
Energies, 1, 3, 27, 28, 31–35, 37, 39–45, 47, 48, 52, 64, 75–77, 79, 81–87, 98, 119–124, 165, 172, 175, 188, 275, 278
Excludable, 192, 194, 196
Expenditures, 4, 110, 181, 182, 184, 185, 196, 207

F

Feminism, 160
Films, 4, 5, 19, 22, 241–243, 246, 247, 259–272, 278
Fins, 5, 107, 108, 135, 182, 207, 224, 269, 277
First aid, 94, 95, 99, 102, 103, 107–108, 111, 131
Foils, 118, 276, 278
Forcing, 36–38
Framework, 12, 16, 21, 22, 29, 38, 69, 87, 88, 90, 158–159, 182, 183, 187, 189, 196, 197, 217, 219, 241, 249, 252–256
Franchise, 225
Free ride, 259–261, 268–270
Frequency, 29, 31–33, 39–41, 44, 48, 68, 79, 82, 106, 131, 133–135, 141, 144, 147

G

Gender, 12, 109, 157–161, 165–170, 172, 175, 176, 216, 234–236, 247, 260, 267, 271
Generations, 28, 36–44, 81, 197, 213, 228, 269, 272, 276, 279
Goods, 2, 18, 51–63, 68, 124, 132, 162, 170, 171, 182, 184, 188–191, 193–195, 197, 212, 224, 226, 230, 231, 233, 235, 265, 268, 279

H

Hawai'i, 3, 5, 14, 56, 60, 262, 265, 271, 277
Hazards, 35, 45, 54, 76, 79, 84, 86–88, 90, 94, 102, 108, 110
He'e nalu, 3, 5, 9–24
Health, 1, 4, 5, 93, 99, 110, 117–150, 171–173, 180, 187, 188, 192, 215, 216
Hegemony, 223, 235, 247

I

Identities, 10–13, 16, 19, 20, 22, 24, 160, 161, 188, 196, 223, 226–229, 236, 248–250, 260
Impacts, 1, 3, 4, 48, 54, 66, 68, 69, 90, 98, 101–103, 108–111, 127, 134, 164, 172, 179, 181–186, 189, 196, 197, 205, 207, 217, 218, 227, 229, 279
Indigenous, 9–21, 23, 24, 165, 171, 181, 267, 271
Industries, 1, 4, 12, 24, 110, 111, 159, 160, 162, 164, 167, 169, 186, 196, 207, 208, 214, 215, 217, 218, 223–236, 243, 244, 247, 252, 261, 271, 272
Injuries, 4, 98, 99, 101–103, 105, 107, 117, 118, 121, 126, 128–150, 172, 175, 192, 211

J

Justice, 157–176

M

Magazines, 3, 99, 118, 241, 243–253, 255, 260, 265, 267, 275
Management, 67–69, 79, 81, 87, 90, 93, 110, 117, 136, 181, 192, 193, 195, 204, 215, 232, 236, 271
Markets, 12, 18, 159, 179, 182–184, 190, 191, 196, 197, 207, 208, 210–212, 217, 224, 226, 227, 230, 232, 235, 243, 247, 253
Modeling, 43, 68, 69
Morphodynamics, 52, 84–87
Music, 3, 31, 165, 211, 228, 243, 259–261, 264, 266, 267, 269, 271

N

News, 245, 252

O

Observations, 34–36, 38, 108, 109, 164–165, 168, 218
Online, 96–99, 105, 109, 111, 130, 245, 246, 271
Origins, 3, 9, 23, 27, 34, 52, 207, 229, 234, 271

P

Pacific, 3, 9, 10, 20, 23, 47, 48, 53, 109, 157
Parks, 95, 167–169, 192, 194, 205, 212, 214, 224, 275–278
People, 3, 4, 9, 13, 17, 18, 21, 23, 52, 76, 77, 79, 80, 84, 89, 90, 93, 94, 96, 98, 100–102, 105, 110, 111, 124, 157, 163–165, 169, 171, 173, 181–185, 188–194, 203, 205, 207, 212, 216, 219, 226, 227, 229, 233, 235, 246, 248, 250, 259, 263, 269, 275
Photos, 2, 6, 35, 55, 57–62, 65–67, 76, 78, 81–83, 85, 88, 89, 122–125, 128, 129, 134, 137, 139, 143, 167, 173, 180, 186, 188, 204, 206, 210, 213, 214, 225, 232, 245, 251, 252, 276–278
Physics, 3, 31, 90
Physiology, 117
Physiotherapy, 131, 136
Points, 5, 11, 13, 28–31, 33, 46, 54, 63, 64, 66, 79, 81, 85, 88, 129, 132, 149, 159, 164, 182, 185, 190, 191, 193, 195, 204, 208, 211, 216, 234, 248, 254, 259, 260, 277
Politics, 20, 260, 263, 266, 270–272
Pools, 51, 89, 121, 194, 275–278
Powers, 1, 3, 5, 11, 22, 23, 76, 117, 120–122, 127–129, 144, 157–160, 170–173, 175, 203, 210, 216, 218, 223, 229–236, 241, 259, 260, 265, 269, 271, 272
Predictions, 52, 255
Prints, 4, 224, 244, 245, 250, 252, 255
Private, 20, 182, 194, 209, 210, 232, 276
Proprioception, 124–126
Public sector, 209, 214–219

R

Radical, 1, 5, 11, 260, 265, 269
Reefs, 3, 52, 54, 56, 64–66, 68, 69, 88, 132, 135, 174, 187, 188, 197, 206, 211, 217, 279
Rehabilitation
Rescues, 4, 89, 90, 93–111
Resorts, 22, 193, 206, 208–210, 212, 214, 218
Resources, 12, 20, 90, 147, 159, 163, 176, 180–182, 184, 187–197, 261
Resuscitation, 100, 103, 106
Rips, 34, 54, 64, 76–87, 90, 94, 95, 97, 98, 100, 102, 105, 162, 224, 231, 236, 263, 271
Risks, 33, 69, 76, 88, 90, 94, 105, 108, 135, 140, 141, 147–149, 192, 217, 219, 229, 235
Rivers, 27, 56, 67, 68, 164, 214, 275, 276
Rocky coasts, 88–90

S

Safety, 17, 51, 75–90, 93, 95, 96, 98–100, 102, 103, 105, 108–111, 195
Sciences, 1–5, 16, 23, 27, 28, 52, 90, 180, 181, 195
Sea, 1, 3, 12–16, 22, 29, 30, 34–37, 40–41, 46, 48, 55, 82, 83, 89, 99, 157, 163, 164, 167, 266, 267, 278, 279
Shore platforms, 88–90
Social media, 4, 96, 98, 99, 165, 212, 213, 245, 246, 248, 249
Socioeconomic, 69, 180, 181, 207, 218, 267
Spectra, 31–32, 36, 44, 53, 64, 194
Standing waves, 33, 275–277
Stereotypes, 165–167, 169
Strength, 3, 77, 85, 90, 109, 120–125, 127–129, 140, 141, 144, 208, 227, 230, 249
Subcultures, 4, 205, 223, 224, 226–231, 233–236, 243, 246–248, 252, 255, 259, 260, 262
Substitution theory, 247, 248
Surfers, 1, 3–5, 11, 15–17, 19–24, 27, 28, 32, 36, 39, 41, 46, 51, 52, 54, 55, 57, 60, 61, 63, 69, 76, 77, 79, 81, 86, 87, 89, 90, 93–111, 117–129, 131–139, 141, 142, 144–150, 158–162, 164–166, 168–175, 179, 180, 182, 184–187, 189–197, 203, 204, 206–213, 215–218, 224, 226, 229–236, 241–249, 251, 260, 262–267, 269, 271, 272, 275–279
Surfs, 1, 2, 4, 5, 9–12, 14, 18–24, 28, 30, 32, 39, 51–54, 60, 64–69, 75–90, 93–96, 98, 99, 101–103, 105–108, 111, 117–150, 157–165, 167–176, 179–181, 183, 184, 186–197, 203–219, 223–236, 241–256, 259–272, 275, 276, 278, 279
Sustainability, 180, 192, 205, 207, 209, 216–218, 231, 263
Swells, 9, 16, 17, 34, 36, 38–41, 47, 51–54, 58, 64, 66–69, 82, 83, 187, 208, 223, 266

T

Techniques, 15, 17, 34–36, 45, 105–109, 117, 146, 161, 184, 190, 191, 197, 249, 260, 269, 276
Training, 96, 97, 99, 100, 102–111, 121, 122, 124, 126–128, 136, 138, 139, 144, 215
Trauma, 133
Travels, 4, 15, 39, 40, 44, 53, 76, 110, 148, 171, 182, 184, 187, 190–192, 203–208, 210–213, 216, 231, 234, 249, 260–263, 265, 267, 271

U

Uses and gratifications (U&G) theory, 247, 248

V

Values, 20, 30, 31, 44, 46, 47, 61, 62, 95, 96, 109–111, 161, 179–197, 207, 215, 217, 223, 226–231, 235, 241, 243, 247, 250, 255, 259, 260, 264–266, 272

W

Wave form, 28–29, 41, 42, 275

Wave peel

Wave riding, 9–13, 15, 18, 123, 124, 126, 181, 224, 235, 276

Web, 45–46, 245

Well-being, 117, 170, 173, 184, 187–189, 192, 196, 212, 213, 217, 234

Y

Youth, 3, 4, 205, 211, 216, 226–228, 233, 234, 248, 259, 260, 262

MIX
Papier aus verantwortungsvollen Quellen
Paper from responsible sources
FSC® C105338

If you have any concerns about our products,
you can contact us on
ProductSafety@springernature.com

In case Publisher is established outside the EU,
the EU authorized representative is:
**Springer Nature Customer Service Center GmbH
Europaplatz 3, 69115 Heidelberg, Germany**

Printed by Libri Plureos GmbH
in Hamburg, Germany